Conference Proceedings and Lecture Notes in Analysis
Volume I

Editors
D. H. Phong
M. Kashiwara

Volume I — *Proceedings of the Conference on Complex Analysis*

Volume II — *Robert Finn - The First Seventy Years*

International Press

International Press Publications

Series in Mathematical Physics
  Essays on Mirror Manifolds, Edited by S.T. Yau
  Quantum Groups, S. Sternberg
  75 Years of Radon Transform, Edited by S. Gindikin and P. Michor
  Perspectives in Mathematical Physics, Edited by R. Penner and S.T. Yau

Series in Geometry and Topology
  L^2 Moduli Spaces on 4-Manifold with Cylindrical Ends, C. Taubes
  The L^2 Moduli Space and a Vanishing Theorem for Donaldson Polynomial Invariants, J. Morgan, T. Mrowka, D. Ruberman
  Lectures on Harmonic Maps, R. Schoen and S.T. Yau
  Lectures on Differential Geometry, R. Schoen and S.T Yau
  Lectures on Low-Dimensional Topology, Edited by K. Johannson
  Chern, A Great Geometer, Edited by S.-T. Yau

Series in Algebraic Geometry
  Proceedings of the International Symposium on Algebraic Geometry and Related Topics, Edited by Y. Namikawa, K. Ueno, and J.-H. Yang

Series in Analysis
  Integrals of Cauchy Type on the Ball, S. Gong
  Proceedings of the Conference on Complex Analysis at the Nankai Institue, Edited by L. Yang

Series in Applied Physics
  Proceedings of the Second International Conference on Applied Physics, Edited by D.H. Feng and T.-Y. Zhang
  Physics of the Quantum Electron Solid, Edited by S.-T. Chui

Collected and Selected Works
  The Collected Works of Freeman Dyson
  The Collected Works of C.B. Morrey
  The Collected Works of P. Griffiths
  Selected Works of V.S. Varadarajan

Proceedings of the Conference on Complex Analysis

Edited by
Zhong Li
Fuyao Ren
Lo Yang
Shunyan Zhang

International Press

Table of Contents

Preface

At the suggestion of Professor S. S. Chern, the Nankai Institute of Mathematics devoted the 1991–1992 academic year to complex analysis, and the International Conference on Complex Analysis was organized at the end of the special year to memory the late Professor Hiong King-lai on the occation of the centenary of his birth.

The special year started from four workshops in September and October 1991. These workshops were concentrated respectively to complex dynamic system, the theory of value-distribution, the quasi-conformal mappings and the geometric theory of functions. Around twenty scholars from the whole country took part in each workshop which lasted about ten days. Many young scholars found that the workshops were very helpful. From March to June 1992, we arranged four courses for graduate students. Namely, Riemann surface, quasi-conformal mappings, complex dynamic system and the theory of value-distribution. Fifteen foreign professors came to Tianjin to give a series of lectures. They are A. Baernstein, B. V. Bojarski, R. L. Devaney, D. Drasin, G. Frank, P. Gauthier, F. W. Gehring, J. Kotus, J. Langley, P. Lappan, C. McMullen, M. Ohtsuka, Lee A. Rubel, M. Shishikura, J. -M. Wu. The special year is very successful to all the participants, especially to young scholars.

The International Conference on Complex Analysis took place between June 19th and 23rd 1992, and around one hundred mathematicians were present. Over one dozen of them were from the outside of the mainland of China. Professors S. S. Chern and F. W. Gehring delivered one-hour lecture in plenary sessions. The one-hour speakers in sections are Jixiu Chen, Chitai Chuang, M. Essen, P. Lappan, Zhong Li, Jianke Lu, D. Minda, Fuyao Ren, Lee A. Rubel, H. M. Srivastava, Yuefei Wang, Y. C. Wong, C. C. Yang, Lo Yang, Jiarong Yu. We would like to thank all of them for their excellent lectures and other participants who gave contributed talks on their own research work.

The conference was supported by the Chinese Educational Committee, the National Natural Science Foundation of China, Tian Yuan Foundation, the Chinese Mathematical Society and Nankai Institute of Mathematics. We are very grateful to these institutions for their support.

This volume contains a collection of papers, contributed to either the special year or the International Conference on Complex Analysis. We would like to thank the contributors for their help and support.

This book was typeset using a LaTeX style file written by Alex Kasman for International Press. Most authors provided their own TeX files, which enabled the publisher to offer a low price text.

Kinglai Hiong:
A Brief Outline of His Life and Works

Lo Yang

Institute of Mathematics, Chinese Academy of Sciences
Beijing, China

Professor Kinglai Hiong, styled Dizhi, was born in Yunnan Province of China on October 20, 1893 (or September 11 by the lunar calender). His father, Mr. Guodong Hiong, was appointed to be the educational director of Zhaozhou perfecture and lived there with Kinglai when he was twelve years old. In 1907, Kinglai went from Zhaozhou to Kunming to enter Yunnan High School and took some special training in French afterwards. In 1913, he got the third in the examination given by Yunnan Province and went to Belgium for the advanced study.

World War I started in the next year. Kinglai moved to France to study at Grenoble University, Paris University, Montpellier University and Marseillee University successively and obtained his master dgree in Montpellier University. When he returned to China, he became a faculty member of Yunnan Industrier School and the School of Road Construction. In 1921, he succeeded Ru Ho as Professor and Chairman of the Department of Mathematics at Nanjing Higher Normal School and Southeast University. In 1926, he was appointed to be Professor and Chairman of the Department of Mathematics, Tsinghua University.

In 1932, Professor Hiong went to Paris again during his sabbatical vacation to do research and obtained Ph. D. of state in 1934. After returning to China, he continued his original job in Tsinghua University. From 1937 to 1949, he was the president of Yunnan University. During that period, Yunnan University became a large national university with five schools and many famous professors.

In 1949, Professor Hiong went to France for the third time and took part in the conference of UNESCO there. Then he was engaged in scientific research at Paris and unfortunately suffered a cerebral haemorrhage. In spite of this, he made great efforts to do research and published many mathematical papers there.

In 1957, in response to the letter of the late Premier Enlai Chou and the expectation of Chinese colleagues, Professor Hiong returned to Beijing to be the research professor and the director of the division of the theory of functions at the Institute of Mathematics, the Chinese Academy of Sciences. He was also a standing member of the National Committee of the Chinese People's Political Consultative Conference. He died in Beijing on February 3, 1969.

Professor Hiong's special interest was complex analysis and his most outstanding contribution was to create the general theory of meromorphic functions of infinite order.

As very well known, the theory of value-distribution is connected with the distribution of the roots of the equation $f(z) = a$ for a meromorphic function

Proceedings of the International Conference on Complex
Analysis at the Nankai Institute of Mathematics, 1992 pp. 3-7
©INTERNATIONAL PRESS 1994

$f(z)$ in the plane and different complex values of a. From 1880 to 1920, Picard, Borel, Valiron and other French mathematicians did systematic researches on entire functions. They paid more attention to the quantity $n(r, f = a)$, which is the number of roots, with due count of multiplicity, of $f(z) = a$ in $|z| \leq r$ and used the maximum modulus $M(r, f)$ as their efficient tool. In 1925, R. Nevanlinna took meromorphic functions as principal object and replaced $n(r, f = a)$ by its integral average

$$N(r, f = a) = \int_0^r \{n(t, f = a)/t\}dt.$$

He introduced the characteristic function

$$T(r, f) = m(r, f) + N(r, f = \infty),$$

where

$$m(r, f) = \frac{1}{2\pi} \int_0^{2\pi} \log^+ |f(re^{i\theta})|d\theta,$$

and

$$\log^+ |f(re^{i\theta})| = \max\{\log |f(re^{i\theta})|, 0\}.$$

The characteristic function gives a description of increasing of $f(z)$. For instance, the order of $f(z)$ is defined by

$$\rho = \limsup_{r \to \infty} \log T(r, f)/\log r.$$

Based on these notations, Nevanlinna created the modern theory of value-distribution of meromorphic functions which contains two fundamental theorems. (cf. [2], [8])

When Professor Hiong went to Paris for the second time, Valiron, Milloux, Bloch, H. Cartan etc. made their efforts to a deeper investigation of the theory of value-distribution. For entire and meromorphic functions of finite order, Valiron [10] introduced the concept of precise order and obtained ideal results. However, the only work for infinite order due to Blumenthal [1] is not so precise and in the case of entire functions. Professor Hiong introduced the type function, the infinite order in his meaning and obtained very beautiful results. To be precise, he proved the following important theorem. [3]

Let $f(z)$ be a meromorphic function of infinite order in the plane. Then there exists a function $\rho(r)$ such that

(a) $\rho(r)$ is a continuous nonnegative nondecreasing function defined on $(0, \infty)$ and tends to infinity with r;

(b) $\lim\limits_{r \to \infty} \{\log U\{r(1 + 1/\log U(r))\}/\log U(r)\} = 1$, where $U(r) = r^{\rho(r)} (r > 0)$;

(c) $\limsup\limits_{r \to \infty} \{\log T(r, f)/\log U(r)\} = 1$.

Using this concept, Professor Hiong [3] himself obtained a series of precise results on the value-distribution of entire and meromorphic functions of infinite order. For instance, he proved the existence of Borel directions in the case of infinite order as follows:

Let $f(z)$ be a meromorphic function of infinite order $\rho(r)$ in the plane. Then there exists a direction arg $z = \theta_0 (0 \leq \theta_0 < 2\pi)$ such that if $n(r, \theta_0, \varepsilon, f = a)$ denotes the number of zeros of $f(z) - a$ in the region $(|z| \leq r) \cap (|argz - \theta_0| < \varepsilon)$, then we have

$$\limsup_{r \to \infty} \{\log n(r, \theta_0, \varepsilon, f = a)/\rho(r) \log r\} = 1$$

for any positive number ε and every complex value a, except two values at most.

Later on, $\rho(r)$ is named as the infinite order of $f(z)$ in the sense of Hiong and becomes a powerful tool in the research of entire and meromorphic functions of infinite order.

Professor Hiong also made the significant contributions to the value distribution of a meromorphic function combined with its derivative. Connecting with Milloux' work, he extended Nevanlinna's fundamental lemma on logarithmic derivatives to the general case [4]. Namely, he obtained the following result.

Let $f(z)$ be meromorphic in $|z| < R (\leq \infty)$ and k a positive integer. If $f(0) \neq 0, \infty$, then we have

$$m(r, f^{(k)}/f) < C_k\{1 + \log^+ \log^+ 1/|f(0)| + \log^+ 1/r$$

$$+ \log^+ 1/(\rho - r) + \log^+ \rho + \log^+ T(\rho, f)\}$$

for any r and ρ with $0 < r < \rho < R$, where C_k is a constant depending only on k.

With the aid of this general lemma, Professor Hiong did several extensions to the Nevanlinna's second fundamental theorem. One simple example of them can be stated as follows [5]:

Let $f(z)$ be meromorphic in the open plane and a, b and c three finite complex numbers with $b \neq 0, c \neq 0$ and $b \neq c$. Then we have

$$T(r, f) < N(r, 1/(f - a)) + N(r, 1/(f^{(k)} - b))$$

$$+ N(r, 1/(f^{(k)} - c)) - N(r, 1/f^{(k+1)}) + S(r, f),$$

where $S(r, f)$ has the properties of the error term in the Nevanlinna theorem.

After the Milloux and Hiong's work, the investigation on the value distribution of a meromorphic function combined with its derivative has been extensively developed. An interesting and important inequality due to W.K. Hayman in 1959, is one of them.

There is a very closed connection between the Nevanlinna theory and the theory of normal families. In the twenties, G.Valiron remarked that the Schottky theorem and the Montel's criterion can be deduced from the Nevanlinna's second fundamental theorem. At the same time, Bloch posed an ingenious idea: there always exists a criterion on normality to correspond every Picard-Liouville type theorem. Based on this idea, Professor Hiong [7] started from his fundamental inequality and proved the Miranda's criterion which says:

Let F be a family of holomorphic functions in the region D. If, for every function $f(z)$ of $F, f(z)$ does not take a finite complex value a and its derivative $f^{(k)}(z)$ does not take a finite nonzero complex value b in D, then F is normal there.

Although this result is known already, Professor Hiong gave a new proof and remarked that the key point to establish a criterion on normality resides in eliminating the so called initial values of the corresponding fundamental inequality. During last thirty years, this method has been adopted by analysts in China, the United States and the United Kingdom.

Besides, Professor Hiong did important work on algebroidal functions, holomorphic and meromorphic functions in the unit disk and the theorem of uniqueness etc. His book "On the meromorphic and algebroidal functions—The extension of one of Nevanlinna theorems" has been highly valued in this field.

Professor Hiong also made the fundamental contributions to the mathematical education in China. In the 1920s, he set up the Departments of Mathematics at both Southeast University in Nanjing and Tsinghua University in Beijing. He is one of forerunners to introduce the modern mathematics to China. As a professor in universities, he wrote and compiled mathematical teaching materials covering more than ten courses. "Advanced Mathematical Analysis" was published afterwards as a textbook and yielded excellent results.

In 1932, Professor Hiong attended the International Congress of Mathematicians at Zurich. In 1935, he took an active part in establishing the Chinese Mathematical Society and together with other mathematicians started the Chinese Journal of Mathematics in next year.

Professor Hiong recognized and trained many outstanding students. Since Hiong realized the backward state of modern science in China during the twenties, he considered as most important thing to train qualified students from the very beginning of his educational career. During the different periods of Southeast University, Tsinghua University and Yunnan University, his teaching was very successful. Many brilliant mathematicians and physicists emerged from his students. Professors S.S. Chern, C.T. Chuang, B.L. Hsu, L.K. Hua, C.Ku, C.C. Lin, T.Z. Ny, S.T. Qian, S.S. Shu, S.F. Tuan, and Z.Y. Zhao etc. are all the outstanding representatives of his students.

Professor Hiong died in 1969 which was during the period of "Gang of Four". His death is certainly a heavy loss to the Chinese mathematical community.

References

[1]   O. Blumental, Principes de la theorie des fonctions entieres d'ordre infini, Paris, 1910.

[2]   W.K. Hayman, Meromorphic functions, Oxford: Clarendon Press, 1964.

[3]   K.L. Hiong, Sur la fonctions entieres et les fonctions meromorphes d'ordre infini, *Journ. Math. Pures et Appl.*, 14(1935), 233–308.

[4]   K.L. Hiong, Sur les fonctions holomorphes dont les derivees admettant une valeur exceptionnelle, *Ann. Ec. Norm, Sup.*, 3e Serie. 72(1955), 165–197.

[5]   K.L. Hiong, Sur la limitation de $T(r, f)$ sans intervention des poles, *Bull. Sci. Math*, 2e Serie, 80(1956), 1–16.

[6]   K.L .Hiong, Sur les fonctions meromorphes et les fonctions algebroides, Memorial Sci. Math., Fasc. 139, Paris, Gauthier-Villar, 1957.

[7] K.L. Hiong, Sur les cycle de Montel-Miranda dans la theorie des familles normales, *Scientia Sinica*, 7(1958), 987–1000.

[8] R. Nevanlinna, Analytic functions, Springer-Verlag, 1970.

[9] G. Valiron, Lectures on the general theory of integral functions, Cambridge, 1923.

[10] G. Valiron, Fonctions entieres et fonctions meromorphes, Mem, Sci. Math., Fasc. 2, Paris, 1925.

On the Beltrami Equation

B.Bojarski

Institute of Mathematics, Polish Academy of Sciences
00-950, Warsaw , ul. Sniadeckich, 8
bojarski@impan.impan.gov.pl

V.Ya.Gutlyanskii

Institute for Applied Mathematics and Mechanics, Ukrainian Academy of
Sciences
340114, Donetsk, ul. Roze Luxemburg, 74,
gut%iamm.donetsk.ua @relay.ussr.eu.net

Abstract: In this paper we give short outline of some fundamental results of
the theory of quasiconformal mappings and consider applications of the
Beltrami equation to some problems of the conformal mapping theory
and partial differential systems.

1   Introduction

New profound connections between theories of conformal and quasiconformal
mappings were discovered during few last decades. Delicate methods of holomorphic function theory allowed to solve many difficult problems for quasiconformal mappings in the complex plane. On the other hand analytical methods
of Beltrami equation theory led up to new approaches to proving fundamental
results in the theory of conformal mappings.

The quasiconformal mappings in the complex plane were introduced by
H.Grotzsch, M.Lavrent'ev and L.Ahlfors (see, [1]–[3]). It was O.Teichmüller
[4] who noticed the deep connection between quasiconformal mappings and
complex function theory and gave the first applications to the Riemann surface theory. New approaches in this field appeared after fundamental works
of M.Lavrent'ev, C.Morrey, L.Ahlfors, L.Bers, P.Belinskii, I.Vekua, F.Gehring,
C.Andreian Cazacu, O.Lehto, S.Kruskal, R.Kuhnau, K.Strebel and many others
(see [5]–[7] and references therein). The progress in the theory of Teichmüller
spaces and deformations of conformal structures on Riemann surfaces gave rise
to a new branch of the theory of univalent analytic functions (see [8]).

In this paper we give short outline of some fundamental results of the
theory of quasiconformal mappings and consider some applications of the Beltrami equation to some problems of the conformal mapping theory and partial

Proceedings of the International Conference on Complex
Analysis at the Nankai Institute of Mathematics, 1992 pp. 8-33

differential equations. The basic tools are the uniqueness and representation theorems for normalized solutions of the Beltrami equation, the fundamental results due to O.Teichmüller, H.Wittich and P.Belinskii on the regularity of a quasiconformal mapping at a point (see [7], p. 54, [5], p. 234) and the ideas of convex analysis related with the concept of invariantly convex sets of complex dilatations.

2   Beltrami differential equation

Let G be a domain in the complex plane $\mathbb{C}$. A continuous complex valued function $f = u + iv$ is said to be *absolutely continuous on lines (ACL)* in a domain G if for each closed rectangle $\{x + iy : a \leq x \leq b, c \leq y \leq d\} \subset G$, the functions $u(x,y)$ and $v(x,y)$ are absolutely continuous on $[a,b]$ for almost all $y \in [c,d]$ and the functions $u(x,y)$ and $v(x,y)$ are absolutely continuous on $[c,d]$ for almost all $x \in [a,b]$. It is a well-known result due to F.Gehring and O.Lehto [21] that if $f \in ACL$ is a local homeomorphism in the domain G then it is differentiable a.e. in G (see also [5], p.128).

A sense-preserving homeomorphism f of a domain G is said to be Q-*quasiconformal* if

1) f is ACL   in   G;

2) $\max_\alpha |\partial_\alpha f(z)| \leq Q \min_\alpha |\partial_\alpha f(z)|$ a.e. in G.

Here $\partial_\alpha f$ is the derivative in the direction α.

Supposing [1] that $|f_z(z)| > 0$ a.e., we can define the function

$$\mu(z) = \frac{f_{\bar{z}}(z)}{f_z(z)}$$

for a.e. $z \in G$. This measurable function μ in G is called *the complex dilatation* of f.

The definition of complex dilatation leads to the *Beltrami differential equation*

$$f_{\bar{z}} = \mu(z) f_z \qquad \text{a.e. in } G. \tag{2.1}$$

where μ is measurable and $\|\mu\|_\infty = \text{ess sup}_D |\mu(z)| \leq k < 1$. Evidently $Q = (1 + k)(1 - k)^{-1}$.

A function f is said to be an L^p-solution of (2.1) in a domain G if f has generalized L^p-derivatives and (2.1) holds a.e. in G. In other words, we consider the generalized solution of (2.1) in the Sobolev class $W^{1,p}(G)$, $p \geq 1$. If f is conformal, μ vanishes identically, and (2.1) reduces to the Cauchy–Riemann equation $f_{\bar{z}} = 0$.

The Beltrami equation has a long history and for details and historical remarks we refer the reader to the monographs L.Ahlfors [3], I.Vekua [6], O.Lehto–K.Virtanen [5] (see also [12]).

---

[1] It is a fact that for Q-quasiconformal mappings the condition $|f_z| > 0$ a.e., for $f \neq const$, is a consequence of 1) and 2) (see [5], [11]).

2.1 Existence and representation theorems

The uniqueness part for the Beltrami equation follows, roughly speaking, from the direct computation of the complex dilatation and the well-known Weyl's theorem.

Theorem A: *Let F and ω be quasiconformal mappings of a domain G whose complex dilatations agree a.e. in G. Then*

$$f = F \circ \omega^{-1} \tag{2.2}$$

is a conformal mapping.

Now we start with the fundamental existence and representation theorems [11].

Theorem B: *Assume μ is measurable in $\mathbb{C}$ with $\|\mu\|_\infty < 1$ and has compact support. Then the Beltrami equation (2.1) in the complex plane $\mathbb{C}$ admits a unique solution of the form*

$$\omega(z) = z - \frac{1}{\pi} \int\!\!\int_{\mathbb{C}} \frac{\rho(\zeta)dm_\zeta}{\zeta - z}, \tag{2.3}$$

where $\rho \in L_p$, $p > 2$, is the unique solution of the singular integral equation

$$\rho - \mu \mathcal{H}_{\mathbb{C}}[\rho] = \mu. \tag{2.4}$$

in the Lebesgue space L_p. Moreover,

$$\omega(z) = z + \sum_{n=0}^{\infty} C_{\mathbb{C}}[(\mu\mathcal{H})^n \mu](z), \tag{2.5}$$

and the series is uniformly convergent in $\mathbb{C}$. The solution (2.3) gives a quasiconformal mapping of the complex plane onto itself.

Here and below for a domain $G \subseteq \mathbb{C}$ the operators

$$C_G[\rho](z) = -\frac{1}{\pi} \int\!\!\int_G \frac{\rho(\zeta)}{\zeta - z} dm_\zeta, \qquad \mathcal{H}_G[\rho](z) = \partial_z C_G[\rho](z),$$

are the complex Cauchy and Hilbert transforms.

The fundamental idea to consider the solution of the Beltrami equation in the form (2.3) is due to L.Ahlfors [14] and I.Vekua [15]. It immediately leads to the existence of the L^2 solution of (2.1). However the further progress on this way was hampered by the fact that if $\rho \in L^2$, where ρ is a solution of the integral equation (2.4), then the mapping (2.3) can be discontinuous and does not lead to a quasiconformal mapping. The crucial further step has been done in [9]–[11] with the idea of using the Calderon–Zygmund theory of singular integral equations in the spaces L^p for $p > 2$ (see [6], Chapter 2). This directly leads to the conclusion that

$$\omega(z) = z - \frac{1}{\pi} \int\!\!\int_{\mathbb{C}} \frac{\rho(\zeta)dm_\zeta}{\zeta - z},$$

where $\rho \in L_p$, $p > 2$, is the unique solution of the singular integral equation (2.4) in the Lebesgue space L_p.

Using the Sobolev imbedding theorems this immediately removes the essential obstacle for continuity in (2.3).

It can be proved rather directly (see, for example, [6]) that a generalized solution of the Beltrami equation whose complex dilatation (or the coefficient μ) is locally Holder continuous is a local diffeomorfism. By a topological argument it follows then that the solution (2.3) is a global diffeomorphism of the complex plane $\mathbb{C}$. By approximation this leads to the quasiconformal solution of the Beltrami equation in the form (2.3).

In general case when the complex dilatation μ is not compactly supported one can use the decomposition of $\mu = \mu_1 + \mu_2$ with $\mu_1 = 0$ and $\mu_2 = 0$ in the neighborhood of the origin and the infinity respectively to prove the existence of the normalized solution of the Beltrami equation (see [16], p. 35, [6], p. 110).

Theorem C: *Let μ be a measurable function in a domain G with $\|\mu\|_\infty < 1$. Then there is a quasiconformal mapping $f : G \to \mathbb{C}$ whose complex dilatation agrees with μ a.e.*

2.2 Dependence of a mapping on its complex dilatation

The Theorem B and the representation formula (2.3)–(2.5) given therein show how a quasiconformal mapping depends on its complex dilatation(see [11], [36], [8], p. 68).

The dependence of normalized solutions of the Beltrami equation (2.1) on auxiliary parameters t is best studied using the inhomogeneous singular integral equation

$$\rho(t) - \mu(t)\mathcal{H}[\rho(t)] = h(t), \tag{2.6}$$

where $\mu(t) \equiv \mu(z, t)$, etc., in the Lebesgue spaces L_p, $1 < p < \infty$, for p sufficiently close to $p = 2$. This means that the admissible values of p are chosen in such a way that the inequality

$$A_p\mu_0 < 1 \tag{2.7}$$

holds, where A_p is the L_p-norm of the Hilbert operator $\mathcal{H}$ and

$$\mu_0 = \sup_t \|\mu(z, t)\|_\infty < 1.$$

Due to Riesz–Thorin convexity theorems A_p is continuous in p at $p = 2$ and $A_2 = 1$. Inequality (2.7) implies the fundamental *a priori* estimate for solutions of (2.6)

$$\|\rho\|_{L_p} \leq \frac{\|h\|_{L_p}}{1 - A_p\mu_0}. \tag{2.8}$$

If $\mu(t)$ is a family of complex dilatations depending on a real or complex parameter t in some neighbourhood of $t_0 = 0$ and $\omega(t)$ the corresponding family of quasiconformal mappings then the representation formulas of type (2.3), (2.5) allow to reduce the study of $\omega(t)$, as functions of the parameter, to the study of one parameter family of the *"densities"* $\rho(t)$.

Proposition 3.1: *Let*

$$\rho(t) - \mu(t)\mathcal{H}[\rho(t)] = h(t)$$

be a parametrized family of integral equations. If $\|\mu(t)\|_\infty \leq \mu_0 < 1$ for all admissible t and $\mu(t)$ and $h(t)$ depend continuously (differentiably) on t, in the sense of Lebesgue spaces L_∞ and L_p respectively, then the solution ρ is continuous in the L_p-sense (differentiable) in t.

Moreover the derivative $\rho'(t)$ may be calculated by solving the integral equation

$$\rho' - \mu(t)\mathcal{H}[\rho'] = \mu'\mathcal{H}[\rho] + h'. \tag{2.9}$$

If the parameter is complex, $\mu(t)$ and $h(t)$ depend holomorphically (real analytic) on t then so does the solution.

We sketch an essential step in the proof of the Proposition 3.1 strictly following the calculations in [11].

Consider, e.g. the differentiability with respect to the real parameter. Denoting by ρ_τ the difference

$$\rho_\tau \equiv \frac{\rho(t+\tau) - \rho(t)}{\tau} - \rho'(t)$$

and analoguously for μ_τ and h_τ we get for ρ_τ the integral equation

$$\rho_\tau - \mu(t)\mathcal{H}[\rho_\tau] - \mu_\tau\mathcal{H}[\rho] = h_\tau.$$

Since $\|\mu_\tau\|_\infty \to 0$ and $\|h_\tau\|_{L_p} \to 0$ for $\tau \to 0$ the estimate (2.7) implies the differentiability of $\rho(t)$ and the equality $\partial\rho/\partial t \equiv \rho'$. For more details see [11]–[13].

As corollaries to the Proposition 3.1 more precise and explicit statements on the dependence of the mapping on the complex dilatation can be obtained. Essentially the proofs are obtained by combining the calculations of the Proposition 3.1 and the integral representation formulas. These calculations lead directly to the explicit formulas for the Gateaux differential of the normalized quasiconformal mapping in its dependence on the infinitesimal variation of the complex dilatation.

Theorem D: *Let μ be a measurable function in the plane with bounded support and $\|\mu\|_\infty < 1$. Let $z \in f(z,t)$ be the quasiconformal mapping of the plane with complex dilatation $t\mu$, normalized with the condition $\lim(f(z,t) - z) = 0$ as $z \to \infty$. Then, for every fixed $z \neq \infty$, the function $t \to f(z,t)$ is holomorphic in the disk $|t| < 1/\|\mu\|_\infty$.*

Let us consider the qausiconformal mappings f of the plane, normalized by the conditions $f(0) = 0$, $f(1) = 1$ and $f(\infty) = \infty$.

Theorem E: *Let*

$$\mu(z,t) = \mu(z) + t\nu(z) + t\varepsilon(z,t),$$

where $\nu, \varepsilon \in L^\infty$ and $\|\varepsilon(z,t)\|_\infty \to 0$ as $t \to 0$. Then

$$\left.\frac{\partial f(z,t)}{\partial t}\right|_{t=0} = -\frac{1}{\pi}\int\int_{\mathbb{C}}\nu(\zeta)\frac{f(z)(f(z)-1)f_\zeta^2(\zeta)}{f(\zeta)(f(\zeta)-1)(f(\zeta)-f(z))}dm_\zeta,$$

uniformly on compact sets.

For the details of the proofs we refer the reader to L.Ahlfors ([16], p. 105), O.Lehto ([8], p.69) and S.Kruskal' ([25], p. 56).

3 On the behavior of a quasiconformal mapping at a point

In this section we give some conditions on μ under which the quasiconformal mapping $f(z)$ is conformal at the prescribed point z_0. Without loss of generality we may assume that the fixed point z_0 is the origin and the mapping is normalized with the condition $f(0) = 0$.

We start with the Belinskii's theorem (see [7], p. 54) which is a natural extension of the one due to O.Teichmüller and H.Wittich [16], [17].

Theorem 3.1: *Let $f(z)$ be a Q-quasiconformal mapping of the finite plane onto itself with $f(0) = 0$ and put*

$$I(r) = \frac{1}{2\pi}\int\int_{|z|<r}\frac{p(z)-1}{|z|^2}dm_z < \infty \quad for \quad r < \infty, \tag{3.1}$$

where

$$p(z) = (1 + |\mu(z)|)/(1 - |\mu(z)|). \tag{3.2}$$

Then f is conformal at $z = 0$ and

$$\left|\frac{f(z)}{z} - f'(0)\right| < |f'(0)|\varepsilon(|z|), \tag{3.3}$$

where the function ε depends only on Q and I and $\varepsilon(|z|) \to 0$ as $|z| \to 0$. The derivative $f'(0)$ satisfies the inequalities

$$\min_{|z|=1}|f(z)|e^{-I(1)} \leq |f'(0)| \leq \max_{|z|=1}|f(z)|e^{I(1)}. \tag{3.4}$$

It's necessary to note that if $I(r) \leq \Delta(r)$, where $\Delta(r) \to 0$ as $r \to 0$, then the conclusions of the theorem hold for a function ε depending only on Q and Δ.

For the details of the proof and the history of the problem we refer the reader to P.Belinskii ([7], p.54) and O.Lehto and K.Virtanen ([5], p. 232). Note that some results related to Theorem 3.1 have been obtained by B.Shabat,

L.Volkovyskij and O.Lehto (see [5], p. 236). The regularity of the mapping $f(z)$ at the point z_0 under the condition

$$\int\int_{|z-z_0|\leq r}\left|\frac{\mu-\mu_0}{z-z_0}\right|^p dm_z < \infty, \qquad p > 2.$$

was demonstrated in [11] by applying the representation Theorem B. O.Lehto has investigated the question when a quasiconformal mapping f with prescribed complex dilatation is regular in the domain D (see [5], p. 235). We shall use this approach in the paragraph 9.

4   On convergence and compactness theorems

This section deals with problems of the behavior of complex dilatations of Q-quasiconformal mappings under locally uniform convergence of such mappings. These questions have been investigated by P.Belinskii, L.Bers, B.Bojarski, F.Gering, O.Lehto, V.Ryazanov, K.Strebel, M.Schiffer and G.Schober, and others (see, for example, [5], [7], [11], [19]–[21]). Our purpose is to give a short outline of some recent important results in this directions. Let us begin with two well-known theorems (see [11], [19] and [20]).

Theorem 4.1: *If $f_n(z) \to f(z)$ locally uniformly and $\mu_n(z) \to \nu(z)$ almost everywhere, then $\mu(z) = \nu(z)$ almost everywhere.*

Moreover, it has been proved that the locally uniform convergence $f_n(z) \to f(z)$ does not yield, in general, the convergence $\mu_n(z) \to \nu(z)$ almost everywhere.

Theorem 4.2: *If $f_n(z) \to f(z)$ locally uniformly, then*

$$|\mu(z)| \leq \limsup_{n\to\infty} |\mu_n(z)|$$

almost everywhere.

In order to generalize the first theorem and to get a refinement of Strebel's result let us introduce some definitions and notations.

Denote by Γ the set of all Mobius transformations of the unit disk D onto itself.

A set M of points in the unit disk $D = \{z : |z| < 1\}$ is said to be *invariantly convex*, if all sets $\gamma(M)$, $\gamma \in \Gamma$ are convex too.

Let $\overline{invco_U}\, M$ denote the smallest with respect to inclusion closed invariantly convex set containing M.

Theorem 4.3: *If a sequence of Q-quasiconformal mappings $f_n(z)$ of a region D with complex dilatation μ_n converges locally uniformly in D to a Q-quasiconformal mapping $f(z)$ with complex dilatations μ, then*

$$\mu(z) \in \overline{invco_U}(Ls_{n\to\infty}\{\mu_n(z)\}) \quad a.e.,$$

where $Ls_{n\to\infty}(M_n) = \cap_{m=1}^{\infty}\left(\overline{\cup_{n\geq m}M_n}\right)$ denotes the topological limit superior of a sequence of sets M_n.

This theorem refines the corresponding result of K.Strebel [20], since

$$\overline{invco_U}(Ls_{n\to\infty}\{\mu_n(z)\}) \subseteq \{z : |\mu| \leq \limsup_{n\to\infty}|\mu_n|\}.$$

Suppose now that the conditions of the Theorem 4.1 are satisfied, i.e. the sequence f_n converges locally uniformly to f, and the μ_n converges a.e. to ν. Then $Ls_{n\to\infty}\{\mu_n(z)\} = \{\nu(z)\}$ and, consequently,

$$\overline{invco_U}(Ls_{n\to\infty}\{\mu_n(z)\}) = \{\nu(z)\}.$$

According to Theorem 4.3, $\mu(z) = \nu(z)$ almost everywhere.

We say that a one-parameter family of nonempty closed planar sets $M(z)$ defined for z in some measurable set E of the extended complex plane $\overline{\mathbb{C}}$, is measurable with respect to the parameter $z \in E$ if for any closed set M_0 the set of points $E_0 = \{z : z \in E, M(z) \subseteq M_0\}$ is measurable with respect to the planar Lebesgue measure.

Let $M(z)$ be an arbitrary family of sets in the disk $D_q = \{z : |z| < q\}$, $q < 1$. Denote by $\mathcal{N}_{M(z)}$ the class of all normalized in a suitable way $Q = (1 + q)(1 - q)^{-1}$-quasiconformal mappings of the plane $\overline{\mathbb{C}}$ whose complex dilatations satisfy the condition

$$\mu(z) \in M(z)$$

for almost all $z \in \mathbb{C}$.

Theorem 4.4: *Let $M(z)$, $z \in \mathbb{C}$ be an arbitrary measurable family of nonempty closed sets in the disk D_q. Then*

$$\overline{\mathcal{N}_{M(z)}} = \mathcal{N}_{\overline{invco}M(z)}$$

in the topology of locally uniform convergence.

As a consequence of Theorems 4.3 and 4.4 follows

Theorem 4.5: *Let $M(z)$, $z \in \mathbb{C}$ be an arbitrary measurable family of nonempty closed sets in the disk D_q. Then $\mathcal{N}_{M(z)}$ is compact in the topology of locally uniform convergence if and only if $M(z)$ is invariantly convex for almost all $z \in \mathbb{C}$.*

These theorems have been proved in [21]. Analogous results hold for families of Q-quasiconformal mappings with restrictions of integral type on the complex or real dilatations (see, for example, [23]).

The simple conclusion that immediately follows from the Theorem 4.5 and the definition of an invariantly convex set is that the compact classes of Q-quasiconformal mappings are generated by the families of complex dilatations which form a convex set. This leads, in particular, to the universal variational procedure over compact families of quasiconformal mappings which we shall describe below.

5   Variational procedure

The methods of the calculus of variations in the theory of quasiconformal mappings has been valuable from at least two points of view. On the one hand, it can be used to solve concrete extremal problems for functionals over a given family of mappings and to determine the extremal functions or the extremal value of the functional. On the other hand, one can obtain existence and representation theorems for solutions of some partial differential equations by constructing appropriate functionals over a compact class of quasiconformal mappings. These equations arise from variational procedure as a necessary condition for the extremal. The first direction has already been investigated in detail (see [24]–[26] and references therein). Concerning the second one we can note only two original papers by M.Schiffer and G.Schober [27], [28] and the paper [29].

In this section a new general variational procedure over some compact families of quasiconformal mappings that are important for applications is presented. It leads to existence theorems for some linear partial differential equations. More importantly, this approach gives representation for the solution in terms of quasiconformal mappings.

5.1   *Variational formulas.*

Let us introduce a special class of quasiconformal self-mappings of the complex plane $\mathbb{C}$.

Let $Q : \mathbb{C} \to [1, \infty]$ be a measurable function such that $\|Q\|_\infty = \text{ess sup}_{\mathbb{C}} Q$ is finite and μ be a complex dilatation in $\mathbb{C}$ satisfying the conditions

$$|\mu(z)| \leq (Q(z) - 1)(Q(z) + 1)^{-1}$$

for almost all $z \in \mathbb{C} \setminus \{\bigcup_{n=1}^{P} K(z_n, \rho_n)\} \setminus \Delta_R$, where $K(z_n, \rho_n) = \{z : |z - z_n| < \rho_n\}$, $\Delta_R = \{z : |z| > R\}$, and $\mu = 0$ for other points [2].

We denote by $w = f(z)$ the homeomorphic generalized solution of the Beltrami equation normalized by $f(\infty) = \infty$ and $f'(\infty) = 1$. This solution realizes a $Q(z)$-quasiconformal mapping of the extended complex plane on itself and the mapping is conformal in the domains $K(z_n, \rho_n)$ and Δ_R. Varying μ, we obtain the class N of all such mappings.

Let us denote by M the set of functions $\mu(z)$ that have the properties stated above.

Theorem 5.1: Let $f \in N$ and let it have the complex dilatation μ and let ν be an arbitrary element of M. Then for sufficiently small values of the parameter $t, t > 0$, the following functions are in the class N:

$$f^*(z) = f(z) + t\frac{1}{\pi} \int \int_{\mathbb{C}} \frac{(\mu(\zeta) - \nu(\zeta)) f_\zeta^2(\zeta) dm_\zeta}{(f(z) - f(\zeta))} + o(t), \qquad (5.1)$$

---

[2] We assume that the discs $\overline{K(z_n, \rho_n)}$ and $\overline{\Delta_R}$ are disjoint

where $t^{-1}o(t) \to 0$ as $t \to 0$, uniformly on compact sets.

Proof: When μ and ν belong to M, the function

$$\mu(z,t) = \mu + t(\nu - \mu)$$

also belongs to M for all $t, 0 \le t \le 1$. Let us denote by f^* the solution of the Beltrami equation with complex dilatation $\mu(z,t)$ that belongs to the class N. The theorem E allows to represent this solution in the form (5.1).   ∎

Note that the analoguous proof of variational formulas holds universally for any class of $Q(z)$-quasiconformal mappings whose complex dilatations form a convex set $M \subseteq B_1 = \{\mu : \|\mu\|_\infty < 1\}$. It consists in choosing admissible variations for μ in the form $\mu_t = \mu + t(\nu - \mu)$, $\nu \in M$, and suitably representing the homeomorphism f_t with complex dilatation μ_t on the basis of Theorems D or E. This approach to the construction of variational formulas was proposed in [30].

5.2   Variational procedure.

Let $\chi : N \to \mathbb{R}$ be an upper semicontinuous functional defined over the class N. By Theorem 4.5 the class N is compact. Then there exists a function $f \in N$ for which

$$\chi(f) = \max_N \chi. \tag{5.2}$$

We also require that χ is Gateaux differentiable, i.e.

$$\chi(f^*) = \chi(f) + tRe \int \int_{\mathbb{C}} g d\kappa + o(t)$$

for every admissible variation $f^* = f + tg + o(t)$ in the class N. Here κ is a finite complex Borel measure with compact support. Furthermore, we suppose that the kernel $(w - f(z))^{-1}$ is locally integrable with respect to the product measure $dm \otimes d\kappa$, where m is the planar Lebesgue measure, and

$$\mathcal{A}(w) = \frac{1}{\pi} \int \int_{\mathbb{C}} \frac{d\kappa}{(w - f(z))} \neq 0$$

almost everywhere.

We formulate necessary conditions for an extremum in terms of the complex dilatation $\mu(z)$ of a mapping $f(z) \in N$ realizing $\max_N \chi$.

Theorem 5.2: *Let μ be the complex dilatation of an extremal mapping $f \in N$. Then under the hypotheses stated above, the extremal function f for the problem (5.2) satisfies the differential relation*

$$f_{\bar{z}}(z) = -k(z) \frac{\overline{\mathcal{A}(f(z))}}{|\mathcal{A}(f(z))|} \overline{f_z(z)} \tag{5.3}$$

a.e. in $\mathbb{C}$, where $k(z) = (Q(z) - 1)(Q(z) + 1)^{-1}$.

Proof: Let μ be the complex dilatation of the extremal function f. Applying the variational formula (5.1), we calculate the variation of the functional

$$\delta\chi = \chi(f^*) - \chi(f) = Re\frac{t}{\pi} \int\int_{\mathbb{C}\times\mathbb{C}} \frac{(\mu(\zeta) - \nu(\zeta))f_\zeta^2 dm_\zeta d\kappa_z}{(f(z) - f(\zeta))}$$
$$+ o(t) \leq 0.$$

Divide by t and let $t \to 0$. Then, by using the Fubini theorem, we may interchange the order of integration and write the necessary condition for the maximum in the form

$$\Re e \int\int_{\mathbb{C}} (\mu(\zeta) - \nu(\zeta))\mathcal{A}(f(\zeta))f_\zeta^2 dm_\zeta \leq 0$$

It follows from this inequality and the arbitrariness of $\nu \in M$ that the characteristic μ of the extremal mapping f for the problem

$$\max_N \chi(f).$$

over the class N gives at the same time an extremum for the problem

$$\min_{\nu \in M} \Re e \int\int_{\mathbb{C}} \nu(\zeta)\mathcal{A}(f(\zeta))f_\zeta^2 dm_\zeta$$

over the class M. So we reduce the initial nonlinear extremal problem over the compact class N of quasiconformal mappings *to a linear* extremal problem over *a convex* class M of their complex dilatations.

Furthermore, in the standard way, we can establish that the dilatation of the extremal mapping has the form

$$\mu(z) = -k(z)\frac{\overline{\mathcal{A}(f(z))}}{|\mathcal{A}(f(z))|}\frac{\overline{f_z}}{f_z}$$

for almost all z. Thus, the extremal function f for the problem (5.2) satisfies the differential relation (5.3). ∎

6   Linear partial differential systems

Let $Q : \mathbb{C} \to [1, \infty)$ be a measurable function such that $\|Q\|_\infty = \text{ess sup}_{\mathbb{C}} Q$ is finite.

We shall consider the partial differential equation

$$\text{div}(Q(z)\,\text{grad}\,U) = 0. \tag{6.1}$$

This equation is of basic importance, in particular, in the theory of steady state fluid flow in an inhomogeneous medium.

By weak solution of the equation $\mathrm{div}(Q\,\mathrm{grad}\,U) = 0$ with singularities at $z_1, \ldots, z_P$ and ∞ we mean a function $U \in W^1_{2,loc}(\mathbb{C} \setminus \{z_1, \ldots, z_P\})$ that admits a conjugate $V \in W^1_{2,loc}(\mathbb{C} \setminus \{z_1, \ldots, z_P\})$ in the sense that U and V satisfy the generalized Cauchy–Riemann equations (see [27])

$$V_x = QU_y, \quad V_y = -QU_x. \tag{6.2}$$

If we define $F(z) = U + \imath V$ then (6.2) is equivalent to the equation

$$F_{\bar{z}} = -k(z)\overline{F_z}. \tag{6.3}$$

with $k(z) = (Q(z) - 1)(Q(z) + 1)^{-1}$.

Theorem 6.1: *Suppose that $Q : \overline{\mathbb{C}} \to [1, \infty)$ is measurable and $\|Q\|_\infty < \infty$. Then for any $x_1, \ldots, x_P \in \mathbb{R}$ there exists a $Q(z)$-quasiconformal mapping $f : \mathbb{C} \to \mathbb{C}$ preserving $0, 1$ and ∞ such that*

$$U(z) + \imath V(z) = \sum_{n=1}^{P} x_n \ln[f(z) - f(z_n)]$$

is a weak solution of the system (6.2) with logarithmic singularities at $z_1, \ldots, z_P$ and ∞.

Proof: Let us consider the extremal problem

$$\max_{N} \Re e \sum_{n,l=1}^{P} x_n x_l \ln \frac{f(z_n) - f(z_l)}{z_n - z_l}$$

over the class N introduced above. It follows then from the Theorem 5.2 that f satisfies the differential relation

$$f_{\bar{z}}(z) = -k(z) \cdot \frac{\overline{\sum_{n=1}^{P} x_n/(f(z) - f(z_n))}}{\sum_{n=1}^{P} x_n/(f(z) - f(z_n))} \overline{f_z(z)}.$$

for almost all $z \in \mathbb{C} \setminus \{\bigcup_{n=1}^{P} K(z_n, \rho_n)\} \setminus \Delta_R$. That is, if

$$F(z) = \sum_{n=1}^{P} x_n \ln \frac{1}{f(z) - f(z_n)}$$

then F satisfies (6.3) for such z. It remains to use the compactness of the class of normalized $Q(z)$-quasiconformal self-mappings of the plane and apply the convergence Theorem 4.1. ∎

Below put for simplicity $P = 1$.

Theorem 6.2: *If additionally*

$$\int\int_{|z-z_1|\leq\delta} \frac{|Q(z) - Q(z_1)|}{|z - z_1|^2} dm_z < \infty, \tag{6.4}$$

and

$$\int\int_{|z|\leq\delta} \frac{|Q(1/z) - Q(\infty)|}{|z|^2} dm_z < \infty, \tag{6.5}$$

then there exist the finite limits

$$\lim_{z\to z_1} \left\{ U(z) - \frac{1}{Q(z_1)} \ln|z - z_1| \right\}, \tag{6.6}$$

$$\lim_{z\to\infty} \left\{ U(z) - \frac{1}{Q(\infty)} \ln|z| \right\}. \tag{6.7}$$

Proof: Let $g(z) = f(z + z_1) - f(z_1)$ and $\phi(w) = w|w|^{Q(z_1)-1}$. Now we can apply the Theorem 3.1 for the composition $\phi \circ g(z)$ because the condition (6.4) implies the convergence of the integral (3.1).

M.Schiffer and G. Schober [27], [28] have proved these theorems using the basic functional

$$\chi = \lim_{r\to 0} \frac{(A(r, z_n, f)/\pi)^{1/2}}{r^{1/Q(z_n)}}$$

over the class of normalized $Q(z)$-quasiconformal mappings of the complex plane $\mathbb{C}$, where $A(r, z_n, f) = \text{Area}(f(|z - z_n| < r))$, and the Dini condition

$$\int_0 \frac{\omega(r)}{r} dr < \infty,$$

on the coefficient $Q(z)$ at the points z_n and ∞, where

$$\omega(r) = \text{ess} \sup_{|z-z_n|\leq r} |Q(z) - Q(z_n)|.$$

The example below shows that O.Teichmüller – H.Wittich – P.Belinskii condition (6.4) does not imply the Dini condition.

Let $z_n = 0$ and $Q(z) = 1 + \alpha(|z|)$ for $z \neq 0$ and $Q(0) = 1$, where $\alpha(r) = 1$ for $r \in [1/n, 1/n + 1/n^3]$, $n = 1, 2, \ldots$ and $\alpha(r) = 0$ for the rest $r > 0$. Then $\omega(r) \equiv 1$ and the Dini condition is violated. However

$$\int\int_{|z|\leq\delta} \frac{|Q(z) - Q(0)|}{|z|^2} dm_z = 2\pi \int_0^\delta \frac{\alpha(r)}{r} dr$$

$$\leq 2\pi \sum_{n=1}^\infty \frac{1}{n^2}.$$

7   On differentiation of one-parameter families of conformal mappings

Now we consider some problems of differentiability with respect to a parameter of the families of univalent analytic functions and give some applications to the geometric function theory.

We need a corollary of the Theorem B.

Corollary 1: *Let ω be a quasiconformal self-mapping of the unit disk D with complex dilatation $\mu = tza(z) + o(t)$, where t is real, $za \in L_\infty$ and $\|o(t)/t\|_\infty \to 0$ as $t \to 0$, normalized with the conditions $\omega(0) = 0$. Then ω admits the representation*

$$\omega = z\left\{1 + tC_D[a](z) - t\overline{C_D[a](1/\bar{z})} + it\gamma\right\} + o(t), \tag{7.1}$$

where γ is a real parameter.

This result follows immediately from the Theorem B if one extends ω to a quasiconformal mapping of the whole plane by the reflection principle and uses the decomposition.

Let now G_t be a family of simply connected domains, $0 \in G_t$, $G_t \neq \mathbb{C}$, depending on a real parameter $t \in T$, $0 \in T$. Let $f(z,t)$ be a conformal mapping of the unit disk D onto G_t with standard normalizations $f(0,t) = 0$, $f'(0,t) > 0$. Suppose that $G_t \to D_0$ as to the kernel. The well-known Caratheodory kernel convergence theorem says that $f(z,t) \to f(z,0)$ as $t \to 0$ locally uniformly in D. We will consider some problems of differentiability of the family $f(z,t)$ with respect to a parameter t.

Historically the first nontrivial result in this directions is due to J.Hadamard. Later, Ch.Lowner, M.Lavrent'ev, M.Schiffer, G.Goluzin, P.Kufarev, W.Hayman, Chr.Pommerenke, N.Lebedev, J.Becker, G.S.Goodman and many others investigated these problems from various view points and proved in this way many fundamental results for the theory of conformal mappings (see [31], p. 156, 183, [32], p. 93 and references therein).

We try to look at this problems from the viewpoint of quasiconformal mapping theory. Let us reformulate the Theorem A in the following way.

Proposition 7.1: *Let $F(z,t)$ be an arbitrary family of quasiconformal mappings of the unit disk D onto G_t and $\omega(z,t)$ be a quasiconformal self-mapping of the disk D with complex dilatation $\mu(z,t) = F_{\bar{z}}(z,t)/F_z(z,t)$. Then the conformal mappings $f(z,t) : D \to G_t$ admit the representation*

$$f(z,t) = F(\bullet, t) \circ \omega^{-1} \circ \phi, \tag{7.2}$$

where ϕ is a suitable Mobius transform of D.

For simplicity let us suppose that $F(0,t) = 0$ and hence ϕ reduces to a rotation. The next simple example explains the main idea of our further considerations.

Let us consider the Jordan domains G_t close to the unit disk D which are bounded by the Jordan starlike curves

$$\Omega(\vartheta, t) = e^{i\vartheta}(1 + tp(e^{i\vartheta})), 0 \leq \vartheta < 2\pi, |t| < t_0,$$

where p is real-valued function and $p \in C^1(\partial D)$. Our problem is to find a conformal mapping $f : D \to G_t$ when t is small and calculate the derivative of f with respect to the parameter t as $t = 0$.

At first one can easily extend p inside the unit disk D from the boundary with the condition that $p \in C^1(\overline{D})$. Indeed, let r_0, $0 < r_0 < 1$, be fixed and put

$$p(z) = \frac{|z| - r_0}{|z|(1 - r_0)} p(e^{i\theta}),$$

if $r_0 \leq |z| \leq 1$, and $p(z) = 0$ for $|z| \leq r_0$. In this way we generate a family of quasiconformal mappings $F : D \to G_t$ for sufficiently small t, which are conformal for $|z| < r_0$ and have the form

$$F(z, t) = z + tzp(z).$$

The complex dilatation of F is

$$\mu(z, t) = \frac{tzp_{\bar{z}}}{1 + tp(z) + tzp_z} = tzp_{\bar{z}} + o(t),$$

where $\|o(t)/t\|_\infty \to 0$ as $t \to 0$.

Using the Corollary 1 we find a quasiconformal self-mapping ω of the unit disk in the form

$$\omega = z \left\{ 1 + tC_D[p_{\bar{z}}](z) - t\overline{C_D[p_{\bar{z}}](1/\bar{z})} + it\gamma \right\} + o(t). \tag{7.3}$$

Then by Proposition 7.1 the conformal mappings $f : D \to G_t$ for sufficiently small t can be presented in the form

$$f(z, t) = F(\omega^{-1}(z, t), t) = z + tz\frac{1}{2\pi i} \int_{\partial D} (\Re ep(\zeta)) \frac{\zeta + z}{\zeta - z} \cdot \frac{d\zeta}{\zeta} + o(t),$$

where $o(t)/t \to 0$ as $t \to 0$ locally uniformly in D.

It is clear that the final result does not depend on the extension of $p(z)$ for $z \in D$.

Theorem 7.1: *The following assertions are equivalent:*

1). There exists a family of quasiconformal mappings $F : D \to G_t$ of the form

$$F(z, t) = f(z, 0) + tzf'(z, 0)p(z) + o(t),$$

where $o(t)/t \to 0$ as $t \to 0$ locally uniformly in U, with complex dilatation

$$\mu(z) = tzp_{\bar{z}}(z) + o(t)$$

where $\|o(t)/t\|_\infty \to 0$ as $t \to 0$;

2). The family $f : U \to D_t$ of the conformal mappings has a derivative with respect to t as $t = 0$

$$\partial f/\partial t|_{t=0} = zf'(z, 0) \left\{ p(z) - C_D[p_{\bar{\zeta}}](z) + \overline{C_D(p_{\bar{\zeta}}](1/\bar{z})} + i\gamma \right\} \tag{7.4}$$

locally uniformly in the disk D, where γ is a suitable real parameter.

Corollary 7.1.1: *Let additionally $p(z) \in C(\overline{D})$ and $p_{\bar{z}} \in C(D)$. Then*

$$\partial f/\partial t|_{t=0} = zf'(z,0)\frac{1}{2\pi i}\int_{\partial D}(\Re ep(\zeta))\frac{\zeta+z}{\zeta-z}\cdot\frac{d\zeta}{\zeta}. \qquad (7.5)$$

Proof: It is easy to calculate that in this case

$$C_{\mathcal{D}}[p_{\bar{\zeta}}](z) = p(z) - \mathcal{K}[p](z)$$

for $z \in D$ and

$$C_{\mathcal{D}}[p_{\bar{\zeta}}](z) = -\mathcal{K}[p](z)$$

for $z \in \mathbb{C} \setminus \overline{D}$, where

$$K[p](z) = \frac{1}{2\pi i}\int_{\partial D}\frac{p(\zeta)}{\zeta-z}d\zeta$$

is the Cauchy integral. Now the formula (7.4) implies (7.5).

Now we give a version of the fundamental Goluzin's variational theorem which is the basic result of the method of interior variations in the class of holomorphic univalent functions in the unit disk ([32], Chapt. 3.4).

Theorem 7.2: *Let f be analytic and univalent in D and $f(0) = 0$. Suppose that*

$$F(z) = f(z) + tzf'(z)p(z) + o(t)$$

where $o(t)/t \to 0$ as $t \to 0$ locally uniformly in D, be holomorphic univalent function for $R = \{r \le |z| < 1\}$ and all sufficiently small t, $|t| < t_0$. Let G_t be the union of $F(R,t)$ and the bounded component of $\mathbb{C} \setminus F(R,t)$. If $f^*(z,t)$, $f^*(0,t) = 0$, maps D onto G_t then

$$f^*(z,t) = f(z) + tzf'(z)\left\{T(z) + \overline{S(1/\bar{z})} + i\gamma\right\} + o(t),$$

where $T(z) + S(z)$ is the Laurent's expansion of $p(z)$ for $r \le |z| < 1$ and $o(t)/t \to 0$ as $t \to 0$ locally uniformly in D.

Proof: For sufficiently small t one can extend F to the quasiconformal mapping of the disk D onto G_t with the complex dilatation $\mu = 0$ for $r \le |z| < 1$ and $\mu = tzp_{\bar{z}}(z) + o(t)$ for other points of D. By the Corollary 1 the quasiconformal self-mapping $\omega(z)$ of the disk D with the same complex dilatation has the representation (7.3) with $a(z) = p_{\bar{z}}(z)$. By Pompeiu's integral formula

$$C_{\mathcal{D}}[p_{\bar{\zeta}}](z) = p(z) - T(z)$$

for $z \in D$ and

$$C_{\mathcal{D}}[p_{\bar{\zeta}}](z) = S(z)$$

for $z \in \mathbb{C} \setminus \overline{D}$. Thus

$$\omega(z,t) = z(1 + t(p(z) - T(z) - \overline{S(1/\bar{z})} + i\gamma)) + o(t).$$

Hence by Proposition 7.1

$$f^*(z,t) = F \circ \omega^{-1} = f(z) + tzf'(z)(T(z) + \overline{S(1/\bar{z})} + i\gamma) + o(t).$$

The Schiffer variation [34] is produced by choosing F in the form $F = \Phi \circ f$ where Φ is a t_0-quasiconformal mapping of the plane $\mathbb{C}$ with complex dilatation concentrating on a small disk near $f(z_0)$.

Another short and remarkable proof of the Goluzin's theorem has been given by Chr.Pommerenke (see [31], p. 186).

8 Asymptotically conformal quasicircles

8.1 Definitions and preliminary results

A Jordan curve $\Gamma \subset \mathbb{C}$ is called a *quasicircle* if it is the image of the unit circle under a quasiconformal mapping of $\mathbb{C}$ (see [5], p.286, [7], 97,98). In 1963 L.Ahlfors [35] gave surprisingly simple geometric characterization of quasicircles. He proved that the curve Γ is quasicircle iff the quantity

$$\gamma \equiv \gamma(w_1, w_2, w) = \frac{|w_1 - w| + |w - w_2|}{|w_1 - w_2|} \tag{8.1}$$

is bounded for all $w_1, w_2 \in \Gamma$ and $w \in \Gamma(w_1, w_2)$, where $\Gamma(w_1, w_2)$ denotes the sub arc of Γ corresponding to $w_1, w_2 \in \Gamma$ with smaller diameter.

Let $\Gamma \subset \mathbb{C}$ be a quasicircle in the complex plane and let $f : D \to \text{int } \Gamma$ denote a conformal mapping of the unit disk $D = \{z : |z| < 1\}$ onto the interior of Γ. By a famous result of L.Ahlfors (see [16], p. 71) f has a quasiconformal extension over the unit circle ∂D. If there exists a quasiconformal extension with complex dilatation $\mu(z)$ such that

$$\text{ess} \sup_{|z| \le t} |\mu(z)| \to 0, \quad t \to 1 + 0, \tag{8.2}$$

then the curve Γ is called *asymptotically conformal* [37], [38]. Chr.Pommerenke and J.Becker proved [37] that this is equivalent to the condition

$$\lim_{|w_1 - w_2| \to 0} \frac{|w_1 - w| + |w - w_2|}{|w_1 - w_2|} = 1 \tag{8.3}$$

uniformly with respect to $w \in \Gamma(w_1, w_2)$.

8.2 On asymptotical homogeneity.

A mapping $f : G \to \mathbb{C}$, $f(0) = 0$, is called conformal at the point $0 \in G$ if

$$f(z) = f'(0)(z + o(|z|)), \quad f'(0) \neq 0.$$

Let the complex dilatation $\mu(z)$ be continuous in the neighborhood of the origin and $\mu(0) = 0$. The quasiconformal mapping

$$f(z) = z(1 - \ln|z|), \quad f(0) = 0,$$

shows that in this case f can be non-conformal at the point $z = 0$. However, as it has been established by Belinskii (see [7], p. 41), in this case

$$f(z) = A(|z|)(z + o(|z|)),$$

where $o(|z|)/|z| \to 0$ as $|z| \to 0$. Moreover, there exists a limit

$$\lim_{\rho \to 0} \frac{A(t\rho)}{A(\rho)} = 1, \quad \forall t > 0.$$

A mapping $f : G \to \mathbb{C}$, $f(0) = 0$, is called *asymptotically homogeneous* at the origin, if

$$\lim_{z \to 0} \frac{f(z\zeta)}{f(z)} = \zeta \tag{8.4}$$

for any $\zeta \in \mathbb{C}$.

One can prove that the asymptotical homogeneity of the quasiconformal mapping f at the origin is equivalent to the conditions

$$f(z) = A(|z|)(z + o(|z|)), \tag{8.5}$$

$$\lim_{\rho \to 0} \frac{A(t\rho)}{A(\rho)} = 1, \quad \forall t > 0, \tag{8.6}$$

where $o(|z|)/|z| \to 0$ as $|z| \to 0$.

If f is asymptotically homogeneous at the origin, then

$$\lim_{r \to 0} \frac{\max_{|z|=r} |f(z)|}{\min_{|z|=r} |f(z)|} = 1,$$

that is the infinitesimal circles centered at the origin are preserved, and as

$$\lim_{|z| \to 0} \frac{|f(z\zeta)|}{|f(z)|} = |\zeta|,$$

therefore the moduli of infinitesimal rings centred at the origin are preserved also. At last, from

$$\lim_{z \to 0} [\arg f(z\zeta) - \arg f(z)] = \arg \zeta$$

for any $\zeta \in \mathbb{C}^* = \mathbb{C} \setminus \{0\}$, it follows that the angles between rays outgoing from the origin in the direction of corresponding points are preserved. And vice versa, if the conditions (8.5), (8.6) hold simultaneously, then the mapping f is asymptotically homogeneous at the origin.

The example

$$f(z) = ze^{i(-\ln|z|)^{1/2}}, \quad f(0) = 0,$$

shows, that under asymptotical homogeneity the radial lines can be transformed to the spirals.

8.3   Some criteria for asymptotical conformality.

In this section we give equivalent conditions of asymptotical conformality for quasicircles.

Theorem 8.1: *A Jordan curve Γ is asymptotically conformal iff there exists a quasiconformal mapping $f : \mathbb{C} \to \mathbb{C}$, $\Gamma = f(\partial D)$, with complex dilatation satisfying one of the following conditions uniformly with respect to $\eta \in \partial D$:*
   1. The dilatation $\mu(z)$ is continuous on ∂D and

$$\lim_{z \to \eta} \mu(z) = 0.$$

   2. On the unit circle $\mu(z)$ is approximately continuous and

$$\mu(\eta) \equiv 0.$$

   3. For all $\eta \in \partial D$

$$\lim_{r \to 0} \frac{1}{r^2} \int\int_{|z-\eta| \leq r} |\mu(z)| dx dy = 0.$$

Corollary 8.1.1: *If for some $\alpha > 0$ there exists the uniform limit with respect to $\eta \in \partial D$*

$$\lim_{t \to 0} \frac{1}{t^2} \int\int_{|z-\eta| \leq t} |\mu(z)|^\alpha dm_z = 0,$$

then Γ is asymptotically conformal.

Corollary 8.1.2: *If for some $\alpha > 0$*

$$\int\int_{|\eta-z| \leq t} \frac{|\mu(z)|^\alpha}{|\eta - z|^2} dm_z$$

converges uniformly with respect to $\eta \in \partial D$ then Γ is asymptotically conformal.

Corollary 8.1.3: *(see [42], p. 111) Let for some $\alpha > 0$*

$$\int_0 k(t)^\alpha \frac{dt}{t} < +\infty,$$

where

$$k(t) = ess \sup_{|1-|z|| \leq t} |\mu(z)|.$$

Then the curve Γ is asymptotically conformal.

Now we give some equivalent conditions of asymptotical conformality in terms of the quasiconformal mapping f.

Theorem 8.2: *A Jordan curve Γ is asymptotically conformal iff there exists a quasiconformal mapping $f : \mathbb{C} \to \mathbb{C}$, $\Gamma = f(\partial D)$ satisfying one of the following conditions uniformly with respect to $\eta \in \partial D$:*

1. *$f(z)$ is asymptotically homogeneous on ∂D.*
2. *If, for each fixed $\delta \geq 0$ $|(z' - \eta)/(z - \eta)| \leq \delta$, then*

$$\lim_{z',z \to \eta} \left\{ \frac{f(z') - f(\eta)}{f(z) - f(\eta)} - \frac{z' - \eta}{z - \eta} \right\} = 0.$$

3. *There exists a limit*

$$\lim_{z \to 0} \frac{f(\eta + z\zeta) - f(\eta)}{f(\eta + z) - f(\eta)} = \zeta, \quad \forall \zeta \in \mathbb{C}.$$

In the Theorem 1 and 2 one can assume in addition that $\mu(z) \equiv 0$ inside or outside D.

The following theorem gives inner and boundary characterization of conformal mappings with asymptotically conformal extensions.

Theorem 8.3: *Let f be a conformal mapping of D onto a Jordan domain $G \subseteq \mathbb{C}$. Then ∂G is the asymptotically conformal curve iff one of the following conditions is fulfilled*

1) f is uniformly conformal in D, that is

$$\lim_{z \to \eta} \frac{f(\eta) - f(z)}{f'(\eta)(\eta - z)} = 1 \tag{8.7}$$

uniformly with respect to $\eta \in D$.

2) Under the condition $|(z' - \eta)/(z - \eta)| \leq \delta$ for each fixed $\delta > 0$,

$$\lim_{z',z \to \eta} \left\{ \frac{f(z') - f(\eta)}{f(z) - f(\eta)} - \frac{z' - \eta}{z - \eta} \right\} = 0;$$

uniformly with respect to $\eta \in D$.

3) Under the condition $|(z' - \eta)/(z - \eta)| \leq \delta$ for each fixed $\delta > 0$

$$\lim_{z',z \to \eta} \left\{ \frac{f(z') - f(\eta)}{f(z) - f(\eta)} - \frac{z' - \eta}{z - \eta} \right\} = 0;$$

as $z, z' \in \overline{D}$ uniformly with respect to $\eta \in \partial D$.

4) Under the condition $|(z' - \eta)/(z - \eta)| \leq 1$

$$\lim_{z',z \to \eta} \left\{ \frac{f(z') - f(\eta)}{f(z) - f(\eta)} - \frac{z' - \eta}{z - \eta} \right\} = 0;$$

as $z, z' \in \partial D$ uniformly with respect to $\eta \in \partial D$.

The conditions of type (8.7) have been investigated before by D.Gaier, Chr.Pommerenke, S.Warschawski and others (see, for example, [38], [39]).

9   Boundary properties of conformal mappings

This section is devoted to the problem of finding sufficient conditions for rectifiability and smoothness of quasicircles. We will formulate such conditions in terms of complex dilatation. We will use the representation theorem and the mentioned above result of O.Teichmüller, H.Wittich and P.Belinskii on the regularity of a quasiconformal mapping at a point.

9.1   Rectifiability of quasicircles.

It's well-known that quasicircles need not be rectifiable, even if they are asymptotically conformal (see [7], p.42). We deal with the problem of finding sufficient conditions for rectifiability of quasicircles.

Let H^1 be the Hardy space of analytic functions in the unit disk D such that

$$\lim_{r \to 1} \int_0^{2\pi} |f(re^{i\theta})|d\theta < \infty. \tag{9.1}$$

By F.Riesz theorem (see [40], p.44) $f' \in H^1$ implies rectifiability of the curve $\Gamma = f(\partial D)$.

Theorem 9.1: *Let $f : D \to \mathbb{C}$ be a conformal mapping. If f has a quasiconformal extension with the complex dilatation $\mu(z)$ such that*

$$\int\int_{1<|z|\leq R} \frac{|\mu(z)|^2}{(1-|z|)^2} dm_z < +\infty \tag{9.2}$$

then $f' \in H^1$ in the unit disk D.

Corollary 9.1.1: *Under the hypothesis of Theorem 9.1 the boundary curve $\Gamma = f(\partial D)$ is rectifiable.*

Corollary 9.1.2: *If for some $0 < \alpha \leq 2$*

$$\int\int_{1<|z|\leq R} \frac{|\mu(z)|^\alpha}{(1-|z|)^2} dm_z < +\infty$$

then Γ is rectifiable.

Proof of the Theorem 9.1. Let $D_R = \{z : |z| < R\}$ and F be the normal solution of Beltrami's equation with the complex dilatation $\nu(z) = \mu(z)$, $z \in D_R$, and $\nu(z) = 0$ for $z \in \mathbb{C} \setminus D_R$ and $\|\nu\|_\infty \leq q < 1$. Inside D_R we have $f(z) = A(F(z))$, where $A(w)$ is some conformal mapping of $F(D_R)$ to $f(D_R)$. Thus, $f' \in H^1$ iff $F' \in H^1$. By Theorem B

$$F_z - 1 = \mathcal{H}\nu + \mathcal{H}(\nu\mathcal{H}\nu) + \ldots = \mathcal{H}(\nu \mathcal{F}_z).$$

Consequently, the condition (9.1) for F is equivalent to

$$\lim_{r \to 1-0} \int_{|z|=r} |\mathcal{H}\nu \mathcal{F}_z)||dz| < +\infty.$$

Further,

$$\|F_z - 1\|_2 \leq \frac{q}{1-q}(\pi(R^2-1))^{1/2} = c \cdot \frac{q}{2} \tag{9.3}$$

because $\mathcal{H} : L^2(\mathbb{C}) \to L^2(\mathbb{C})$ and $\|\mathcal{H}(\phi)\|_2 = \|\phi\|_2$ for every $\phi \in L^2(\mathbb{C})$.

The above integral equals

$$I = \frac{r}{\pi} \int_0^{2\pi} d\theta \left| \int \int_{D_R} \frac{\mu(\zeta) F_\zeta(\zeta)}{(\zeta - z)^2} dm_\zeta \right|$$

$$\leq \frac{r}{\pi} \int \int_{D_R} |\mu(\zeta) F_\zeta(\zeta)| dm_\zeta \int_0^{2\pi} \frac{d\theta}{|z - \zeta|^2}.$$

Evaluating the Poison integral by the inequality (9.3) and using Holder inequality we derive

$$I \leq 2 \int \int_{D_R} \frac{|\mu(\zeta) F_\zeta(\zeta)|}{|\zeta| - 1} dm_\zeta \leq c \left(\int \int_{D_R} \frac{|\mu(\zeta)|^2}{(1 - |\zeta|)^2} dm_\zeta \right)^{1/2}.$$

Now the condition (9.1) implies $f' \in H^1$.

It is possible to derive from Theorem 9.1 some results on asymptotically conformal selfmappings, say of D, which is due to L.Carleson [41].

Corollary 9.1.3: *Let ω be a quasiconformal mapping of D onto itself with the complex dilatation $\mu(z)$ satisfying one of the conditions (9.1). Then the boundary values $\omega(\eta)$, $\eta \in \partial D$, yield, for $0 < \alpha \leq 2$, an absolutely continuous homeomorphism of the unit circle.*

9.2 Smoothness of quasicircles.

This section is devoted to the problem of finding sufficient conditions for smoothness of quasicircles.

Theorem 9.2: *Let $f : D \to \mathbb{C}$ be a conformal mapping. If f has a quasiconformal extension with the complex dilatation $\mu(z)$ such that*

$$\int \int_{|\eta - z| \leq r} \frac{|\mu(z)|}{|\eta - z|^2} dm_z \tag{9.4}$$

converges uniformly in the unit circle ∂D, then $g = \ln f'$ is continuous in $\overline{D}$.

Corollary 9.2.1: *Under the hypothesis of Theorem 9.2 the boundary curve $\Gamma = f(\partial D)$ is smooth.*

Corollary 9.2.2: *If for some $0 < \alpha \leq 1$*

$$\int \int_{|\eta - z| \leq r} \frac{|\mu(z)|^\alpha}{|\eta - z|^2} dm_z \tag{9.5}$$

converges uniformly in the unit circle ∂D, then Γ is smooth.

Corollary 9.2.3: *If for some $r > 1$*

$$\int \int_{1 < |z| \leq r} \frac{|\mu(z)|}{(1 - |z|)^2} dm_z < +\infty \tag{9.6}$$

then Γ is smooth.

Corollary 9.2.4: *If for some $0 < \alpha \leq 1$*

$$\int\int_{1<|z|\leq r} \frac{|\mu(z)|^\alpha}{(1-|z|)^2} dm_z \tag{9.7}$$

converges, then Γ is smooth.

Corollary 9.2.5: *(see [42], p. 112) Let for some $0 < \alpha \leq 1$*

$$\int_0 k(t)^\alpha \frac{dt}{t} < +\infty,$$

where

$$k(t) = ess \sup_{1<|z|\leq 1+t} |\mu(z)|. \tag{9.8}$$

Then the curve Γ is smooth.

Corollary 9.2.6: *Let for some $\delta > 0$*

$$\mu(z) = O((1-|z|)^\delta) \quad as \quad |z| \to 1+. \tag{9.9}$$

Then Γ is smooth.

Note that J.Anderson and A.Hinkkanen have proved rectifiability of Γ under the same condition on the complex dilatation.

Proof of the Theorem 9.2. By definition of the uniform convergence of the integral there exists a function $\Delta(r) \to 0$, as $r \to 0$ such that

$$\int\int_{|\eta-z|\leq r} \frac{|\mu(z)|}{|\eta-z|^2} dm_z \leq \Delta(r), \forall \eta \in \mathbb{C} : |\eta| = 1.$$

If

$$w(z) = w_\eta(z) = f(z+\eta) - f(\eta),$$

then $w(0) = 0$, and by Theorem 3.1

$$|w(z)/z - w'(0)| < |w'(0)|\varepsilon(|z|),$$

where $\varepsilon(|z|)$ depends only on Q, Δ and $\varepsilon(|z|) \to 0$ as $|z| \to 0$. Moreover

$$|w'(0)| \leq \max_{|z|=1} |w(z)|e^{\Delta^*(1)},$$

where $\Delta^*(1) = (1+Q)\Delta(1)/\pi$.
Since

$$\max_{|z|=1} |w_\eta(z)| \leq 2\max_{|z|\leq 2} |f(z)| = c_1,$$

for any $|\eta| = 1$, then

$$\left| \frac{f(\eta + z) - f(\eta)}{z} - f'(\eta) \right| \leq c_2 \varepsilon(|z|),$$

where $c_2 = c_1 e^{\Delta^*(1)}$, and c_2 not depends on η, $|\eta| = 1$.

Hence $f'(\eta)$ is continuous on $|\eta| = 1$, because the quotients

$$\frac{f(\eta + z) - f(\eta)}{z}$$

are continuous.

References

1. H. Grötzsch Über die Verzerrung bei schlichten nichtkonformen Abbildungen und über eine damit zusammenhangende Erweiterung des Picardschen Satzes *Ber. Verh. Sachs. Acad. Wiss. Leipzig* 80 1928

2. M.A. Lavrentieff Sur une classe de representations continues, *Rec. Math.*, 48, 1935

3. L. Ahlfors On quasiconformal mappings *J. Analyse Math.* 3 1954 pp. 1–58 and 207–208

4. O. Teichmüller Untersuchungen über konforme und quasikonforme Abbildung *Deutsche Mathematik* 3 1938 pp. 621–678

5. O. Lehto, K. Virtanen *Quasikonforme Abbildungen Berlin etc.: Springer–Verlag 1972*

6. I.N. Vekua *Generalized analytic functions Fizmatgiz Moskva 1959 Russian*

7. P. P. Belinskii *General properties of quasiconformal mappings Nauka Novosibirsk 1974 Russian*

8. O. Lehto *Univalent Functions and Teichmüller Spaces, Graduate texts in mathematics; 109 Springer–Verlag 1987*

9. B. Bojarski Homeomorphic solutions of the Beltrami systems *Doklady Acad. Nauk USSR* 102, N 4 1955 pp. 661–664 Russian

10. B. Bojarski On solutions of elliptic systems in the plane *Doklady Acad. Nauk USSR* 102, N 5 1955 pp. 871–874 Russian

11. B. Bojarski Generalized solutions of the PD system of the first order and elliptic type with discontinuous coefficients *Matem. Sbornik* 43(85) 1957 pp. 451–503

12. B. Bojarski General properties of solutions of elliptic systems in the plane *Confer. of function theory, Moscow 1957 Russian*

13. B. Bojarski Old and new on Beltrami equation *Functional Analytic Methods in Complex Analysis and Applications to Partial Differential Equations World Scientific Singapore–New Jersey–London–Hong Kong 1990*

14. L. Ahlfors Conformality with respect to Riemannian metrics *Ann. Acad. Sci. Fenn. Ser. A. 1 206 1955* pp. 1–22

15. I.V. Vekua The problem of reducing differential forms of elliptic type to canonical form and the generalized Cauchy–Riemann System *Doklady Acad. Nauk USSR* 100 1955 pp. 197–200 Russian

16. L. Ahlfors *Lectures on quasiconformal mappings Princeton, N.,J. Van Nostrand 1966*

17. O. Teichmüller Untersuchungen uber konforme und quasikonforme Abbildung *Deutsche Math.* 3 1938 pp. 621–678

18. H. Wittich Zum Beweis eines Satzes über quasikonforme Abbildungen *Math. Zeitschr.* 51 1948 pp. 275–288

19. L. Bers On a theorem of Mori and the definition of quasiconformality *Trans. Amer. Math Soc.* 84 1957 pp. 78–84

20. K. Strebel Ein Konvergenzsatz für Folgen quasikonforme Abbildungen *Comment. Math. Helv* 44 1969 pp. 469–475

21. V.I. Ryazanov Some questions of convergence and compactness for quasiconformal mappings *Amer. Math. Soc. Trasl. (2)* 131 1986 pp. 7–19

22. F. Gehring, O. Lehto On the total differentiability of functions of a complex variable *Ann. Acad. Sci. Fenn. A 1* 272 1959 pp.

23. V.Ya. Gutlyanskii, V.I. Ryazanov On quasiconformal mappings with restrictions of integral type on the Lavrent'ev characteristic *Soviet Math. Dokl.* 36, N 3 1988 pp. 456–459

24. G. Schober *Univalent functions – Selected Topics Springer Berlin etc.: 1975*

25. S. L. Kruskal *Quasiconformal mappings and Riemannian surfaces Nauka Novosibirsk 1975*

26. S. L. Kruskal, R. Kuhnau *Quasiconformal mappings–new methods and applications Nauka Novosibirsk 1984 Russian*

27. M. Schiffer, G. Schober Representation of fundamental solutions for generalized Cauchy–Riemann equations by quasiconformal mappings *Ann. Acad. Sci. Fenn. Ser. A 1* 2 1976 pp. 501–531

28. M. Schiffer, G. Schober A variational method for general families of quasiconformal mappings *Journal D'analyse Math.* 34 1978 pp. 240–264

29. V.Ya. Gutlyanskii, V.I. Ryazanov On representation of the solutions with singularities for generalized Cauchy–Riemann systems *Dokl. Akad. Nauk SSSR* 44, N 2 1992 pp. 456–459

30. V.Ya. Gutlyanskii The product of conformal radii of nonoverlapping domains. *Doklady Akademii Nauk Ukr.SSR, Ser. A* 4 1977 pp. 44–48 See: Ten papers on complex analysis, AMS Translations, Ser.2 ,Vol. 122 (1984),p. 65-69.

31. Ch. Pommerenke *Univalent functions Vandenhoeck u. Ruprecht Gottingen 1975*

32. G.M. Goluzin *Geometric function theory of complex variable Nauka Moscow 1966 Russian*

33. Lowner Ch. *Charles Loewner, Collected Papers Boston – Basel Birkhauser 1988*

34. M. Schiffer A method of variations within the family of simple functions *Proc. London Math. Soc. (2)* 44 1938 pp. 432–449

35. L. Ahlfors Quasiconformal reflections *Acta Math.* 109 1963 pp. p. 291–301

36. L.Ahlfors, L. Bers Riemann's mapping theorem for variable metrics *Ann. Math.* 72, N 2 1960 pp. p.385–404

37. J. Becker, Chr.Pommerenke Über die quasikonforme Fortsetzung schlichten Funktionen *Math. Zeit.* 161, N1 1978 pp. p. 69–80

38. Chr. Pommerenke On univalent functions, Bloch functions and VMOA *Math. Ann.* 236 1978 pp. p. 199–208

39. Chr. Pommerenke, S.E. Warschawski On the quantitative boundary behavior

of conformal maps *Comment. Math. Helv.* 57 1982 pp. p. 107–129

40. P.L. Duren *Theory of H^p spaces Academic Press New York 1970*

41. L. Carleson On mappings conformal at the boundary *J. Analyse Math.* 19 1967 pp. p. 1–13

42. J. Becker On asymptotically conformal extension of univalent functions *Complex Variables* 9 1987 pp. p. 109–120

Normality Critereon and Singular Directions[1]

Huai-hui Chen

Department of Mathematics, Nanjing Normal University
Nanjing 210024, Jiangsu Province, P. R. CHINA

Xin-hou Hua

Institute and Department of Mathematics, Nanjing University
Nanjing 210008, Jiangsu Province, P. R. CHINA

Abstract: We give a criterion for the normality of a family of meromorphic functions or holomorphic functions which generalized a theorem due to P. Lappan and, as its applications, prove some theorems on singular directions of a meromorphic function or an entire function which strengthen or complement the properties of classical Julia direction and Borel direction so that a function f not only assumes every complex number b near the direction infinitely often, but also has comparatively big derivatives at zeros of $f - b$, the number of exceptional values b depends on the order of the involved derivatives.

1 Introduction

From the second fundamental inequality of Nevanlinna one can derive the following theorem (see [2], p. 35).

Theorem A: *A meromorphic function f in the plane reduces to a constant if f has $a_i(i = 1, ..., [\frac{2}{k}] + 3)$, finite or infinite, as its multiple values of order at least $k + 1$. Here $k = 1, 2$ or 3, a_i are mutually distinct. For an entire function f, the number $[\frac{2}{k}] + 3$ is replaced by $[\frac{1}{k}] + 2$.*

When $k = 1$ in Theorem A, we have

Theorem B: *A meromorphic function f in the plane reduces to a constant if f has five distinct multiple values.*

In 1974, Lappan [1] proved a criterion for normality of a function in the unit disk, which has a similar version for a family of functions: *a family of meromorphic functions is normal if every function in the family has five distinct multiple values which is independent of the functions.* A slight modification of

---

[1]Research was supported by the National Natural Science Foundation of China.

Proceedings of the International Conference on Complex
Analysis at the Nankai Institute of Mathematics, 1992 pp. 34-40

his proof provides an improvement of this criterion, which he presented in a lecture recently.

Theorem C: *A family of meromorphic functions is normal if there exist five distinct complex numbers $a_i(i = 1, ..., 5)$, finite or infinite, and a positive number M such that for every function f in the family, $f(z) = a_i(1 \leq i \leq 5)$ implies $f^{\#}(z) \leq M$, where $f^{\#}$ is the spherical derivative of f.*

The proof of Theorem C is based on Theorem B and a theorem of Zalcman [3] formulated as follows:

Theorem D: *If a family of meromorphic functions on a domain D is not normal at a point $z_{\circ} \in D$, then there exist a sequence $f_n \subset F$, a sequence $z_n \to z_{\circ}$ and a positive sequence $\rho_n \to 0$ such that $f_n(z_n + \rho_n \zeta)$ converges spherically and uniformly to a nonconstant meromorphic function $g(\zeta)$.*

In this paper, we give a generalization of Theorem C and, as its applications, prove some theorems which strengthen or complement the properties of classical Julia direction and Borel direction.

2 A generalization of Theorem C and its applications

Theorem 1: *A family of meromorphic functions is normal if there exist distinct values $a_i(i = 1, ..., [\frac{2}{k}] + 3)$, finite or infinite, and a positive number M such that, for every function f in the family, $f(z) = a_i \neq \infty$ implies $|f^{(j)}(z)| \leq M(j = 1, ..., k)$, and $f(z) = a_i = \infty$ implies $|(\frac{1}{f})^{(j)}(z)| \leq M(j = 1, ..., k)$. For a family of holomorphic functions, the number $[\frac{2}{k}] + 3$ can be replaced by $[\frac{1}{k}] + 2$. Here $k = 1, 2,$ or 3.*

Proof: Suppose that the family is not normal at a point z_0. Then Theorem D asserts the existence of sequences z_n, ρ_n and f_n, such that $g_n(\zeta) = f_n(z_n + \rho_n \zeta)$ converges to a nonconstant meromorphic function $g(\zeta)$. If $g(\zeta_{\circ}) = a_i \neq \infty$, by Hurwitz theorem there exist a sequence ζ_n such that $g_n(\zeta_n) = f_n(z_n + \rho_n \zeta_n) = a_i$ $(n = 1, 2, ...)$, $\zeta_n \to \zeta_{\circ}$. So,

$$g^{(j)}(\zeta_{\circ}) = \lim_{n \to \infty} g_n^{(j)}(\zeta_n) = \lim_{n \to \infty} \rho_n^j f_n^{(j)}(z_n + \rho_n \zeta_n) = 0, \quad j = 1, ..., k,$$

for $|f_n^{(j)}(z_n + \rho_n \zeta_n)| \leq M$. When $g(\zeta_{\circ}) = a_i = \infty$, we have $(\frac{1}{g})^{(j)}(\zeta_{\circ}) = 0(j = 1, ..., k)$ by the same reasoning. Hence, $g(\zeta)$ has $a_i(i = 1, ..., [\frac{2}{k}] + 3)$ as its multiple values of order at least $k + 1$. In virtue of Theorem A, $g(\zeta)$ reduces to a constant, a contradiction. This proves Theorem 1. ∎

Theorem C is the special case "$k = 1$" of Theorem 1.

Remark 1: *The number $[\frac{2}{k}] + 3$ in Theorem 1 is best possible.*

Consider the Weierstrass elliptic function which satisfying the following equation

$$\wp'(z) = 4\wp^3(z) - 28\wp(z) - 24 = 4(\wp(z) + 1)(\wp(z) + 2)(\wp(z) - 3).$$

If $\wp(z) = -1$ or $\wp(z) = -2$ or $\wp(z) = 3$, then $\wp'(z) = 0$. Thus, $-1, -2$ and 3 are 2-multivalues of $\wp(z)$. Furthermore, if $z_\circ$ is a pole of $\wp(z)$ with order m, then $2(m+1) = 3m$, which results in m=2. Thus, ∞ is also a 2-multivalue of $\wp(z)$. Let

$$f_n(z) = \wp(nz), \quad n = 1, 2, \ldots$$

From the above discussion we know that $-1, -2, 3$ and ∞ are multivalues of $f_n(z)$ for all n. Let $F = \{f_n(z)\}$, then F satisfies all the assumptions of Theorem 1 for $a_1 = -1$, $a_2 = -2$, $a_3 = 3$, $a_4 = \infty$, $k = 1$ and $M = 0$. However, the family F is not normal because $\wp(z)$ is of order 2.

The function which conformally maps an equilateral triangle onto the upper half plane is an example for $k = 2$, and $(\wp + 1)^2$ is an example for $k = 3$.

The first application of Theorem 1 is an improvement of Theorem A.

Theorem 2: *If $f(z)$ is a meromorphic function in the plane which has the property that*

$$\varlimsup_{r \to \infty} \frac{T(r, f)}{(\log r)^2} = \infty, \tag{1}$$

$k = 1, 2,$ or 3, then for every complex number a (finite or infinite) with $[\frac{2}{k}] + 2$ possible exceptions, there exists a sequence $z_n \to \infty$ such that

$$f(z_n) = a \quad (n = 1, 2, \ldots) \tag{2}$$

$$\sum_{j=1}^{k} \lim_{n \to \infty} |z_n|^j |f^{(j)}(z_n)| = \infty \tag{3}$$

or, when $a = \infty$,

$$\sum_{j=1}^{k} \lim_{n \to \infty} |z_n|^j |(\frac{1}{f})^{(j)}(z_n)| = \infty. \tag{4}$$

Proof: Set $f_n(z) = f(nz)$, $n = 1, 2, \ldots$. It is well-known that under the hypothesis (1), this family is not normal in $|z| > 1$. Suppose that there are $[\frac{2}{k}] + 3$ exceptions, namely $a_i(i = 1, \ldots, [\frac{2}{k}] + 3)$, then there exists $M > 0$ such that $f(z) = a_i(1 \leq i \leq [\frac{2}{k}] + 3)$ imply $|z|^j |f^{(j)}(z)| \leq M$ or, when $a_i = \infty$, $|z|^j |(\frac{1}{f})^{(j)}(z)| \leq M(j = 1, \ldots, k)$. Hence, $|z| > 1$ and $f_n(z) = a_i$ implies

$$|f_n^{(j)}(z)| = n^j |f^{(j)}(nz)| \leq |nz|^j |f^{(j)}(nz)| \leq M$$

or, when $a_i = \infty$, $|(\frac{1}{f_n})^{(j)}(z)| \leq M(j = 1, \ldots, k)$. Therefore, the family $f_n(z)$ $(n = 1, \ldots, |z| > 1)$ satisfies the assumption of Theorem 1, consequently, it is normal. This is impossible. The theorem is proved.   ∎

In fact, $f(nz)(n = 1, 2, ...)$ is not normal at a point $z_o = r_o e^{i\theta_o}$, so we have proved actually that the sequence z_n can be choosed near the half line $\arg z = \theta_o$.

Theorem 3: *For a meromorphic function satisfying* (1), *there exists a half line* $arg z = \theta_o$, *such that for every complex number a (finite or infinite) with* $[\frac{2}{k}] + 2$ *possible exceptions, there is a sequence* $z_n \to \infty$, *such that* $\arg z_n \to \theta_o$, *and* (2), (3) *(or* (4), *if* $a = \infty$) *are satisfied. Here,* $k = 1, 2$ *and* 3 *and* θ_o *is independent of* k.

The half line $\arg z = \theta_o$ is the classical Julia direction. Let $k = 3$ in Theorem 3, then only 2 exceptions are possible. So, Theorem 3 is a strengthening of the property of a Julia direction.

We know that $f_n(z) = f(nz)$ $(n = 1, 2, ...)$ is not normal in $|z| > 1$ for transcendental entire function f(z). So, the same reasoning proves the following theorem.

Theorem 4: *For a transcendental entire function* $f(z)$, *there exists a half line* $\arg z = \theta_o$, *such that for every finite complex number a, with* $[\frac{1}{k}] + 1$ *possible exceptions, there is a sequence* $z_n \to \infty$, *such that* $arg z_n \to \theta_o$, *and* (2) *and* (3) *are satisfied. Here,* $k = 1$ *and* 2, *and* θ_o *is independent of* k.

When $k = 2$, Theorem 4 strengthens the property of the Julia directions of a transcendental entire function.

3 Further results

Now we consider a meromorphic function whose characteristic function has more rapid growth than (1) or has a finite and positive order.

Theorem 5: *If* $f(z)$ *is a meromorphic function in the plane which has the property that*

$$\overline{\lim_{r \to \infty}} \frac{T(r, f)}{(\log r)^{2+\alpha}} = \infty, \quad \alpha > 0, \tag{5}$$

then there exists a half line $\arg z = \theta_o$ *such that for every complex number (finite or infinite), with* $[\frac{2}{k}] + 2$ *possible exceptions, there is a sequence* $z_n \to \infty$ *such that* $\arg z_n \to \theta_o$, $f(z_n) = a(n = 1, 2, ...)$, *and*

$$\sum_{j=1}^{k} \lim_{n \to \infty} \{ \frac{|z_n|}{(\log |z_n|)^{\frac{\alpha}{2}}} \}^j |f^{(j)}(z_n)| = \infty$$

or, when $a = \infty$,

$$\sum_{j=1}^{k} \lim_{n \to \infty} \{ \frac{|z_n|}{(\log |z_n|)^{\frac{\alpha}{2}}} \}^j |(\frac{1}{f})^{(j)}(z_n)| = \infty.$$

Here, $k = 1, 2$ and 3, and $\theta_\circ$ is independent of k.

Proof: There exists a sequence $z_n' \to \infty$ such that

$$\frac{|z_n'|}{(\log |z_n'|)^{\frac{\alpha}{2}}} f^{\#}(z_n') \to \infty. \tag{6}$$

Otherwise, we have

$$f^{\#}(z) \le M (\log |z|)^{\frac{\alpha}{2}} \frac{1}{|z|}, \quad |z| \ge r_\circ,$$

$$\begin{aligned}
S(r) &= \frac{1}{\pi} \int_{|z| \le r} f^{\#}(z)^2 \, dx \, dy \\
&= \frac{1}{\pi} \int_0^r \int_0^{2\pi} f^{\#}(\rho e^{i\theta})^2 \rho \, d\rho \, d\theta \\
&\le 2M^2 \int_{r_\circ}^r \frac{(\log \rho)^\alpha}{\rho} \, d\rho + S_\circ, \\
&= S_\circ' + \frac{2M^2}{1+\alpha} (\log r)^{1+\alpha},
\end{aligned}$$

$$\begin{aligned}
T(r, f) &= \int_0^{r_\circ} \frac{S(\rho)}{\rho} \, d\log \rho \\
&\le T_\circ + \int_{r_\circ}^r \frac{2M^2}{1+\alpha} (\log \rho)^{1+\alpha} \, d\log \rho \\
&= T_\circ' + \frac{2M^2}{(1+\alpha)(2+\alpha)} (\log r)^{2+\alpha},
\end{aligned}$$

this contradicts (5).

Let $\arg z_n' \to \theta_\circ$, we conclude that $\arg z = \theta_\circ$ is the half line we want to find. To prove this, suppose that there exist $\epsilon > 0$, $M > 0$, and $a_i (i = 1, ..., [\frac{2}{k}] + 3)$ such that $|\arg z - \theta| \le \epsilon$, $f(z) = a_i$ $(1 \le i \le [\frac{2}{k}] + 3)$ implies

$$\{\frac{|z|}{(\log |z|)^{\frac{\alpha}{2}}}\}^j |f^{(j)}(z)| \le M \quad (j = 1, ..., k) \tag{7}$$

or, when $a_i = \infty$,

$$\{\frac{|z|}{(\log |z|)^{\frac{\alpha}{2}}}\}^j |(\frac{1}{f})^{(j)}(z)| \le M \quad (j = 1, ..., k). \tag{8}$$

Set

$$\rho_n' = \frac{|z_n'|}{(\log |z_n'|)^{\frac{\alpha}{2}}},$$

$$f_n(\zeta) = f(z_n' + \rho_n' \zeta), \quad |\zeta| < 1, \quad n = 1, 2, ...$$

Because of (6),
$$f_n^{\#}(0) = \rho_n' f^{\#}(z_n') \to \infty,$$
consequently, by Marty's criterion, $f_n(\zeta)$ $(n = 1, 2, ...)$ is not normal at $\zeta = 0$. On the other hand, note that

$$\rho_n'/|z_n'| \to 0,$$

it follows from (7) or (8) that $|\zeta| < 1$ and $f(\zeta) = a_i$ $(0 \le i \le [\frac{2}{k}] + 3)$ imply

$$|f_n^{(j)}(\zeta)| = \rho_n'^j |f^{(j)}(z_n' + \rho_n'\zeta)| \le M' \quad (j = 1, ..., k)$$

or, when $a_i = \infty$,

$$|(\frac{1}{f_n})^{(j)}(\zeta)| = \rho_n'^j |(\frac{1}{f})^{(j)}(z_n' + \rho_n'\zeta)| \le M' \quad (j = 1, ..., k)$$

for a sufficiently large n. So, Theorem 1 asserts the normality of the family $f_n(\zeta)(n = 1, 2, ...)$, a contradiction. This completes the proof of Theorem 5. ∎

As to functions of positive and finite order, we have

Theorem 6: *If $f(z)$ is a meromorphic functiopn of order λ $(0 < \lambda < \infty)$, then there exists a half line $\arg z = \theta_\circ$ such that for every complex number a (finite or infinite), with $[\frac{2}{k}] + 2$ possible exceptions, there is a sequence $z_n \to \infty$ with the property that $\arg z_n \to \theta_\circ$, $f(z_n) = a(n = 1, 2, ...)$, and*

$$\sum_{j=1}^{k} \lim_{n\to\infty} |z_n|^{j(1-\frac{\lambda}{2})+\epsilon_n} |f^{(j)}(z_n)| \to \infty$$

or, when $a = \infty$,

$$\sum_{j=1}^{k} \lim_{n\to\infty} |z_n|^{j(1-\frac{\lambda}{2})+\epsilon_n} |(\frac{1}{f})^{(j)}(z_n)| \to \infty.$$

Here, $k = 1, 2$ and 3, $\theta_\circ$ is independent of k, $\epsilon_n \to 0$ is independent of a.

The proof of Theorem 6 is the same as Theorem 5. At this moment, there exists a sequence $z_n' \to \infty$ such that

$$|z_n'|^{1-\frac{\lambda}{2}+\epsilon_n'} f^{\#}(z_n') \to \infty, \tag{9}$$

where $\epsilon_n' \to 0$. Then, by considering the family of functions

$$f_n(\zeta) = f(z_n' + |z_n'|^{1-\frac{\lambda}{2}+\epsilon_n'}\zeta), \quad |\zeta| < 1, \quad n = 1, 2, ...,$$

we can completes the proof of Theorem 6. It is easy to see that for a Borel direction $\arg z = \theta_\circ$ of the function f, one can find a sequence z_n' satisfying (9) and $\arg z_n' \to \theta_\circ$. Therefore a Borel direction possesses the property in Theorem 6.

Remark 2: *The examples in Remark 1 also indicate the number $[\frac{2}{k}]+2$ of exceptions in above theorems is best possible. For a transcendental entire function the number $[\frac{1}{k}]+1$ of exceptions in Theorem 4 is also best as shown by e^z and $\sin z$.*

Remark 3: *Weierstrass elliptic function $\wp(z)$ has order 2, and $\wp'(z)$ has same value at points where $\wp$ assumes a fixed number. This means that the exponent $j(1-\frac{\lambda}{2})$ in Theorem 6 is sharp. However, the functions e^z and $\sin z$ inspire us to ask: Can the index $j(1 - \frac{\lambda}{2})$ in Theorem 6 be replaced by $j(1 - \lambda)$ for an entire function of order λ ? There is another interesting question: May we expect that the sequence z_n in Theorem 6 has index of convergence λ?*

References

1. Lappan, P. A criterion for a meromorphic function to be normal *Comment. Math. Helv.* 49 1974 pp. 492-495

2. Yang, L. Value distribution theory and its new researches *Science Press Beijing 1982*

3. Zalcman L. A heuristic principle in complex function theory *Amer. Math. Monthly* 82 1975 pp. 813-817

Keywords: Meromorphic function, normal family, singular direction.
AMS classification: 30D45, 30D35

On the Dynamics of Analytic Endomorphisms of the Circle [1]

Guizhen Cui

Department of Mathmatics, Peking University
Beijing 100871, P. R. China

Abstract: By the linearization representation of the analytic endomorphisms of the circle, we prove that a topological conjugacy between two analytic endomorphisms of the circle can be modified to real analytic if they have the equal eigenvalues along corresponding cycles

Suppose f is an analytic endomorphism of the unit circle S^1 with no critical point and has degree $d \geq 2$. f is called expanding if there is a continuous measure $\varphi(\theta)d\theta$ on S^1 such that the norm of the tangent map of f is bigger than one, i.e., $f'(\theta)\frac{\varphi(f(\theta))}{\varphi(\theta)} > 1$. For an expanding map $f : S^1 \to S^1$, there is an isomorphism $S : R \to R$ satisfies

$$S^d(t) = \delta S(\frac{t}{\delta})$$

which is formally identical to the Cvitanovic-Feigenbaum equation(CFE), such that the analytic conjugate class of f uniquely determines the affine conjugate class of S. In this paper, a necessary and sufficient condition was given for which solution of CFE corresponds to an expanding map, which is a question proposed in *[2]*.

D. Sullivan and M. Shub *[4]* has given an analytic classification for expanding maps, i.e., two expanding maps are analytically conjugated if they have the same eigenvalues along corresponding cycles. Applying the relation of f and the solution S of CFE. I gave an analytic classification for non-critical-point maps.

1 Lineariztion of the Endomorphisms

Suppose $F : R \to R$ is an isomorphism satisfies condition $(*)$:

$(*)$ 1. $F(0) = 0, F'(0) = \delta > 1$;

    2. $F(t + 1) = F(t) + d$ and

    3. $F^i(t) = t + 1$ has unique solution $t_i \in (0, 1)$
      for any $i > 0$ with $\{t_i\}$ decreasing to zero.

[1]Research was supported by National NSFC

Proceedings of the International Conference on Complex
Analysis at the Nankai Institute of Mathematics, 1992 pp. 41-44
©INTERNATIONAL PRESS 1994

Such isomorphisms F_1 and F_2 are called analytic conjugated if there is an isomorphism $\Phi : R \to R$ satisfies:

$$\Phi(t+1) = \Phi(t) + 1 \quad \text{and}$$

$$F_1\Phi = \Phi F_2$$

Lemma 1: *For non-critical-point map f, there is an isomorphism $F : R \to R$ satisfies $(*)$ such that F is a lifting of f^n for some $n > 0$ (by f^n we denote the d-fold composition of f).*

For an expanding map f, we can choose $n = 1$.

This lemma comes from a proposition of R. Mãné[3] which is said that f has no wandering interval and f has finite non-repelling periodic poits. An interval $I \subset S^1$ is called wandreing if $f^i(I) \cap f^j(I) = \phi$ for any $i \neq j$; $i, j > 0$.

Lemma 2: *Suppose $F : R \to R$ is an isomorphism satisfies condition $(*)$. Then there is an isomorphism $\gamma : R \to R$ such that*

$$\gamma F \gamma^{-1}(t) = \delta t$$

This lemma can be derived by a little modifying for the proof of lemma 2.2 of *[1]*.

Set $S(t) = \gamma(\gamma^{-1}(t) + 1)$, S satisfies the equation

$$S^d(t) = \delta S(\frac{t}{\delta}).$$

One can easily verified that the analytically conjugate class of F uniquely determines the affine conjugate class of S.

2   Expressing the expanding property

Suppose $f : S^1 \to S^1$ is an expanding map. Then we have an isomorphic lifting $F : R \to R$ of f satisfies $(*)$, there is an isomorphism $\gamma : R \to R$ such that $\gamma F \gamma^{-1}(t) = \delta t$ and $S(t) = \gamma(\gamma^{-1}(t) + 1)$ satisfied CFE. So we can induce a isomorphic solution of CFE from any expanding map. Inversely, what kind of isomorphic solutions of CFE are induced by expanding maps?

Theorem A: *Suppose S is an isomorphic solution of CFE. Then S is induced by an expanding map iff S has an invariant neighborhood U of R in C with smooth boundary consisted of two simple curves such that $\frac{1}{\delta}(U) \subset U$.*

Proof: If S is induced by an expanding map f, i.e., there is an isomorphism $\gamma : R \to R$ such that $\gamma F \gamma^{-1} = \delta$ and $S(t) = \gamma(\gamma-1(t) + 1)$ where $F : R \to R$ is a lifting of f satisfies $(*)$, then the univalent convergence domain of γ contains a band domain B such that $\overline{F^{-1}(B)} \subset B$. let $\gamma(B) = U$, then U satisfies the condition.

To prove the sufficiency, we map the simple connected invariant convergence region to a band domain B by Riemann mapping theorem with the origin and the infinity fixed, and $S(0)$ to one by choosing a suitable width of band domain. The equation implies that S has no fixed point on R. So the infinity is the only parabolic fixed point of the conjugation of S. This means that $\gamma S \gamma^{-1}(z) = z + 1$ where γ is the Riemann mapping. Set $F = \gamma S \gamma^{-1}$. Since $\frac{1}{\delta}(U) \subset U$, $\overline{F^{-1}(B)} \subset B$, i.e., the projection f of F is expanding. ◻

3   Analyticity of conjugacies

From the relation of F and S, we can derive the relations of eigenvalues of F and S.

Lemma 3: *Let $F : R \to R$ be an isomorphism satisfies $(*)$, $\gamma F \gamma^{-1}(t) = \delta t$ and $S(t) = \gamma(\gamma^{-1}(t) + 1)$. Then $S(x) = \delta^i x$ has unique solutions $x_i = \gamma(t_i)$ and*

$$(F^i)'(t_i) = \frac{\delta^i}{S'(x_i)}.$$

Theorem B: *Suppose f_1 and f_2 are topologically conjugated. Then f_1 is analytically conjugated to f_2 iff f_1 and f_2 have equal eigenvalues along corresponding periodic points.*

This theorem is a direct conseqence of the following lemmas and lemma 3.

Lemma 4: *Let S_1 and S_2 be isomorphic solutions of CFE. If the solutions x_i, y_i of the functional equations $S_1(x) = \delta^i x, S_2(y) = \delta^i y$ uniquely exist and $\{x_i\}, \{y_i\}$ are strictly decreasing to zero with $S_1'(x_i) = S_2'(y_i)$, then S_1 is equivalent to S_2.*

Proof: Consider

$$R_1(x) = \frac{x}{S_1(x)}, \qquad R_2(y) = \frac{y}{S_2(y)}.$$

Since $S_1(0) \neq 0 \neq S_2(0)$, S_1 and S_2 are isomorphisms of R, we have that R_1 and R_2 are analytic near the origin and their inverse function exist there. Then $y = R_2^{-1} R_1(x) = \phi(x)$ is also analytic in a neighborhood W of the origin satisfying

$$x S_2(\phi(x)) = S_1(x)\phi(x);$$

hence

$$\phi(x_i) = y_i.$$

On calculating the derivatives of the two side, we have

$$x\phi'(x) = \phi(x)\frac{[\frac{S_1(x)}{x}] - S_1'(x)}{[\frac{S_1(x)}{x}] - S_2'(\phi(x))}.$$

Applying the condition $S_1'(x_i) = S_2'(y_i)$, we get

$$x_i \phi'(x_i) = \phi(x_i).$$

The analytic function $x\phi'(x) - \phi(x)$ has x_i as its zero points, but $\{x_i\}$ tends to zero. So

$$\phi(x) = x\phi'(x), \quad \text{and} \quad \phi(x) = cx.$$

This implies that S_1 is equivalent to S_2. ∎

Lemma 5: *If f_1^n is analytically conjugated to f_2^n, then f_1 is also analytically conjugated to f_2.*

Proof: Let $\varphi f_1^n = f_2^n \varphi$. Then φ maps the repelling periodic points of f_1 is infinite, i.e., $\varphi f_1 \varphi^{-1} - f_2$ has infinite zero point. So $\varphi f_1 = f_2 \varphi$. ∎

References

[1] G. Cui**On Real Analytic Expanding Endomorphisms of S^1 and Cvitanovic-Feigenbaum Eqation,***to apear in Acta Math. Sinica*

[2] A. Douady and J. Hubburd**On the Dynamics of Polynomial -like Mappings,** *Ann. Sci. Ecole Norm. Sup. vol. (4)18(1985), 287-343*

[3] R. Mane, **Hyperbolicity, Sinks and Measure in One Dimentional Dynamics,** *Commun. Math. Phys. vol. 100(1985),495-524*

[4] M. Shub and D. Sullivan **Expanding Endomorphisms of the Circle Revisited,***Ergodic Theory and Dynamical Systems(1985)*

Keywords: linearization, expanding
1980 Mathematics subject classification: 30C99

Coefficients of Bounded Univalent Functions

Daniel Dreibelbis[1], James Rovnyak[1]
University of Virginia, Mathematics-Astronomy Building
Charlottesville, VA 22903, USA

Kin Y. Li[2]
Hong Kong University of Science and Technology
Clear Water Bay, Kowloon, Hong Kong

1 Coefficients of Bounded Holomorphic Functions

Let D be the open unit disk in the complex plane. For any holomorphic function $B : D \to \overline{D}$, the coefficients of the power series for $B(z)$ can be easily characterized by operator techniques. Let M_B denote the multiplication operator $f \mapsto fB$ on the Hardy space $H^2(D)$ of square summable power series on D, where the inner product is the usual inner product of the coefficient sequences. Since the operator norm of M_B is equal to the sup-norm of B on D (see *[9]*), the following theorem gives a necessary and sufficient condition on the coefficient sequence of B.

Theorem 1: *$B(z) = b_0 + b_1 z + b_2 z^2 + \cdots$ represents a holomorphic function from D to $\overline{D}$ if and only if $\langle fB, fB \rangle_{H^2(D)} \le \langle f, f \rangle_{H^2(D)}$ for every $f \in H^2(D)$. This is equivalent to*

$$\left\| \begin{pmatrix} b_0 & 0 & 0 & 0 & 0 & 0 & \cdots \\ b_1 & b_0 & 0 & 0 & 0 & 0 & \cdots \\ b_2 & b_1 & b_0 & 0 & 0 & 0 & \cdots \\ b_3 & b_2 & b_1 & b_0 & 0 & 0 & \cdots \\ \vdots & \vdots & \vdots & \vdots & \vdots & \vdots & \ddots \end{pmatrix} \right\| \le 1.$$

The matrix in the theorem is the matrix of M_B with respect to the orthonormal basis $\{z^n\}_{n=0}^{\infty}$ on $H^2(D)$. In principle, when one has to check if a sequence $\{b_n\}$ satisfies the condition in the theorem or not, one only has to check that $I - M_B^* M_B \ge 0$ by examining the determinants of the principal minors. The matrix of M_B^* is, of course, the conjugate transpose of the matrix above.

---

[1]Supported by the National Science Foundation.
[2]Supported by a HKUST Direct Allocation Grant.

Proceedings of the International Conference on Complex
Analysis at the Nankai Institute of Mathematics, 1992 pp. 45-58
©INTERNATIONAL PRESS 1994

Closely related to this is the Carathéodory-Fejér problem. It consists of finding conditions on $b_0, b_1, \ldots, b_r$ so that $b_0 + b_1 z + \cdots + b_r z^r$ can be extended to a power series of a holomorphic function from D to $\overline{D}$. The operator theoretic approach to the problem is to restrict the multiplication operator to the subspace spanned by $\{1, z, z^2, \ldots, z^r\}$.

Theorem 2: *[Schur [11]] $b_0 + b_1 z + \cdots + b_r z^r$ can be extended to a power series of a holomorphic function from D to $\overline{D}$ if and only if*

$$\left\| \begin{pmatrix} b_0 & 0 & 0 & 0 & \cdots & 0 \\ b_1 & b_0 & 0 & 0 & \cdots & 0 \\ b_2 & b_1 & b_0 & 0 & \cdots & 0 \\ b_3 & b_2 & b_1 & b_0 & \cdots & 0 \\ \vdots & \vdots & \vdots & \vdots & \ddots & \vdots \\ b_r & b_{r-1} & b_{r-2} & b_{r-3} & \cdots & b_0 \end{pmatrix} \right\| \leq 1.$$

For a proof, see *[9]*.

2   Coefficients of Bounded Univalent Functions

For convenience, we define a *normalized Riemann mapping* to be a univalent, holomorphic function $B(z)$ from D to D such that $B(0) = 0$ and $B'(0) > 0$. In *[5]*, L. de Branges gave a characterization of the coefficients of the power series for a normalized Riemann mapping. Analogous to the situation in the previous section, he also considered spaces of square summable power series (but with weights). In place of the multiplication operator, he studied the substitution operator $C_B : f \mapsto f \circ B$, which is more suitable for dealing with univalent functions.

For a real $\nu, \sigma = \{\sigma_i\}$ a sequence of nonnegative numbers, let the *Grunsky space* $\mathcal{G}_\sigma^\nu$ be the set of all generalized power series of the form

$$f(z) = \sum_{n=1}^{\infty} a_n z^{\nu+n}$$

such that

$$\sum_{n=1}^{\infty} |(\nu + n)\sigma_n a_n^2| < \infty.$$

We identify two series $f(z) = \sum_{n=1}^{\infty} a_n z^{\nu+n}$ and $g(z) = \sum_{n=1}^{\infty} c_n z^{\nu+n}$ if $a_n = c_n$ for all n such that $(\nu + n)\sigma_n \neq 0$. Further, we can introduce a (possibly indefinite) inner product on $\mathcal{G}_\sigma^\nu$ by defining

$$\langle f, g \rangle_{\mathcal{G}_\sigma^\nu} = \sum_{n=1}^{\infty} (\nu + n)\sigma_n a_n \overline{c_n}.$$

This makes $\mathcal{G}_\sigma^\nu$ a Krein space *[1]*. In case $\sigma = \{\sigma_i | \sigma_i \equiv 1 \text{ for all } i\}$, we write $\mathcal{G}^\nu$ for $\mathcal{G}_\sigma^\nu$. Similarly, in case $\sigma = \{\sigma_i | \sigma_i = 1 \text{ for } 1 \leq i \leq r; \ \sigma_i = 0 \text{ for } i > r\}$, we write $\mathcal{G}_r^\nu$ for $\mathcal{G}_\sigma^\nu$.

For real $\nu, B(z) = b_1 z + b_2 z^2 + \cdots$ with $0 < b_1 < 1$, define

$$B(z)^\nu = (b_1 + b_2 z + \cdots)^\nu z^\nu = b_1^\nu z^\nu + \nu b_1^{\nu-1} b_2 z^{\nu+1} + \cdots.$$

If $g(z) = \sum_{n=1}^\infty c_n z^{\nu+n} \in \mathcal{G}^\nu$, then

$$(g \circ B)(z) = \sum_{n=1}^\infty c_n B(z)^{\nu+n} = (c_1 b_1^{\nu+1}) z^{\nu+1} + ((\nu+1) c_1 b_1^\nu b_2 + c_2 b_1^{\nu+2}) z^{\nu+2} + \cdots.$$

Theorem 3: *[de Branges]* $B(z) = b_1 z + b_2 z^2 + \cdots$ *is a normalized Riemann mapping from D to D if and only if $\langle g \circ B, g \circ B \rangle_{\mathcal{G}^\nu} \leq \langle g, g \rangle_{\mathcal{G}^\nu}$ for every nonpositive integer ν and every $g \in \mathcal{G}^\nu$.*

If $C_B : g \in \mathcal{G}^\nu \mapsto g \circ B \in \mathcal{G}^\nu$, the contractive inequality in the theorem is equivalent to $J(I - C_B^* C_B) \geq 0$. Here C_B^* denotes the Krein adjoint of C_B and J is the canonical symmetry (see *[1]*). Again for a given sequence $\{b_n\}$, we can check the contractivity of C_B by examining determinants.

In *[10]*, J. Rovnyak has generalized the *only if* part of the theorem to all $\mathcal{G}_\sigma^\nu$ spaces with $\sigma_1 \geq \sigma_2 \geq \cdots \geq 0$ and ν real. In *[3],[4]*, de Branges made the following conjecture.

Conjecture $B(z) = b_1 z + b_2 z^2 + \cdots + b_r z^r$ *can be extended to a power series of a normalized Riemann mapping from D to D if and only if*

$$\langle g \circ B, g \circ B \rangle_{\mathcal{G}_r^\nu} \leq \langle g, g \rangle_{\mathcal{G}_r^\nu}$$

for all real ν and $g \in \mathcal{G}_r^\nu$ (cf. [7]).

In *[8]*, it was shown that contractivity of substitution by $B(z)$ on $\mathcal{G}_r^\nu$ is equivalent to contractivity of substitution by $B(z)$ on $\mathcal{G}_r^{-\nu-r-1}$. In *[7],[8]*, the conjecture for $r \leq 2$ was proved. Notice that de Branges' theorem can be viewed as saying that the conjecture is true in the limit as r tends to infinity. Quite recently, we found the conjecture to be false when $r = 3$. This will be discussed in the last section.

3   Admissible Families of Grunsky Spaces

For a family of Grunsky spaces, de Branges introduced the concept of admissibility. It is a sufficient condition for substitution by normalized Riemann mappings to be contractive on certain pairs of elements in a one-parameter family of Grunsky spaces.

Specifically, for a real ν, let $\mathcal{G}_{\sigma(t)}^\nu, t \geq 1$, be a family of Grunsky spaces such that $\sigma_1(t), \sigma_2(t), \ldots$ are absolutely continuous and $\sigma_n(t) \equiv 0$ for sufficiently large n. The family $\mathcal{G}_{\sigma(t)}^\nu, t \geq 1$ is *admissible* if

(i)   for all n,

$$(2\nu + n)\sigma_n(t) + t\sigma_n'(t) = (2\nu + n)\sigma_{n+1}(t) - \frac{2\nu + n}{n+1} t\sigma_{n+1}'(t)$$

a.e. for $t \geq 1$, and

(ii)   for all n,

$$\tau_n(t) = (\nu + n)[\sigma_n(t) - \sigma_{n+1}(t) + \frac{1}{n+1}t\sigma'_{n+1}(t)]$$

is nonnegative *a.e.* for $t \geq 1$.

For any normalized Riemann mapping $B(z)$, admissibility of $\mathcal{G}^\nu_{\sigma(t)}, t \geq 1$ implies (see *[10]*)

$$\langle g \circ B, g \circ B \rangle_{\mathcal{G}^\nu_{\sigma(a)}} \leq \langle g, g \rangle_{\mathcal{G}^\nu_{\sigma(b)}}$$

and

$$\left\langle \frac{B(z)^\nu - B'(0)^\nu z^\nu}{\nu} + g(B(z)), \frac{B(z)^\nu - B'(0)^\nu z^\nu}{\nu} + g(B(z)) \right\rangle_{\mathcal{G}^\nu_{\sigma(a)}} - \langle g, g \rangle_{\mathcal{G}^\nu_{\sigma(b)}}$$

$$\leq 4 \sum_{n=1}^\infty (\nu + n) \left[\frac{(2\nu + 1)_{n-1}}{n!} \right]^2 \left(\left(\frac{a}{b} \right)^{2\nu} \sigma_n(a) - \sigma_n(b) \right) \tag{1}$$

for $a, b \geq 1$ such that $a = B'(0)b$. When $\nu = 0$, the inner product on $\mathcal{G}^\nu_{\sigma(t)}$ is positive definite and the inequality (1) is to be interpreted as

$$\left\| \log \frac{B(z)}{zB'(0)} + g(B(z)) \right\|^2_{\mathcal{G}^0_{\sigma(a)}} - \|g\|^2_{\mathcal{G}^0_{\sigma(b)}} \leq 4 \sum_{n=1}^\infty \frac{\sigma_n(a) - \sigma_n(b)}{n},$$

which was used by de Branges to solve the Bieberbach Conjecture *[2]*.

Following *[8]*, we make the following definition. We say that given numbers $b_1, b_2, \ldots, b_r (b_1 > 0)$ satisfy the *admissible families condition* if $B(z) = b_1 z + \cdots + b_r z^r$ satifies inequality (1) for every admissible family $\mathcal{G}^\nu_{\sigma(t)}, t \geq 1$, such that $\sigma_n(t) \equiv 0$ for $n \geq r$.

For $r = 2$, the conjecture in Section 2 is true (see *[8]*).

Theorem 4: *For numbers $b_1, b_2 (b_1 > 0), B(z) = b_1 z + b_2 z^2$, the following are equivalent:*

   (i) $\langle g \circ B, g \circ B \rangle_{\mathcal{G}^\nu_2} \leq \langle g, g \rangle_{\mathcal{G}^\nu_2}$ *for all real ν and all g in $\mathcal{G}^\nu_2$;*

  (ii) b_1, b_2 *satisfy the admissible families condition;*

 (iii) $b_1 \leq 1, |b_2| \leq 2b_1(1 - b_1);$

 (iv) $B(z)$ *can be extended to a power series of a Riemann mapping from D to D.*

For $r = 3$, conditions *(i)* and *(ii)* are not equivalent. In *[8]*, it was shown that

$$\frac{1}{3}z + \frac{1}{4}e^{i\pi/4}z^2 + \frac{i}{3}z^3$$

satisfies condition *(ii)* , but not condition *(i)* (hence not condition *(iv)* either.)

4 Extension of Contractive Inequalities

Suppose $B(z) = b_1 z + \cdots + b_r z^r$ satisfies $\langle g \circ B, g \circ B \rangle_{\mathcal{G}_r^\nu} \leq \langle g, g \rangle_{\mathcal{G}_r^\nu}$ for all $g \in \mathcal{G}_r^\nu$ and $b_1 < 1$. In $[6]$, the condition on b_{r+1} so that $C(z) = B(z) + b_{r+1} z^{r+1}$ will satisfy $\langle g \circ C, g \circ C \rangle_{\mathcal{G}_{r+1}^\nu} \leq \langle g, g \rangle_{\mathcal{G}_{r+1}^\nu}$ for all $g \in \mathcal{G}_{r+1}^\nu$ was identified. If this one-step extension is possible for all r, then Theorem 3 will imply $B(z)$ can be extended to a power series of a Riemann mapping from D to D.

The condition $\langle g \circ B, g \circ B \rangle_{\mathcal{G}_r^\nu} \leq \langle g, g \rangle_{\mathcal{G}_r^\nu}$ allows us to discuss complementation on $\mathcal{G}_r^\nu$ (see $[5]$). The space $\mathcal{G}_r^\nu(B)$ is defined to consist of all $f(z) \in \mathcal{G}_r^\nu$ such that

$$\|f\|_{\mathcal{G}_r^\nu(B)}^2 = \sup_{g \in \mathcal{G}_r^\nu} \left(\langle f + g \circ B, f + g \circ B \rangle_{\mathcal{G}_r^\nu} - \langle g, g \rangle_{\mathcal{G}_r^\nu} \right) < \infty.$$

In $[5],[10]$, it was shown that $\mathcal{G}_r^\nu(B)$ is a Hilbert space with the above norm.

Theorem 5: *Assume $r \geq 2$. When $\nu = 0$ or $-r$, substitution by $C(z) = B(z) + b_{r+1} z^{r+1}$ is contractive on $\mathcal{G}_{r+1}^\nu$ for any b_{r+1}. When $\nu \neq 0$ and $-r$, substitution by $C(z)$ is contractive on $\mathcal{G}_{r+1}^\nu$ if and only if*

$$\left| \frac{Q_r(\nu)}{\nu B'(0)^{\nu+r}} + \left\langle \frac{B\left(\frac{1}{z}\right)^\nu - B'(0)^\nu \left(\frac{1}{z}\right)^\nu}{\nu}, \frac{B^*(z)^\mu - B'(0)^\mu z^\mu}{\mu} \right\rangle_{\mathcal{G}_{r-1}^\mu(B^*)} \right|^2$$

$$\leq \left(\frac{1 - B'(0)^{2\nu}}{\nu} - \left\| \frac{B(z)^\nu - B'(0)^\nu z^\nu}{\nu} \right\|_{\mathcal{G}_{r-1}^\nu(B)}^2 \right) \times$$

$$\times \left(\frac{1 - B'(0)^{2\mu}}{\mu} - \left\| \frac{B^*(z)^\mu - B'(0)^\mu z^\mu}{\mu} \right\|_{\mathcal{G}_{r-1}^\mu(B^*)}^2 \right), \qquad (2)$$

where $\mu = -\nu - r$, $Q_r(\nu)$ is the coefficient of $z^{\nu+r}$ in $C(z)^\nu$ and $B^*(z) = \overline{b_1} z + \overline{b_2} z^2 + \cdots$.

Since $Q_r(\nu) = \nu b_1^{\nu-1} b_{r+1} + \phi(\nu, b_1, \cdots, b_r)$, the inequality can be put in the form

$$|b_{r+1} - C(\nu, b_1, \cdots, b_r)| \leq R(\nu, b_1, \cdots, b_r). \qquad (3)$$

That is, for every ν, the condition on b_{r+1} is that it lies in the closed disk with center $C(\nu, b_1, \cdots, b_r)$ and radius $R(\nu, b_1, \cdots, b_r)$. For a proof of Theorem 5, see $[6]$.

In the case $r = 2$, this gives (see $[6]$)

$$\left| b_3 + \left(\frac{\nu - 1}{2b_1} + \frac{(\nu+1)b_1^{2\nu+1}}{1 - b_1^{2\nu+2}} \right) b_2^2 \right|^2 \leq b_1^{2\nu} \left(\frac{\nu+1}{1 - b_1^{2\nu+2}} \right)^2$$

$$\times \left(\frac{(1 - b_1^{2\nu})(1 - b_1^{2\nu+2})}{\nu(\nu+1)b_1^{2\nu-2}} - |b_2|^2 \right) \left(\frac{(1 - b_1^{2\nu+2})(1 - b_1^{2\nu+4})}{(\nu+1)(\nu+2)b_1^{2\nu}} - |b_2|^2 \right) \quad (4)$$

Given b_1, b_2 satisfying condition *(iii)* of Theorem 4, in $[6]$ it was shown that the choice

$$b_3 = \frac{3 - 5b_1}{4b_1(1 - b_1)} b_2^2$$

satisfies the inequality for all real ν. So extension from $\mathcal{G}_2^\nu$ to $\mathcal{G}_3^\nu$ is always possible.

5 Impossibility of Extension

In this section we will show that substitution by

$$B(z) = \frac{1}{6}z + \frac{1}{4}z^2 + \left(\frac{4}{15} + \frac{i}{18}\right)z^3$$

is contractive on $\mathcal{G}_3^\nu$ for all ν, but there is no b_4 so that substitution by $C(z) = B(z) + b_4 z^4$ is contractive on $\mathcal{G}_4^{-1/2}$ and $\mathcal{G}_4^{-3/2}$ simultaneously. This implies by Rovnyak's generalization of Theorem 3 that the conjecture in Section 2 is false when $r = 3$.

The example was found while analyzing data from some computer experiments. The first author was working with the second author in an NSF supported undergraduate research project to study contractivity of polynomials on Grunsky spaces via computers. The above example was suggested after examining some graphic outputs. Subsequently, three independent verifications were made by the authors. Two of these verifications were made using C++ programming and the computer algebra program MAPLE. Below we shall present the mathematical details of a more direct verification which only needs at one stage a simple PASCAL program. The details of the undergraduate research project can be obtained by writing to one of the first two authors.

Throughout this section, let

$$b_1 = \frac{1}{6}, \quad b_2 = \frac{1}{4}, \quad \text{and } b_3 = \frac{4}{15} + \frac{i}{18}.$$

We first show that there is no b_4 so that substitution by $C(z)$ is contractive on $\mathcal{G}_4^{-1/2}$ and on $\mathcal{G}_4^{-3/2}$. This is done by identifying the disk in the inequality (2) in Theorem 5 for $r = 3, \nu = -\frac{1}{2}$ and $\nu = -\frac{3}{2}$, respectively, and see that the disks are disjoint.

Using the definition of $Q_r(\nu)$, we find that

$$\frac{Q_3(\nu)}{\nu b_1^{\nu+3}} = b_1^{-4}b_4 + (\nu - 1)b_1^{-5}b_2 b_3 + \frac{(\nu - 1)(\nu - 2)}{6}b_1^{-6}b_2^3.$$

To deal with the inner product in $\mathcal{G}_2^\nu(B)$, we first find the norm of an arbitrary element in $\mathcal{G}_2^\nu(B)$, then apply the polarization identity. Notice that b_1, b_2 satisfy condition (iii) of Theorem 4 and hence condition (i) of Theorem 4. For arbitrary complex numbers c, d, we have, by the definition of the norm,

$$\left\| cz^{\nu+1} + dz^{\nu+2} \right\|_{\mathcal{G}_2^\nu(B)}^2$$

$$= \sup_{a_1,a_2} \left(\left\| cz^{\nu+1} + dz^{\nu+2} + a_1 B(z)^{\nu+1} + a_2 B(z)^{\nu+2} \right\|_{\mathcal{G}_2^\nu}^2 - \left\| a_1 z^{\nu+1} + a_2 z^{\nu+2} \right\|_{\mathcal{G}_2^\nu}^2 \right)$$

$$= \sup_{a_1,a_2} \left((\nu + 1)\left| a_1 b_1^{\nu+1} + c \right|^2 + (\nu + 2)\left| (\nu + 1)a_1 b_1^\nu b_2 + a_2 b_1^{\nu+2} + d \right|^2 \right.$$

$$-(\nu+1)|a_1|^2 - (\nu+2)|a_2|^2)$$

$$= \sup_{a_1} \left((\nu+1)\left|a_1 b_1^{\nu+1} + c\right|^2 + (\nu+2)\frac{|(\nu+1)a_1 b_1^{\nu} b_2 + d|^2}{1 - b_1^{2\nu+4}} - (\nu+1)|a_1|^2 \right)$$

$$= \sup_{a_1} \left(A|a_1|^2 + B|a_1| + C \right)$$

$$= \frac{(\nu+1)(1 - b_1^{2\nu+4})|c|^2 + (\nu+2)|d|^2 - (\nu+2)b_1^{2\nu}|(\nu+1)b_2 c - b_1 d|^2}{(1 - b_1^{2\nu+2})(1 - b_1^{2\nu+4}) - (\nu+1)(\nu+2)b_1^{2\nu}|b_2|^2}$$

(In the third equality above, to eliminate a_2 we applied the formula

$$\sup_{z} \left(|p + qz|^2 - |z|^2 \right) = \frac{|p|^2}{1 - |q|^2}$$

when $\nu+2$ is positive (in which case $|q| = |b_1^{\nu+2}| < 1$), and the formula

$$\inf_{z} \left(|p + qz|^2 - |z|^2 \right) = \frac{|p|^2}{1 - |q|^2}$$

when $\nu+2$ is negative (in which case $|q| = |b_1^{\nu+2}| > 1$). These were shown in [7]. Then we expanded the expression and found it to be of the form shown in the fourth equality, where

$$A = \frac{(\nu+1)^2(\nu+2)b_1^{2\nu}|b_2|^2 - (\nu+1)(1 - b_1^{2\nu+4})(1 - b_1^{2\nu+2})}{1 - b_1^{2\nu+4}}$$

$$B = \frac{2|(\nu+1)(1 - b_1^{2\nu+4})b_1^{\nu+1}\bar{c} + (\nu+2)(\nu+1)b_1^{\nu} b_2 \bar{d}|}{1 - b_1^{2\nu+4}}$$

$$C = \frac{(\nu+1)(1 - b_1^{2\nu+4})|c|^2 + (\nu+2)|d|^2}{1 - b_1^{2\nu+4}}.$$

It follows from [7] that $A < 0$. Then we applied

$$\max_{x} \left(Ax^2 + Bx + C \right) = \frac{4AC - B^2}{4A}$$

and simplified to get the last equality.)

For convenience, we will set

$$D = \frac{(1 - b_1^{2\nu+2})(1 - b_1^{2\nu+4})}{(\nu+1)(\nu+2)b_1^{2\nu}} - |b_2|^2.$$

Note that if ν is replaced by $\mu = -\nu - 3$, D remains the same.

In the case $c = 1$ and $d = 0$, we get

$$\left\| z^{\nu+1} \right\|_{\mathcal{G}_2^{\nu}(B)}^2 = \frac{1}{D} \left(\frac{1 - b_1^{2\nu+4}}{(\nu+2)b_1^{2\nu}} - (\nu+1)|b_2|^2 \right). \tag{5}$$

In the case $c = 0$ and $d = 1$, we get

$$\left\| z^{\nu+2} \right\|_{\mathcal{G}_2^{\nu}(B)}^2 = \frac{1}{D} \frac{1 - b_1^{2\nu+2}}{(\nu+1)b_1^{2\nu}}. \tag{6}$$

Applying the polarization identity, we get

$$\langle z^{\nu+1}, z^{\nu+2}\rangle_{\mathcal{G}_2^\nu(B)} = \frac{b_1 b_2}{D}. \tag{7}$$

Since

$$\frac{B(z)^\nu - B'(0)^\nu z^\nu}{\nu} = b_1^{\nu-1}b_2 z^{\nu+1} + \left(b_1^{\nu-1}b_3 + \frac{\nu-1}{2}b_1^{\nu-2}b_2^2\right) z^{\nu+2}$$
$$+ \left(b_1^{\nu-1}b_4 + (\nu-1)b_1^{\nu-2}b_2 b_3 + \frac{(\nu-1)(\nu-2)}{6}b_1^{\nu-3}b_2^3\right) z^{\nu+3}$$
$$+ \cdots,$$

we can now identify the disk inequalities by using equations (5), (6), (7). For $\nu = -\frac{1}{2}$, the disk inequality is

$$\left| b_4 - C\left(-\frac{1}{2}, b_1, b_2, b_3\right)\right| \le R\left(-\frac{1}{2}, b_1, b_2, b_3\right),$$

where

$$C\left(-\frac{1}{2}, b_1, b_2, b_3\right) \approx .220339 + .093461i \text{ and } R\left(-\frac{1}{2}, b_1, b_2, b_3\right) \approx .010587.$$

For $\nu = -\frac{3}{2}$, the disk inequality is

$$\left| b_4 - C\left(-\frac{3}{2}, b_1, b_2, b_3\right)\right| \le R\left(-\frac{3}{2}, b_1, b_2, b_3\right),$$

where

$$C\left(-\frac{3}{2}, b_1, b_2, b_3\right) \approx .233051 + .106579i \text{ and } R\left(-\frac{3}{2}, b_1, b_2, b_3\right) \approx .004667.$$

Since

$$\left| C\left(-\frac{1}{2}, b_1, b_2, b_3\right) - C\left(-\frac{3}{2}, b_1, b_2, b_3\right)\right| > .016$$
$$> R\left(-\frac{1}{2}, b_1, b_2, b_3\right) + R\left(-\frac{3}{2}, b_1, b_2, b_3\right),$$

the disks are disjoint.

Next we will show that substitution by $B(z)$ is contractive on $\mathcal{G}_3^\nu$ for all ν. In (4), if we set $\alpha = \nu + 1$, then we will get

$$\frac{1}{18^2} + \frac{3^2}{16^2}\left(\alpha - \frac{26}{45} + \frac{2\alpha}{36^\alpha - 1}\right)^2 \le 36^{\alpha+1}\left(\frac{1}{36^{\alpha+1}}\frac{36^{\alpha-1} - 1}{\alpha - 1} - \frac{1}{16}\frac{\alpha}{36^\alpha - 1}\right)$$
$$\times \left(\frac{1}{36^{\alpha+2}}\frac{36^{\alpha+1} - 1}{\alpha + 1} - \frac{1}{16}\frac{\alpha}{36^\alpha - 1}\right). \tag{8}$$

Note that this inequality remains the same if α is replaced by $-\alpha$. This can be checked directly or one can apply the statement made following the conjecture. Thus, it suffices to prove the inequality for $\alpha \geq 0$.

In the case $\alpha \geq 2$, we have

$$-\frac{37}{35}\alpha \leq \alpha - \frac{26}{45} + \frac{2\alpha}{36^\alpha - 1} = \alpha\left(\frac{36^\alpha + 1}{36^\alpha - 1}\right) - \frac{26}{45} \leq \frac{37}{35}\alpha,$$

$$36\alpha(\alpha + 1) \leq 36^\alpha - 1,$$

$$600\alpha(\alpha - 1) \leq 36^\alpha - 1,$$

$$\frac{1}{40(\alpha + 1)} + \frac{1}{16}\frac{\alpha}{36^\alpha - 1} \leq \frac{1}{40(\alpha + 1)} + \frac{1}{16 \times 36(\alpha + 1)}$$

$$\leq \frac{1}{36(\alpha + 1)}\left(\frac{35}{36}\right)$$

$$\leq \frac{1}{36(\alpha + 1)}\left(\frac{36^{\alpha+1} - 1}{36^{\alpha+1}}\right)$$

and

$$\frac{1}{40^2(\alpha - 1)} + \frac{1}{16}\frac{\alpha}{36^\alpha - 1} \leq \frac{1}{40^2(\alpha - 1)} + \frac{1}{16 \times 600(\alpha - 1)}$$

$$\leq \frac{1}{36^2(\alpha - 1)}\left(\frac{35}{36}\right)$$

$$\leq \frac{1}{36^2(\alpha - 1)}\left(\frac{36^{\alpha-1} - 1}{36^{\alpha-1}}\right).$$

Therefore,

$$\frac{1}{18^2} + \frac{3^2}{16^2}\left(\alpha - \frac{26}{45} + \frac{2\alpha}{36^\alpha - 1}\right)^2 \leq \frac{1}{18^2} + \left(\frac{3}{16}\right)^2\left(\frac{37}{35}\right)^2\alpha^2$$

$$\leq .04007\alpha^2 \qquad \leq \frac{36^{\alpha-1}}{50\alpha^2}$$

$$\leq 36^{\alpha+1}\frac{1}{40^2(\alpha - 1)}\frac{1}{40(\alpha + 1)}$$

$$\leq 36^{\alpha+1}\left(\frac{1}{36^{\alpha+1}}\frac{36^{\alpha-1} - 1}{\alpha - 1} - \frac{1}{16}\frac{\alpha}{36^\alpha - 1}\right)$$

$$\times \left(\frac{1}{36^{\alpha+2}}\frac{36^{\alpha+1} - 1}{\alpha + 1} - \frac{1}{16}\frac{\alpha}{36^\alpha - 1}\right).$$

It only remains to check (8) when $0 \leq \alpha < 2$. The difference of the right side of (8) minus the left side of (8) can be plotted with any computer graphics package over the finite interval $0 \leq \alpha < 2$, and the graph make it clear that the desired inequality is true in this range. We present a different procedure that places less dependence on computer graphics. First expand both sides of (8) to get

$$\frac{1}{18^2} + \left(\frac{3}{16}\right)^2 \left(\alpha - \frac{26}{45}\right)^2 + \frac{3}{8}\left(\alpha - \frac{26}{45}\right)\left(\frac{3}{8}\frac{\alpha}{36^\alpha - 1}\right) + \left(\frac{3}{8}\frac{\alpha}{36^\alpha - 1}\right)^2$$

$$\leq \frac{1}{36^{\alpha+2}}\left(\frac{36^{\alpha-1} - 1}{\alpha - 1}\right)\left(\frac{36^{\alpha+1} - 1}{\alpha + 1}\right)$$

$$-\frac{1}{6}\left(\frac{36^{\alpha-1} - 1}{\alpha - 1} + \frac{1}{36}\frac{36^{\alpha+1} - 1}{\alpha + 1}\right)\left(\frac{3}{8}\frac{\alpha}{36^\alpha - 1}\right) + 36^\alpha\left(\frac{3}{8}\frac{\alpha}{36^\alpha - 1}\right)^2.$$

(Here the cases $\alpha = 0$ and $\alpha = 1$ are to be taken in the limit sense.) Grouping common factors and arranging terms, we get

$$\frac{1}{18^2} + \left(\frac{3}{16}\right)^2 \left(\alpha - \frac{26}{45}\right)^2$$

$$+\frac{3}{8}\frac{\alpha}{36^\alpha - 1}\left[\frac{3}{8}\left(\alpha - \frac{26}{45}\right) + \frac{1}{6}\frac{36^{\alpha-1} - 1}{\alpha - 1} + \frac{1}{216}\frac{36^{\alpha+1} - 1}{\alpha + 1} - \frac{3}{8}\alpha\right]$$

$$\leq \frac{1}{36^{\alpha+2}}\left(\frac{36^{\alpha-1} - 1}{\alpha - 1}\right)\left(\frac{36^{\alpha+1} - 1}{\alpha + 1}\right).$$

Multiplying both sides by $36^{\alpha+2}(36^\alpha - 1)/\alpha$, we get

$$\left[4 + \left(\frac{27}{4}\right)^2 \left(\alpha - \frac{26}{45}\right)^2\right] 36^\alpha\frac{36^\alpha - 1}{\alpha}$$

$$+81 \times 36^\alpha\left(\frac{36^{\alpha-1} - 1}{\alpha - 1}\right) + \frac{81}{36}36^\alpha\left(\frac{36^{\alpha+1} - 1}{\alpha + 1}\right)$$

$$\leq \left(\frac{36^{\alpha-1} - 1}{\alpha - 1}\right)\left(\frac{36^\alpha - 1}{\alpha}\right)\left(\frac{36^{\alpha+1} - 1}{\alpha + 1}\right) + \frac{81 \times 13}{10}36^\alpha.$$

If we expand out $\left(\alpha - \frac{26}{45}\right)^2 = \alpha^2 - \frac{52}{45}\alpha + \left(\frac{26}{45}\right)^2$ and transpose the negative term to the right, we get

$$\left[\left(\frac{27}{4}\right)^2 \alpha^2 + \frac{1921}{100}\right] 36^\alpha\frac{36^\alpha - 1}{\alpha} + 81 \times 36^\alpha\left(\frac{36^{\alpha-1} - 1}{\alpha - 1}\right) + \frac{81}{36}36^\alpha\left(\frac{36^{\alpha+1} - 1}{\alpha + 1}\right)$$

$$\leq \left(\frac{36^{\alpha-1} - 1}{\alpha - 1}\right)\left(\frac{36^\alpha - 1}{\alpha}\right)\left(\frac{36^{\alpha+1} - 1}{\alpha + 1}\right) + \frac{1053}{20}36^\alpha(36^\alpha + 1). \qquad (9)$$

Call the left side $F(\alpha)$ and the right side $H(\alpha)$. Observe that each factor in each term of $F(\alpha)$ and $H(\alpha)$ is positive, nondecreasing and convex on $[0, 2]$, so both $F(\alpha)$ and $H(\alpha)$ are convex on $[0, 2]$.

To establish inequality (9) on $[0, 2]$, we divide up the interval into subintervals $[a_n, b_n]$ such that there is a tangent line $L_n(\alpha)$ to $H(\alpha)$ at some $\alpha = c_n \in (a_n, b_n)$, satisfying

$$F(a_n) < L_n(a_n) \quad \text{and} \quad F(b_n) < L_n(b_n).$$

By convexity of $F(\alpha)$ and $H(\alpha)$, it follows that for all $\alpha \in [a_n, b_n]$,

$$F(\alpha) < L_n(\alpha) < H(\alpha).$$

The tangent lines can be found by running a simple computer program. We provide one such PASCAL program below.

```
program      inequality(output);
var      ln36, left, right, slope, value, x:real;
function      g(x:real):real;
begin
    if x = 0 then g:= ln36 else g:= (exp(x * ln36) - 1)/x
end;
function      dg(x:real):real;
begin
    if x = 0
    then dg:=  sqr(ln36)/2.0
    else dg:= (exp(x * ln36) * (ln36 * x - 1) + 1)/ sqr(x)
end;
procedure      tangent(x:real;   var slope, value:real);
var      exp36:real;
begin
    exp36 := exp(x * ln36);
    slope :=  dg(x-1)* g(x)* g(x+1)+ g(x-1)* dg(x)* g(x+1)
            + g(x - 1) * g(x) * dg(x + 1)
            +52.65 * ( sqr(exp36) * 2 * ln36 + exp36 * ln36);
    value :=  g(x-1)* g(x)* g(x+1)+52.65*( sqr(exp36)+exp36)
end;
function      f(x:real):real;
begin
    f:= ((45.5625 *  sqr(x) + 19.21) *  g(x) + 81 *  g(x - 1)
            +2.25 *  g(x + 1)) * exp(x * ln36)
end;
begin
    ln36 :=  ln(36);      right := 2.0;
    repeat
        left := 0;
        x := (left + right)/2.0;      tangent(x, slope, value);
        while slope * (right - x) + value <  f(right) do
            begin x := (x + right)/2.0;      tangent(x, slope, value)
        end;
        while slope * (left - x) + value <  f(left) do left := (x +
    left)/2.0;
        writeln(' On [',left : 11 : 9,',', right : 11 : 9,'], the tangent to
y = H(x)
            at ', x : 11 : 9,' has equation');
        writeln(' L(x)=',slope : 15 : 7,'(x-',x : 11 : 9,')+',value : 12 :
    5,'.');
```

$$\text{writeln(' F(',}left : 11 : 9,\text{')=', f(}left\text{): 14 : 7,'} < \text{L(',}left : 11 : 9,\text{')=',}$$
$$slope * (left - x) + value : 14 : 7);$$
$$\text{writeln(' F(',}right : 11 : 9,\text{')=',f(}right\text{): 14 : 7,'} < \text{L(',}right : 11 : 9,\text{')=',}$$
$$slope * (right - x) + value : 14 : 7);$$
$$\text{writeln; writeln;}$$
$$right := left$$
$$\text{until } right = 0$$
$$\text{end.}$$

The program produced an output of 52 tangent lines. Here we will list a few of them. The complete list can be obtained by running the above program.

On $[1.816406250, 2.000000000]$, the tangent to $y = H(x)$ at 1.875000000 has equation
$$L(x) = 1281728431.0445077(x - 1.875000000) + 151149355.16656.$$
$$F(1.816406250) = 56123805.8756633 < L(1.816406250) = 76048079.9100473$$
$$F(2.000000000) = 218079993.6000000 < L(2.000000000) = 311365409.0471249$$
On $[1.649665833, 1.816406250]$, the tangent to $y = H(x)$ at 1.702880859 has equation
$$L(x) = 295425260.6824139(x - 1.702880859) + 35844248.88537.$$
$$F(1.649665833) = 16402147.4634448 < L(1.649665833) = 20123185.7043692$$
$$F(1.816406250) = 56123805.8756633 < L(1.816406250) = 69382517.0048312$$
On $[1.498231664, 1.649665833]$, the tangent to $y = H(x)$ at 1.546561718 has equation
$$L(x) = 80685866.5102746(x - 1.546561718) + 10049999.60617.$$
$$F(1.498231664) = 5384682.3888346 < L(1.498231664) = 6150447.3459398$$
$$F(1.649665833) = 16402147.4634448 < L(1.649665833) = 18369044.4279982$$
On $[1.360698679, 1.498231664]$, the tangent to $y = H(x)$ at 1.404592185 has equation
$$L(x) = 25547293.8286764(x - 1.404592185) + 3258189.84941.$$
$$F(1.360698679) = 1966103.4253539 < L(1.360698679) = 2136829.5598246$$
$$F(1.498231664) = 5384682.3888346 < L(1.498231664) = 5650425.1338695$$

$$\vdots$$

$$\vdots$$

On $[0.066590697, 0.086975604]$, the tangent to $y = H(x)$ at 0.076103653 has equation
$$L(x) = 1924.8187828(x - 0.076103653) + 344.80344.$$
$$F(0.066590697) = 325.7352117 < L(0.066590697) = 326.4927201$$
$$F(0.086975604) = 365.0225367 < L(0.086975604) = 365.7299722$$
On $[0.043700145, 0.066590697]$, the tangent to $y = H(x)$ at 0.058266859 has equation
$$L(x) = 1728.0567129(x - 0.058266859) + 312.25921.$$
$$F(0.043700145) = 287.0245426 < L(0.043700145) = 287.0871027$$
$$F(0.066590697) = 325.7352117 < L(0.066590697) = 326.6432746$$

On $[0.024581331, 0.043700145]$, the tangent to $y = H(x)$ at 0.032775108 has equation

$$L(x) = 1483.1955575(x - 0.032775108) + 271.41374.$$

$$F(0.024581331) = 258.5270060 < L(0.024581331) = 259.2607690$$
$$F(0.043700145) = 287.0245426 < L(0.043700145) = 287.6177079$$

On $[0.000000000, 0.024581331]$, the tangent to $y = H(x)$ at 0.012290666 has equation

$$L(x) = 1313.3126610(x - 0.012290666) + 242.80899.$$

$$F(0.000000000) = 226.3393988 < L(0.000000000) = 226.6675051$$
$$F(0.024581331) = 258.5270060 < L(0.024581331) = 258.9504788$$

Finally, we will also show that b_1, b_2, b_3 satisfy the admissible families condition. Thus the contractive substitution and admissible families conditions together still are not sufficient to characterize the initial coefficients of the power series of normalized Riemann mappings. From *[8]*, we have the following theorem.

Theorem 6: *Given numbers $b_1, b_2, b_3 (b_1 > 0)$ which are not $1, 0, 0$ satisfy the admissible families condition if and only if $b_1 < 1, |b_2| \leq 2b_1(1 - b_1)$ and*

$$\left| b_3 + \frac{1}{2}\left(\frac{\nu - 1}{b_1} + \frac{1}{1 - b_1}\right) b_2^2 \right|^2 \leq \frac{2b_1^2(1 - b_1)^2(b_1^{-1} + 1) - |b_2|^2}{2b_1(1 - b_1)}$$

$$\times \left[8(\nu + 1)^2 b_1^2(1 - b_1) - (2\nu + 3)(2\nu + 1)b_1^2(1 - b_1^2) - \frac{2\nu + 2 - (2\nu + 1)b_1}{2(1 - b_1)}|b_2|^2 \right]$$

for all real ν.

It is easy to see that for our numbers $b_1, b_2, b_3, 0 < b_1 < 1, |b_2| \leq 2b_1(1 - b_1)$. Now substituting the numbers into the long inequality in the theorem, we get the quadratic inequality

$$\frac{4199}{186624}\nu^2 + \frac{11599}{466560}\nu + \frac{5063}{518400} \geq 0.$$

Since the discriminant is

$$\left(\frac{11599}{466560}\right)^2 - 4 \times \frac{4199}{186624} \times \frac{5063}{518400} = -\frac{7099879}{27209779200},$$

the inequality holds for all real ν.

References

[1]   J. Bognár, *Indefinite Inner Product Spaces*, Springer-Verlag, Berlin-New York, 1974.

[2]   L. de Branges, "A proof of the Bieberbach conjecture," Acta Math. 154 (1985), 137-152.

[3]   L. de Branges, "Unitary linear systems whose transfer functions are Riemann mapping functions," Integral Equations and Operator Theory 19 (1986), 105-124.

[4] L. de Branges, "Underlying concepts in the proof of the Bieberbach conjecture," Proceedings of the International Congress of Mathematicians 1986, pp. 25-42, Berkeley, California, 1986.

[5] L. de Branges, *Square Summable Power Series*, in preparation.

[6] G. Christner, K. Y. Li, and J. Rovnyak, "Julia operators and coefficient problems," Oper. Theory: Adv. Appl., to appear.

[7] K. Y. Li, "de Branges' conjecture on bounded Riemann mapping coefficients," J. Analyse Math. 54 (1990), 289-296.

[8] K. Y. Li and J. Rovnyak, "On the coefficients of Riemann mappings of the unit disk into itself," Oper. Theory: Adv. Appl., 62 (1993), 145-163.

[9] M. Rosenblum and J. Rovnyak, *Hardy Classes and Operator Theory*, Oxford University Press, New York, 1985.

[10] J. Rovnyak, "Coefficient estimates for Riemann mapping functions," J. Analyse Math. 52 (1989), 53-93.

[11] I. Schur, "Über Pitenzreihen, die im Innern des Einheitskreises beschränkt sing, I, II", J. reine Angew. Math. 147 (1917), 205-232; ibid. 148 (1918), 122-145.

On Minimal Thinness, Boundary Behavior of Positive Harmonic Functions and Quasiadditivity of Capacity

Matts Essen

Uppsala University, Department of Mathematics
P.O. Box 480, S–751 06 Uppsala, Sweden

0 Introduction

This is an expanded version of a survey of recent results in two related areas, presented at the "International Conference on Complex Analysis", organized by Nankai Institute of Mathematics in Tianjin, China, in June 1992. The first three sections are concerned with the connection between minimal thinness, reduced functions and certain results on the boundary behavior of positive harmonic functions going back to A. Beurling. It turns out that a key observation in the proof of Theorem 1 in Section 1 is that capacity is not only subadditive but can also in certain cases be "quasiadditive", i.e. almost additive. A number of interesting examples where this new concept can be used have recently been given by H. Aikawa: in Section 4, we present some of these results. No proofs are given in the first four sections. The main references are the papers *[4]*, *[1]* and *[2]*. For simplicity, we have restricted our discussion in the first three sections to the plane. However, it is natural also to consider problems of this type in higher dimensions. As an example, we give a proof of the minimum principle of Beurling in higher dimensions in Section 5.

1 Minimal thinness and boundary behavior of positive harmonic functions

Let Ω be an open connected subset of the unit disc U in the plane which is regular with respect to the Dirichlet problem. The complement $E = U \setminus \Omega$ is relatively closed in U. Let $SH(U)$ denote the class of nonnegative superharmonic functions in U. If h is a harmonic function in U, we consider the reduced function

$$R_h^E(z) = \inf\{u(z) : u \in SH(U), \, u \geq h \quad \text{on } E\},$$

and the lower regularization

$$\widehat{R}_h^E(z) = \liminf R_h^E(w), \, w \to z, \, w \in U,$$

which is superharmonic in U (cf. Helms *[13]*).

If we choose $h = 1$ and assume that E has a positive distance to the unit circle T, $\widehat{R}_1^E$ will be harmonic in Ω with boundary values 1 on $\partial E \cap U$ and 0

Proceedings of the International Conference on Complex
Analysis at the Nankai Institute of Mathematics, 1992 pp. 59-68
©INTERNATIONAL PRESS 1994

on T. It is easy to see that $\widehat{R}_1^E$ is a Green potential. On the other hand, if we choose E as the set $\{z \in U : |z - 1| \leq 1/2\}, \widehat{R}_1^E$ can not be a Green potential since it is 1 on $T \cap \{|z - 1| \leq 1/2\}$. Can we characterize sets E which are such that $\widehat{R}_1^E$ is a Green potential?

Definition: A relatively closed subset E of a domain A is minimally thin in A at a point τ on the Martin boundary ∂A if there exists $z_0 \in A \setminus E$ such that $\widehat{R}_h^E(z_0) < h(z_0)$, where h is the (essentially unique) minimal harmonic function in A with pole at τ.

When $A = U$ and $\tau \in T, h$ is the Poisson kernel P_τ in U, where

$$P_\tau(z) = P(\tau, z) = (1 - |z|^2)|z - \tau|^{-2}.$$

As another example, we choose $A = \{\Re z > 0\}$. In this case, h is the Poisson kernel in the half–plane at finite boundary points and $\Re z$ if $\tau = \infty$.

We quote a special case of a very general result of Naim *[15]*(Section 26).

Theorem A: *Let* $h(z) = \int_T P(\tau, z)d\mu(\tau)$ *be a positive harmonic function in* U. *Then*

$$\widehat{R}_h^E(z) = \int_N P(\tau, z)d\mu(\tau) + \{\text{a Green potential}\}$$

where $N = \{\tau \in T : E \text{ is not minimally thin at } \tau\}$.

An immediate corollary is that $\widehat{R}_h^E$ is a Green potential if and only if $\mu(N) = 0$. If $h = 1$, we have $d\mu(e^{i\theta}) = d\theta/2\pi$ and we see that $\widehat{R}_1^E$ will be a Green potential if and only if E is minimally thin a.e. on T.

To formulate a new criterion for a set E to be minimally thin at a boundary point $\tau \in T$ in U, we let $\{Q\}$ be a Whitney decomposition of U into approximate "squares" with centres $z(Q)$ and

$$\text{diam.} Q \leq \text{dist.} \{z(Q), T\}/3 = t/3,$$

(cf. Stein *[16]*). If $c(E)$ denotes the capacity of E, we consider

$$W(\tau) = W(\tau, E) = \sum((1 - |z(Q_k)|)/|1 - \tau\overline{z(Q_k)}|)^2 (\log(4t_k/c(E_k)))^{-1},$$

where $E_k = E \cap Q_k$.

Theorem 1: *E is minimally thin at τ if and only if $W(\tau) < \infty$.*

Corollary: *$\widehat{R}_1^E$ is a Green potential if and only if $W(\tau) < \infty$ a.e. on T.*

We consider also

$$W_0(\tau) = W_0(\tau, E) = \sum_{E \cap Q_k \neq \emptyset} ((1 - |z(Q_k)|)/|1 - \tau\overline{z(Q_k)}|)^2.$$

The functions W and W_0 have recently appeared in two papers. In the problem considered by Hayman and Lyons *[12]* (cf. also Bonsall and Walsh

[6] and Bonsall *[5]*), the crucial condition is that we must have $W_0(\tau) = \infty$ everywhere on T. In *[17]*, Volberg considers "boundary layers": a necessary condition is that $W(\tau) < \infty$ everywhere on T.

What is the relation between the functions W or W_0, minimal thinness, certain known results on the boundary behavior of harmonic functions going back to Beurling *[4]* and the Hayman–Lyons and Volberg results? If we start from Naim's Theorem A and Theorem 1, we can give simple proofs of all these results. There are analogues in higher dimensions and in more general domains (cf. Aikawa *[1]*(Theorem 4 and Remark) and Maz'ya *[14]*). We shall discuss these results as well as "quasiadditivity" of capacities, a concept introduced and investigated by Aikawa *[1]*, *[2]*; a special case was considered already in Essén *[9]*.

We begin with the result of Beurling. Let $S = \{z_n\}$ be a sequence of points in U and let

$$S(\delta) = \cup_1^\infty \{z \in U : \rho(z, z_n) < \delta\},$$

where ρ is the hyperbolic metric in U. Beurling defines S to be an equivalence sequence at $\tau \in T$ if for each positive harmonic function h in U, the inequalities

$$h(z) \geq P(\tau, z), \ z \in S,$$

imply that

$$h(z) \geq P(\tau, z), \ z \in U.$$

In our language, the result of Beurling *[4]* can be stated as

Theorem B: *S is an equivalence sequence at τ if and only if $S(\delta)$ is not minimally thin at τ for some positive δ, or equivalently, that $W_0(\tau, S) = \infty$.*

To give an idea of the proof, we note first that we can assume that the hyperbolic discs in $S(\delta)$ are disjoint (cf. the proof of Theorem B in *[9]*). It follows that

$$W(\tau, S(\delta)) \approx W_0(\tau, S(\delta)).$$

This depends on the fact that for the set $S(\delta)$, all terms containing capacity will be essentially constant for all indices k. If $S(\delta)$ is not minimally thin at τ, we have $W_0(\tau) = \infty$. Mapping U onto $\{\Re z > 0\}$ with τ going to ∞, we find that

$$\sum X_k^2/(1 + Y_k^2) = \infty,$$

where $Z_k = X_k + iY_k$ is the image of z_k. This is the condition used by Beurling.

Beurling's theorem has been generalized to elliptic partial differential equations by Maz'ya *[14]* and to harmonic functions in higher dimensions by Dahlberg *[7]*. Since Maz'ya's result deserves more attention than it has received, we have included it in Section 6.

We note that Theorem B holds also in balls or half–spaces in higher dimensions (cf. Section 5) or even in $C^{1,\alpha}$–domains: it is natural to state results of this type in terms of minimal thinness.

Beurling's theorem is an important tool in the work of Hayman and Lyons. A second tool is given by

Lemma 1:   *E is not minimally thin at $\tau \in T$ if and only if*

$$P(\tau, z) = \int_\Gamma P(\tau, \zeta) w(z, d\zeta)$$

where $\Gamma = \partial\Omega \cap U$ and we integrate with respect to harmonic measure in Ω.

In the proof, we use only the definition of minimal thinness.

2   The results of Hayman--Lyons and Volberg

To describe the problem of Hayman and Lyons, we let B be a subclass of the class $C^+(T)$ of non–negative continuous functions on T. B is a basis for $C^+(T)$ if the positive cone spanned by B is dense in $C^+(T)$ (we use the uniform norm). A subset S of U will be called basic if the set $B = \{P(\cdot, z) : z \in S\}$ is a basis for $C^+(T)$.

Theorem C: *(Hayman and Lyons [12]) A set $S \subset U$ is basic if and only if*

$$W_0(\tau) = W_0(\tau, S) = \infty \quad \text{for all } \tau \in T.$$

Equivalently, S is basic if and only if the set $S(\delta)$ is not minimally thin at any point of T for some positive δ.

Volberg defines $\Omega = U \setminus E$ to be a boundary layer if there exists a constant $c > 0$ such that $w(0, I) \geq c|I|$ for all arcs $I \subset T$ where $w(0, I)$ is harmonic measure of the arc I in the domain Ω evaluated at the origin. A necessary and sufficient condition for Ω to be a boundary layer is that

$$P^w(\tau) := \int_\Gamma P(\tau, \zeta) w(0, d\zeta) \leq 1 - c \quad \text{for all } \tau \in T,$$

where $\Gamma = \partial\Omega \cap U$ (cf. *[17]*). We note that if there exists one point $\tau \in T$ where E is not minimally thin, we can use Lemma 1 to conclude that

$$1 = P(\tau, 0) = \int_\Gamma P(\tau, \zeta) w(0, d\zeta).$$

Consequently, Ω is not a boundary layer. A necessary (but not a sufficient) condition for Ω to be a boundary layer is that E is minimally thin everywhere on T. If we assume also that the series defining $W(\cdot, E)$ is uniformly convergent on T, Ω will be a boundary layer.

3   A sketch of the proof of Theorem 1

To give an idea of the proof of Theorem 1, we start with minimal thinness at infinity in the half–plane $D = \{\Re z > 0\}$. We write $z = x + iy$. A relatively closed subset E of D is minimally thin at infinity in D if there exists $z_0 = x_0 + iy_0 \in D \setminus E$ such that $\widehat{R}_x^E(z_0) < x_0$. Such a set can be characterized by a criterion of Wiener type. Let us for a moment assume that E is a compact subset of D. The fundamental distribution on E is a measure λ_E supported by E such that the Green potential

$$G\lambda_E(z) = \int G_D(z,\zeta)d\lambda_E(\zeta) = \widehat{R}_x^E(z), \ z \in D,$$

where $G_D(z,\zeta)$ is Green's function in D with pole at ζ. The Green energy of L is defined as

$$\gamma(E) = \int G\lambda_E(z)d\lambda_E(z);$$

it can be extended to an outer capacity defined on subsets of D (cf. *[10]*).

Returning to the general case, we can prove that a relatively closed subset E of D is minimally thin at infinity in D if and only if

$$\sum_1^\infty \gamma(E^{(n)})2^{-n} < \infty,$$

where $E^{(n)} = E \cap \{2^n \le |z| < 2^{n+1}\}, n = 1, 2, \dots$ For a discussion of the history of these results, we refer to *[10]*(Section 3).

We wish to replace the Green energy $\gamma(\cdot)$ in this criterion by ordinary capacity $c(\cdot)$. Let $\{Q\}$ be a Whitney decomposition of D as in *[11]*(Section 5). To a subset $\widetilde{E}$ of D and a Whitney square Q_k, we associate $\widetilde{E}_k = \widetilde{E} \cap Q_k$ and the numbers $\{t_k, R_k, \theta_k\}$ where t_k is the distance from the centre $z(Q_k)$ of Q_k to ∂D, R_k is the distance from $z(Q_k)$ to the origin and $\theta_k = \arccos(t_k/R_k)$.

Theorem D: *(cf. Essén [8](Theorem 2)). $\widetilde{E}$ is minimally thin at infinity in D if and only if*

$$\sum (\cos\theta_k)^2(\log(3t_k/c(\widetilde{E}_k)))^{-1} < \infty$$

where we sum over all Whitney squares in a neighborhood of infinity.

To prove Theorem 1 starting from Theorem D, we consider the function $F(z) = (\tau + z)/(\tau - z)$ which maps U conformally onto D with τ going to infinity. Then E will be minimally thin at $\tau \in T$ in U if and only if $\widetilde{E} = F(E)$ is minimally thin at infinity in D. Applying Theorem D and going back to U, we obtain Theorem 1.

Since minimally thin sets are preserved by inversion maps, it is easy to deduce an analogue of Theorem D at a finite boundary point $z_0 \in \partial D$. For simplicity, we take $z_0 = 0$.

Theorem D':   *A set $\widetilde{E}$ is minimally thin at $O \in \partial D$ in D if and only if (i) in Theorem D holds: this time, we sum over all Whitney squares in a neighborhood of the origin.*

4 Quasiadditivity of capacity

It is well–known that capacity is sub–additive: if $E = \cup E_k$, then we have

$$c(E) \leq \sum c(E_k).$$

We say that a capacity $c(\cdot)$ is quasiadditive with respect to a decomposition $\{Q_k\}$ of a domain Ω if there exists a constant A such that

$$\sum c(E \cap Q_k) \leq Ac(E)$$

for all subsets E of Ω.

It turns out that Green capacity $\gamma(\cdot)$ is quasi–additive with respect to Whitney decompositions of a half–plane (or a half–space) D: this is a key fact in the proof of Theorem D (cf. *[8]*).

However, it was observed by Aikawa that many different capacities have this property: the word "quasi–additive" was introduced by him. In *[1]*, he proved that Riesz capacity is quasi–additive with respect to Whitney decompositions of domains which are components of complements of closed sets without interior points in $\mathbb{R}^n$ such as $m-$ dimensional affine subspaces with $m < n$. If $m = n-1$, we get half–spaces (the case $n = 2$ was discussed above). If $m = 0$, the closed set contains one point only, and we consider decompositions of $\mathbb{R}^n \setminus \{0\}$ into spherical shells (for details, we refer to *[1]*).

The following discussion leading up to Theorem E is taken from Aikawa *[2]*.

Definition: Let $\{Q_j\}$ and $\{Q_j^*\}$ be families of Borel subsets of $X \subset \mathbb{R}^n$ such that

(i) $Q_j \subset Q_j^*$

(ii) $X = \cup Q_j$

(iii) $\sum \chi_{Q_j^*} \leq$ Const. Then we say that $\{Q_j, Q_j^*\}$ is a quasidisjoint decomposition of X.

Remark. The third condition above says that the enlarged sets $\{Q_j^*\}$ do not overlap too much. Natural examples are given by Whitney decompositions.

If f and g are two positive quantities, we say that $f \approx g$ if there exists a positive constant M such that $M^{-1} \leq f/g \leq M$.

Definition: Let σ be a measure on X. We say that σ is comparable to a capacity $C(\cdot)$ with respect to a decomposition $\{Q_j\}$ if

(i) $\sigma(Q_j) \approx C(Q_j)$ for all j.

(ii) $\sigma(E) \leq$ Const.$C(E)$ for all Borel sets E.

Let k be a nonnegative, lower semicontinuous function on $X \times X$. We consider the capacity C_k defined by

$$C_k(E) = \inf\{\|\mu\| : k(\cdot, \mu) \geq 1 \quad \text{on } E\}$$

where $k(\cdot, \mu) = \int_X k(\cdot, y)d\mu(y)$.

We say that such a kernel has the Harnack property with respect to $\{Q_j, Q_j^*\}$ if

$$k(x,y) \approx k(x',y) \quad \text{for } x, x' \in Q_j \quad \text{and } y \in X \setminus Q_j^*,$$

where the constant of comparison M is independent of j.

Theorem E: *Let $\{Q_j, Q_j^*\}$ be a quasidisjoint decomposition of X. If there exists a measure σ comparable to $C_k(\cdot)$, then for every $E \subset X$,*

$$C_k(E) \approx \sum_j C_k(E \cap Q_j).$$

This result is of interest when we can find such a measure σ. In *[2]*, Aikawa uses "Hardy's inequality" (cf. Ancona *[3]*) which is valid in "uniformly $\triangle$–regular" domains to construct σ. He uses this result to study minimally thin sets in non–tangentially accessible domains and deduces generalizations of the above results of Beurling, Maz'ya and Dahlberg (and many other authors). For details, we refer to *[2]* and the references in *[2]*.

In the case of Riesz capacity (cf. *[1]*), Aikawa uses a different tool: a weighted mean inequality of Muckenhoupt type.

5 Beurling's minimum principle in higher dimensions

Let $S = \{z_n\}_1^\infty$ be a sequence in the half–space $D = \{(x,y) : x \in \mathbb{R}^{n-1}, y > 0\}$ and let $S(\delta) = \cup_1^\infty \{z \in D : \rho(z, z_n) < \delta\}$ where ρ is the hyperbolic metric in the half–space D. We say that S is an equivalence sequence at $\tau \in \partial D$ if for each positive harmonic function h in D, the inequalities

$$h(z) \geq P(\tau, z), \ z \in S,$$

imply that

$$h(z) \geq P(\tau, z), \ z \in D,$$

where $P(\tau, z)$ is the Poisson kernel in D with pole at τ. If $\tau = \infty, P(\tau, z) = y$.

Theorem B': *Let D be as above with $n \geq 3$. S is an equivalence sequence at τ if and only if $S(\delta)$ is not minimally thin at τ in D for some positive δ.*

In the proof, we need the following analogue of Theorems D and D': we use notation given in connection with Theorem D (cf. Essén *[8]*(Theorem 2) and *[9]*).

Theorem F: *Let D be as above with $n \geq 3$. A set $\widetilde{E} \subset D$ is minimally thin at infinity in D or at $0 \in \partial D$ if and only if*

(i) $$\sum (\cos \theta_k)^2 c(E_k) R_k^{2-n} < \infty$$

where we sum over all Whitney cubes in D near infinity or near 0; $c(\cdot)$ denotes Newtonian capacity.

Without loss of generality, we can assume that $S(\delta)$ is a union of non–intersecting hyperbolic balls (cf. the proof of Theorem B in *[9]*). The capacity of the ball centered at $z_k = (x_k, y_k)$ is of the order of magnitude $(\delta y_k)^{n-2}$ (the radius of the ball is roughly δy_k). There is essentially at most one such ball in each Whitney cube and we conclude that $c(E_k) \approx y_k^{n-2}, E \cap Q_k \neq \emptyset$. We note that $\cos \theta_k = y_k / R_k$. Thus a set of type $S(\delta)$ is minimally thin at $\tau = \infty$ if and only if

$$\sum_{E \cap Q_k \neq \emptyset} (y_k / R_k)^n < \infty. \qquad\qquad *$$

An equivalent condition is

$$\sum_{E \cap Q_k \neq \emptyset} (y_k / |x_k|)^n < \infty.$$

If $S(\delta)$ is minimally thin at ∞, we define

$$P(z) = \sum y y_k^n (|x - x_k|^2 + y^2)^{-n/2}$$

which is a harmonic function in D with

$$P(z_k) > y_k,$$

and

$$P((0,y))/y = \sum y_k^n (|x_k|^2 + y^2)^{-n/2} \to 0, y \to \infty.$$

It follows that S is not an equivalence sequence at infinity.

Conversely, assume that $S(\delta)$ is not minimally thin at infinity for some and thus all positive δ (cf. $(*)$!). If

$$h(z) \geq y, \quad z \in S,$$

we can use Harnack's theorem to conlude that

$$h(z) \geq C(\delta)y \quad \rho(z, z_k) \leq \delta, k = 1, 2, \dots$$

Since $S(\delta)$ is not minimally thin at infinity, we conclude that

$$h(z) \geq C(\delta)y, z \in D.$$

This inequality holds for all positive δ. Since $C(\delta) \to 1$ as $\delta \to 0$, we have

$$h(z) \geq y, \quad z \in D.$$

Thus S is an equivalence sequence at infinity in D.

We note that as soon as we have proved that condition $(*)$ is necessary and sufficient for $S(\delta)$ to be minimally thin at infinity, we can use essentially the same argument as in the case $n = 2$. (cf. the proof of Theorem B in *[9]*).

6 Maz'ya's generalization of Beurling's minimum principle for positive harmonic functions

To describe the work of Maz'ya from *[14]*, we let Ω be a domain in $\mathbb{R}^n$ with compact closure $\overline{\Omega}$ and boundary $\partial\Omega$ of class $C^{1,\alpha}, 0 < \alpha < 1$, and consider the operator

$$Lu = -(a_{ij}(z)u_i)_j,$$

in Ω, where $u_i = \partial u/\partial z_i, a_{ij} = a_{ji} \in C^\alpha(\overline{\Omega})$ and there exists a constant $a \in (0, 1]$ such that

$$a|\xi|^2 \le a_{ij}(z)\xi_i\xi_j \le a^{-1}|\xi|^2.$$

We shall study boundary behavior of solutions of the equation $Lu = 0$ near a point of $\partial\Omega$ which we assume to be 0. Without loss of generality, we suppose that $a_{ij}(0) = \delta_i^j$. The expression $(a_{ij}(z)u_i)_j$ should be understood in a weak sense.

We call the sequence $S = \{z_m\}_1^\infty$ converging to 0 in Ω determining if for every positive $L-$harmonic function u in Ω the inequalities

$$u(z_m) \ge \lambda(\partial G/\partial\nu)(z_m, 0), m = 1, 2, ...; \lambda > 0,$$

imply

$$u(z) \ge \lambda(\partial G/\partial\nu)(z, 0) \quad \text{for all } z \in \Omega.$$

Theorem: *The sequence S is determining if and only if it contains a subsequence $\{z_{m_\nu}\}_{=1}^\infty$ such that*

$$\inf_{\nu\neq\mu} |z_{m_\nu} - z_{m_\nu}|/dist.(z_{m_\nu}, \partial\Omega) > 0,$$

and

$$\sum_1^\infty [dist.(z_{m_\nu}, \partial\Omega)/|z_{m_\nu}|]^n = \infty.$$

Near $0, \partial\Omega$ can be considered to be a part of R^{n-1}. In fact, since $\partial\Omega \in C^{1,\alpha}$, we can locally map $\partial\Omega$ into R^{n-1} preserving our assumption that near 0, the coefficients of the differential operator are in $C^\alpha(\overline{D})$, where $D = \{(x,y) : x \in R^{n-1}, y > 0\}$. Now the conditions of the theorem can be rewritten as

(i) $\inf_{\nu\neq\mu} |z_{m_\nu} - z_{m_\nu}|/|z_{m_\nu} - \overline{z_{m_\nu}}| > 0$, with $z = (x, y), x \in R^{n-1}, y > 0$ and $\overline{z} = (z, -y)$,

(ii) $\sum(y_{m_\nu}/|z_{m_\nu}|)^n = \infty$.

Remark. We note that Maz'ya's condition (ii) says that the set $S(\delta)$ is not minimally thin at 0 in D (cf. condition $(*)$ in Section 5).

References

[1]   H. Aikawa, Quasiadditivity of Riesz capacity, Math. Scand. 69 (1991), 15–30.

[2]   H. Aikawa, Quasiadditivity of capacity and minimal thinness. To appear, Ann. Acad. Sci. Fenn.

[3] A. Ancona, On strong barriers and an inequality of Hardy for domains in $\mathbb{R}^n$, J. London Math. Soc. (2) 34 (1986), 274–290.

[4] A. Beurling, A minimum principle for positive harmonic functions. Annal.Acad. Scient.Fennicae Ser. AI 372 (1965).

[5] F.F. Bonsall, Domination of the supremum of a bounded harmonic function by its supremum over a countable subset. Proc. of the Edinburgh Math.Soc. 30 (1987), 471–477.

[6] F.F. Bonsall and D. Walsh, Vanishing ℓ^1 –sums of the Poisson kernel, and sums with positive coefficients. Proc. of the Edinburgh Math.Soc. 32 (1989), 431–447.

[7] B. Dahlberg, A minimum principle for positive harmonic functions. Proc. of the London Math. Soc. XXXIII (1976), 238–250.

[8] M. Essén, On Wiener conditions for minimally thin and rarefied sets. In ”Complex Analysis”, ed. J. Hersch and A. Huber, Birkhuser Verlag, Basel 1988, 41–50.

[9] M. Essén, On minimal thinness, reduced functions and Green potentials, to appear, Proc. of the Edinburgh Math. Soc.

[10] M. Essén and H. Jackson, On the covering properties of certain exceptional sets in a half–space, Hiroshima Math. J. 10 (1980), 233–262.

[11] M. Essén, H. Jackson and P.J. Rippon, On minimally thin and rarefied sets in $\mathbb{R}^p, p \geq 2$. Hiroshima Math. J. 15 (1985), 393–410.

[12] W.K. Hayman and T.J. Lyons, Bases for positive continuous functions, J. London Math. Soc. (2) 42 (1990), 292–308.

[13] L.L. Helms, Introduction to potential theory. Wiley-Interscience, New York-London 1969.

[14] V.G. Maz'ya, Beurling's theorem on a minimum principle for positive harmonic functions. (Russian), Zapiski Nauchnykh Seminarov LOMI 30 (1972), 76–90; (English translation), J. Soviet Math. 4 (1975), 367–379.

[15] L. Naim, Sur le rôle de la frontière de R.S. Martin dans la théorie du potentiel. Ann. Inst. Fourier 7 (1957), 183–281.

[16] E.M. Stein, Singular integrals and differentiability of functions. Princeton University Press 1970.

[17] A.L. Volberg, A criterion on a subdomain of the disc to have its harmonic measure comparable with Lebesgue measure. USSR Academy of Sciences. Steklov Mathematical Institute, Leningrad department, LOMI preprints E-2-89; Proc. Amer. Math. Soc. 112 (1991), 153–162.

On Homeomorphisms with Integrable Dilatation[1]

Ainong Fang

Department of Mathematics, Shanghai Jiao Tong University
Shanghai, 200030, P. R. China.

Abstract: A modulus inequality, ACL^n property and N-condition are given for homeomorphism f with sphere dilatation $H(x, f) \in L(D)$.

1 Introduction

Suppose that Γ is a family of nonconstant curves γ in $\overline{R}^n$. Let $adm(\Gamma)$ be the family of all nonnegative Borel functions $\rho : \quad R^n \to R^1$ such that

$$\int_\gamma \rho ds \geq 1$$

for every locally rectifiable curve $\gamma \in \Gamma$. We call

$$(1.1) \qquad\qquad M(\Gamma) = \inf_{\rho \in adm(\Gamma)} \int_{R^n} \rho^n dm$$

the modulus of Γ. Let D and D' be domains in $\overline{R}^n, y = f(x) : D \to D'$ be a sense-preserving homeomorphism and $\Gamma' = f(\Gamma)$. f is K-quasiconformal iff

$$(1.2) \qquad\qquad K^{-1}M(\Gamma) \leq M(\Gamma') \leq KM(\Gamma)$$

for every curve family Γ in D *[1, 11, 5, 13]*.

We call

$$(1.3) \qquad\qquad H(x, f) = \limsup_{r \to 0} \left[\frac{max_{|y-x|=r}|f(y) - f(x)|}{min_{|y-x|=r}|f(y) - f(x)|} \right]$$

the sphere dilatation.

It is well known that for every K-quasiconformal mapping f, $K \geq 1$ and $n \geq 2$ there is a constant $H = K^{\frac{2}{n}}$ such that $H(x, f) \leq H$. Conversely, for each constant H with $H(x, f) \leq H$, there exists a $K = H^{n-1}$ such that f is K-quasiconformal. So we can define the K-quasiconformal mapping by K or H *[6][13]*(34.2).

In 1928 H. Grötzsch introduced the concept of quasiconformal mapping, since then its theories and applications are springing up vigorously *[12]*. It is

---

[1]Project supported by the National Natural Science Foundation of China and JTU.

Proceedings of the International Conference on Complex
Analysis at the Nankai Institute of Mathematics, 1992 pp. 69-78
©INTERNATIONAL PRESS 1994

also well known that if $D \subset \overline{R}^2$, then D is K-quasiconformally equivalent to the unit disk B iff $C(D)$ the complement of D in $\overline{R}^2$, is a nondegenerate continuum. When $D \subset \overline{R}^n (n \geq 3)$, if $f : D \to B$ (the unit ball) is K-quasiconformal and $C(D) \cap \overline{B(x_0, b)}$ has at least two components which meet $\overline{B(x_0, a)}$, then $K \geq c(\log \frac{b}{a})^{n-1}$ where $x_0 \in R^n, 0 < a < b < \infty$, and c is a positive constant which depends only on n [5]. Therefore, it is reasonable to consider reducing the requirement on the dilatation $H(x, f)$. This is not new thinking and some pioneer works have be done by F. W. Gehring, J. Väisälä, O. Lehto, C. Q. He, T. Iwance, etc. [6, 13, 11, 8, 9]. they studied respectively the K-quasiconformal mappings with the exceptional sets or removable point sets or Riemannian integrable dilatation $H(x, f)$. T. Iwance and V. Šveràk established Stoilow's type factorization of mappings in 2-dimension with integrable dilatation

$$(1.4) \qquad K(z, f) = \begin{cases} |Df(z)|^2/J(z, f), & \text{if} \quad J(z, f) \neq 0, \\ 1, & \text{if} \quad J(z, f) = 0, \end{cases}$$

where $|Df(z)|$ denotes the norm of the differential $Df(z)$. We have studied this problem as well.

Definition. *A sense-preserving homeomorphism $y = f(x) : D \to D'$ is said to have integrable dilatation, denoted by HID, if it satisfies the following conditions:*
    (1) f is ACL,
    (2) f is differentiable a.e.,
    (3) $H(x, f) \in L(D)$.

We prove that the sphere dilatation $H(x, f)$ is upper semicontinuous and that if $H(x, f) \in L(D)$ then f is ACL [2, 3]. In this paper, we establish a modulus inequality and ACL^n property of HID and prove HID to satisfy the N-condition. In another paper, we give a distortion theorem, equicontinuity, compactness, convergent theorem, ..., and so on [4].

2 ACL^n-property and N-condition

We know that K-quasiconformal mappings are ACL^n and satisfy the N-condition. In this section we show that HID mappings are also ACL^n and satisfy the N-condition under some conditions.

Theorem 2.1: *Let $y = f(x) : D \to D'$ be a HID mapping and $H(x, f) \in L^{n-1}(D)$ and $H(y, f^{-1}) \in L^{n-1}(D')$. We have*

$$(2.1) \qquad \int_D \left| \frac{\partial f}{\partial x_i} \right|^n dv_x \leq \int_D |f'(x)|^n dv_x \leq \int_{D'} H(y, f^{-1})^{n-1} dv_y,$$

$$(2.2) \qquad \int_{D'} \left| \frac{\partial f^{-1}}{\partial y_i} \right|^n dv_y \leq \int_{D'} |f^{-1'}(y)|^n dv_y \leq \int_D H(x, f)^{n-1} dv_x.$$

where $|f'(x)| = \max_{|h|=1} |f'(x)h|$.

Proof:

According to (1.3) and the paper *[13]*(34.1), we have

$$\left|\frac{\partial f}{\partial x_i}\right|^n \leq |f'(x)|^n \leq H(x,f)^{n-1} J(x,f).$$

Thus

$$\int_D \left|\frac{\partial f}{\partial x_i}\right|^n dv_x \leq \int_D |f'(x)|^n dv_x \leq \int_D H(x,f)^{n-1} J(x,f) dv_x.$$

Because *[2]* $H(x,f) = H(y,f^{-1})$, where $y = f(x)$ and f^{-1} is the inverse of f, it follows from the paper *[13]*(24.5)

$$\int_D H(x,f)^{n-1} J(x,f) dv_x. \leq \int_{D'} H(y,f^{-1})^{n-1} dv_y.$$

Therefore (2.1) is true.

(2.2) can be proved similarly.   ■

T. Iwance and V. Šverák *[9]* studied the above problem, and obtained

$$\int_{D'} |f^{-1'}(y)|^{n-1} dv_y \leq \int_D K^{n-1}(x,f) dv_x.$$

The Jacobian function occurs in many different context, such as geometric theory of measure and integration, mapping degree theory, quasiconformal analysis, nonlinear elasticity, etc. There arises now a natural question: Under what conditions on f is the Jacobian function locally integrable. T. Iwance and C. Sbordone studied the integrability of the Jacobian under $f \in D^n log^{-1} D(D, R^n)$ which is Orlicz-Sobolev class *[10]*. Theorem 2.1 also gives such a condition. In fact, we have

Proposition 2.2: *Under the condition of Theorem 2.1, $J(x,f) \in L(D)$.*

Now we have that the *HID* mappings satisfy the *N*-condition *[13]* and

Theorem 2.3: *Under the condition of Theorem 2.1, if E is measurable, then $f(E)$ is also measurable and*

$$(2.3) \qquad\qquad m(f(E)) = \int_E J(x,f) dv_x$$

and $J(x,f) \neq 0$ a.e.

Proof: First, we prove that if E is open and $\overline{E} \subset D$, then Theorem 2.3 is true. Because $H(x, f) \in L^{n-1}(D), H(x, f)$ is finite a.e. on E. Set

$$A_N = \{x | H(x, f) > N\} \cap E.$$

Clearly

$$(2.4) \qquad\qquad m(A_N) \to 0 \quad as \quad N \to \infty.$$

If $m(A_N) \neq 0$, we take an open set $E_N \subset E$ such that $E_N \supset A_N$ and

$$(2.5) \qquad\qquad m(E_N \setminus A_N) < m(A_N).$$

If $m(A_N) = 0$, we take an open set E_N such that $E_N \supset A_N$ and

$$(2.6) \qquad\qquad m(E_N) < \frac{1}{N}.$$

Since $y = f(x)$ is a homeomorphism, it follows from (2.4), (2.5) and (2.6)

$$(2.7) \qquad\qquad m(f(E_N)) \to 0 \quad as \quad N \to \infty.$$

Obviously, $H(x, f) \leq N$ on $E \setminus E_N$. According to the paper *[13]*(34.1, 33.3), we have

$$(2.8) \qquad m(f(E)) - m(f(E_N))$$

$$= \int_{f(E) \setminus f(E_N)} dv_y$$

$$= \int_{E \setminus E_N} J(x, f) dv_x$$

$$= \int_E J(x, f) dv_x - \int_{E_N} J(x, f) dv_x$$

Using the absolute continuity of integral, (2.7) and (2.8), we obtain

$$(2.9) \qquad\qquad m(f(E)) = \int_E J(x, f) dv_x.$$

Now we prove that Theorem 2.3 is also true for each Borel set $E \subset D$ as *[11]*(6.6). Since $y = f(x)$ is ACL and $H(x, f)$ is finite a.e. According to the paper *[13]*(32.1, 24.1, 24.2)

$$(2.10) \qquad\qquad m(f(E)) \geq \int_E J(x, f) dv_x.$$

Without loss of generality, we can assume $\overline{E} \subset D$. For $\epsilon > 0$ there is an open set $E_\epsilon \supset E$ and $\overline{E}_\epsilon \subset D$ such that

$$(2.11) \qquad\qquad m(E_\epsilon \setminus E) < \epsilon.$$

Applying the absolute continuity of integral, (2.9) and (2.11), we obtain

$$(2.12) \qquad \int_E J(x,f)dv_x = lim_{\epsilon \to 0} \int_{E_\epsilon} J(x,f)dv_x$$

$$= \lim_{\epsilon \to 0} m(f(E_\epsilon)) \geq m(f(E))$$

Joining (2.10) and (2.12), we obtain

$$m(f(E)) = \int_E J(x,f)dv_x$$

where E is any Borel set in D.

3 Modulus inequality

The modulus inequality (1.2) plays an important role in K-quasiconformal mappings. It is very unfortunate that (1.2) is not true for HID mappings. Therefore we have to find its substitute. We fortunately get it in this section

We recall some definitions. Suppose that Γ is a family of nonconstant curves γ in $\overline{R}^n$. Let $F(\Gamma)$ be the family of all nonnegative Borel functions $\rho : R^n \to R^1$ such that $V(\rho) = \int_{R^n} \rho^n dv \neq 0, \infty$. For such a ρ, set

$$L_\gamma(\rho) = \begin{cases} \int_\gamma \rho ds, & \text{if} \rho \in L(\gamma), \\ \infty, & \text{otherwise}, \end{cases}$$

$$L(\rho) = \inf_{\gamma \in \Gamma} L_\gamma(\rho).$$

We call

$$(3.1) \qquad \Lambda(\Gamma) = \sup_{\rho \in F(\Gamma)} \frac{L(\rho)^n}{V(\rho)}$$

the extremal length of Γ.

It is well known that

$$(3.2) \qquad \Lambda(\Gamma) = \frac{1}{M(\Gamma)}.$$

Suppose that S is a family of nonconstant surface s in $\overline{R}^n$. Let $adm(S)$ be the family of all nonnegative Borel functions $\rho : R^n \to R^1$ such that $V(\rho) = \int_{R^n} \rho^n dv \neq 0, \infty$. For such a ρ, set

$$A_s(\rho) = \begin{cases} \int_s \rho^{n-1} d\sigma, & \text{if } \rho \in L^{n-1}(S), \\ \infty, & \text{otherwise}, \end{cases}$$

$$A(\rho) = \inf_{s \in S} A_s(\rho).$$

We call

$$(3.3) \qquad \beta(S) = \sup_{\rho \in adm(S)} \frac{A(\rho)^{\frac{n}{n-1}}}{V(\rho)}$$

the extremal area of S.

A ring R is a domain whose complement consists of a bounded and an unbounded component which we denote by C_0 and C_1, respectively. Let $B_0 = \partial C_0$ and $B_1 = \partial C_1$, these are the components of ∂R. We define conformal capacity of R as

$$Cap(R) = \inf_u \int_R |\nabla u|^n d\omega,$$

where the infimun is taken over all continuous functions u which are ACL in the ring R with boundary values 0 on B_0 and 1 on B_1.

F. W. Gehring [7](1) proved that (1): if S is a family of nonconstant closed surfaces in R which separate C_0 and C_1, then

$$(3.4) \qquad Cap(R) = \beta(S).$$

(2): If $\Gamma = \Delta(C_0, C_1, R)$ denote the family of all curves γ which join C_0 and C_1 in R, then

$$(3.5) \qquad Cap(R) = M(\Gamma).$$

Following Gehring [5] we define modulus of R by

$$(3.6) \qquad mod(R) = \left[\frac{\omega_{n-1}}{Cap(R)}\right]^{\frac{1}{n-1}},$$

where ω_{n-1} denotes the surface area of the unit sphere in R^n.

Theorem 3.1: *Suppose that f is HID with $H(y, f^{-1}) \in L^{n-1}(D')$ and $H(x, f) \in L^{(n-1)^2}(D)$. Then we have*

$$(3.7) \qquad \left[\omega_{n-1} \int_\alpha^\beta (r^{n-1} \int_{s(0,r)} H(x,f)^{(n-1)^2} ds_\omega)^{-\frac{1}{n-1}} dr\right]^{\frac{1}{n-1}}$$

$$\leq mod(f \circ R)$$

$$\leq \omega_{n-1}^{\frac{1}{n-1}} \left[\int_{s(0,1)} (\int_\alpha^\beta \frac{H(x,f)}{r} dr)^{-(n-1)} ds_\omega\right]^{-\frac{1}{n-1}},$$

where $R = \{x \in R^n : \alpha < |x - x_0| < \beta\} \subset D, x_0 \in D$ and ds_ω is the surface area element of $s(0,1)$, $s(0,r) = \{x \in R^n : |x| = r\}$.

Proof:

We prove firstly the right hand of (3.7) by the extremal length. Denote by $\Gamma = \{\gamma\}$ the family of curves γ in $f \circ R$ which join the two boundary components of $f \circ R$. Let $\Gamma_0 = \{\gamma_\omega\}$, where $f^{-1}(\gamma_\omega) = x_0 + [\alpha + r(\beta - \alpha)]\omega, \omega \in s(0,1), 0 \leq r \leq 1$. Obviously $\Gamma_0 \subset \Gamma$. Let $z \in f^{-1}(\gamma_\omega), dr = |dz|, \rho \in F(\Gamma)$ and

$$(3.8) \qquad |f'(z)| = \max_{|h|=1} |f'(z)h|.$$

Noting the ACL property of $f(z)$, we have

$$L_{\gamma_\omega}(\rho) = \int_{\gamma_\omega} \rho(w)|dw| \leq \int_\alpha^\beta \rho \circ f(z)|f'(z)|dr$$

$$= \int_\alpha^\beta \rho \circ f(z)J(z,f)^{\frac{1}{n}}|f'(z)|J(z,f)^{-\frac{1}{n}}dr$$

where $J(x,f)$ is the Jacobian of f in x. According to $[13](34.1,2)$ we have

$$(3.9) \qquad L_{\gamma_\omega}(\rho) \leq \int_\alpha^\beta \rho \circ f(z)J(z,f)^{\frac{1}{n}}r^{\frac{n-1}{n}}H(z,f)^{\frac{n-1}{n}}r^{-\frac{n-1}{n}}dr.$$

Applying Hölder inequality we obtain

$$L_{\gamma_\omega}(\rho)^n \leq \int_\alpha^\beta [\rho \circ f(z)]^n J(z,f)r^{n-1}dr \left(\int_\alpha^\beta \frac{H(z,f)}{r}dr \right)^{n-1}$$

Thus

$$L_{\gamma_\omega}(\rho)^n \int_{s(0,1)} \left(\int_\alpha^\beta \frac{H(z,f)}{r}dr \right)^{-(n-1)} ds_\omega$$

$$\leq \int_{s(0,1)} \int_\alpha^\beta (\rho \circ f(z))^n J(z,f)r^{n-1}dr ds_\omega$$

$$= \int_R (\rho \circ f(z))^n J(z,f)dv_z$$

$$= \int_{f \circ R} \rho^n dv_w = V(\rho).$$

The following inequality can be obtained from the equalities (3.1), (3,2), (3.5), and (3.6),

$$(3.10) \qquad mod(f \circ R)$$

$$\leq \left\{ \omega_{n-1} \left[\int_{s(0,1)} \left(\int_\alpha^\beta \frac{H(z,f)}{r}dr \right)^{-(n-1)} ds_\omega \right]^{-1} \right\}^{\frac{1}{n-1}}.$$

Now we prove the left hand of (3.7) by the extremal area. Denote by $S = \{s\}$ the family of nonconstant closed surfaces s in $f \circ R$ which separate the two boundary components of $f \circ R$. Obviously the images of the spheres $f^{-1}(s_r) = \{|y - x_0| = r, \alpha < r < \beta\} \subset D, x_0 \in D$ under the mapping $f(y)$ form a subfamily of S. Let $z \in s_r, \rho \in adm(S)$ and ds be the area element of $(n-1)$-dimension surface s_r. It follows from (3.3) and (3.8)

$$A_{s_r}(\rho) = \int_{s_r} \rho(z)^{n-1} ds_z$$

$$\leq \int_{s(x_0,r)} [\rho \circ f(y)]^{n-1} |f'(y)|^{n-1} ds_y$$

$$= \int_{s(x_0,r)} [\rho \circ f(y)]^{n-1} J(y,f)^{\frac{n-1}{n}} |f'(y)|^{n-1} J(y,f)^{-\frac{n-1}{n}} ds_y.$$

Using the Hölder inequality we have

(3.11)
$$A_s(\rho)^{\frac{n}{n-1}} \leq \int_{s(x_0,r)} [\rho \circ f(y)]^n J(y,f) ds_y \times$$

$$\times \left(\int_{s(x_0,r)} H(y,f)^{(n-1)^2} ds_y \right)^{\frac{1}{n-1}}.$$

Hence

$$A_s(\rho)^{\frac{n}{n-1}} \left(\int_{s(x_0,r)} H(y,f)^{(n-1)^2} ds_y \right)^{-\frac{1}{n-1}}$$

$$\leq \int_{s(x_0,r)} [\rho \circ f(y)]^n J(y,f) ds_y.$$

Integrating both sides of this inequality with respect to r over $[\alpha, \beta]$, it follows from Fubini theorem, (3.3), (3.4) and (3.6) that

(3.12)
$$\left[\omega_{n-1} \int_\alpha^\beta \left(r^{n-1} \int_{s(x_0,r)} H(x,f)^{(n-1)^2} ds_\omega \right)^{-\frac{1}{n-1}} dr \right]^{\frac{1}{n-1}}$$

$$\leq mod(f \circ R).$$

Combining (3.10) and (3.12), we complete the proof of theorem 3.1.   ∎

If we replace $H(x,f)$ by $K(x,f)$, we obtain easily the following proposition from (1.4), (3.9) and (3.11).

Proposition 3.2: *Suppose that a sense-preserving homeomophism $y = f(x) : D \to D'$ satisfies the following condition: (1) f is ACL, (2) f is differentiable a.e.,*

(3) $K(x,f) \in L^{n-1}(D)$. Then we have

$$(3.13) \qquad \left[\omega_{n-1} \int_\alpha^\beta \left(r^{n-1} \int_{s(x_0,r)} K(x,f)^{n-1} ds_\omega \right)^{-\frac{1}{n-1}} dr \right]^{\frac{1}{n-1}}$$

$$\leq mod(f \circ R)$$

$$\leq \omega_{n-1}^{\frac{1}{n-1}} \left[\int_{s(0,1)} \left(\int_\alpha^\beta K(x,f)^{\frac{1}{n-1}} r^{-1} dr \right)^{-(n-1)} ds_\omega \right]^{-\frac{1}{n-1}}.$$

The inequality (3.13) and its proof are an n-dimensional version of the paper *[8]*.

Acknowledgement: *On June 19, 1992, I reported some results of this paper and [3, 4] at the international conference of the complex analysis, Tianjin, China. After that meeting, I had a helpful discussion with Professor F. W. Gehring. After he returned to U. S. A. , he sent quickly the papers [9, 10] to me. I would like to thank him for his helpful suggestions and sending me those papers.*

References

[1]   Ahlfors, L(1966). Lectures on quasiconformal mappings, Van Nostrand.

[2]   Fang, A. N. On local characteristic of homeonorphism, to appear.

[3]   Fang, A. N. On *ACL*-properties of homeomorphism under weak hypothesis, to appear.

[4]   Fang, A. N. Compactness of homeomorphism with integrable dilatation, to appear.

[5]   Gehring, F. W (1976), Complex analysis and its applications 2, 213-268, Intcr. Atomic Agency, Vienna.

[6]   Gehring, F. W (1962) , Rings and quasiconformal mappings in space, Trans. Amer. Math. Soc. 103, 353-393.

[7]   Gehring, F. W (1962), Extremal length definitions for the conformal capacity of rings in space, Michigan Math. J. , 9, 137-150.

[8]   He, C. Q. (1987) On the existence theorem of quasiconformal homeomorphism, Proceedings of the Symposium on complex analysis, Xian, China.

[9]   Iwance, T. and Šverák, V. On mappings with integrable dilatation, to appear.

[10]   Iwance, T. and Sbordone, C. On the integrability of the Jacobian under minimal hypothesis, to appear.

[11]   Lehto, O. and Virtanen, K. I. (1973). Quasiconformal mappings in the plane, Springer-Verlag.

[12]   Li, Z. (1988). Quasiconformal mappings and their applications in Riemann surfaces, Science Press, China.

[13]   Väisälä, J. (1971). Lectures on n-dimensional quasiconformal mappings, Springer-Verlag.

Keywords: Quasiconformal   mappings,   Condition(N),

Modulus   inequality.

1991 Mathematics Subject Classification. Primary 30C62.

Measurable Dynamics of Some Holomorphic Maps

Liping Fang

Department of Applied Mathematics, Beijing Institute of Technology
Beijing, 100081, China

Abstract: We study the measurable dynamics of holomorphic self–maps on D, where $D = \mathbb{C}$ or $\mathbb{C}^*$. We show the finite–type holomorphic self–map on D is recurrent. $F(z) = z^m e^{P(z)+Q(\frac{1}{z})}$ and $f(z) = mz + P(e^z) + Q(e^{-z})$ are not ergodic on their Julia sets, where $P(z)$ and $Q(z)$ are polynomials with the same degree.

0 Introduction

In this note, we study the measurable dynamics. There is current interest in this subject. Julia sets of some maps have positive measure or whose Hausdorff dimension is greater than one. Their structure is very complicated. Sullivan *[13]* posed a question about the ergodicity for the exponential map whose Julia set is $\mathbb{C}$. He pointed out that there are several possible ways of defining ergodicity, and it's not clear that the exponential map is not ergodic in every sense. Under a larger equivalence relation, E. Ghys, L. Goldberg and D. Sullivan *[6]* showed that e^z is recurrent, and C. McMullen *[11]* showed any member in sine family is not ergodic on its Julia set. In contrast, M. Rees *[12]* and M.Yu. Lyubich *[8]* independently proved that e^z is not recurrent and not ergodic under another definition.

We'll describe the measurable dynamics of finite type holomorphic self–maps on $\mathbb{C}^*$ and $\mathbb{C}$ in this note.

Let $g : \mathbb{C} \to \mathbb{C}$ be an arbitrary measurable map. Two points z_1 and z_2 are equivalent under g if there exist $n, m \geq 0$ such that $g^n(z_1) = g^m(z_2)$. The equivalence relation on $\mathbb{C}$ is called the g–relation, and an equivalence class of points is a large orbit of g.

Definition 1. *A measurable set $X \subset \mathbb{C}$ is a cross section to the g–relation if X contains at most one point in each large orbit.*

Definition 2. *We say g is recurrent (or conservative) if there are no positive measure cross sections to g–relation.*

Definition 3. *We say g is ergodic if any measurable set made up of large orbits has zero or full measure.*

Proposition. If the g–relation is ergodic, it is recurrent.

Proceedings of the International Conference on Complex
Analysis at the Nankai Institute of Mathematics, 1992 pp. 79-89

We'll show the inverse proposition is not correct.

Theorem 1: *Let $f(z) : \mathbb{C}^* \to \mathbb{C}^*$ be a finite type holomorphic self–map. Then $f(z)$ is recurrent on its Julia set.*

Corollary 1: *Suppose $f(z) = z^m e^{P(z)+Q(\frac{1}{z})}$, where P and Q are polynomials with the same degree and $m \in \mathbb{Z}$, then f is recurrent on its Julia set.*

If $J(f)$ has zero measure, it is meaningless to discuss the recurrence and ergodicity. According to Theorem 1 in [5], for $f(z)$ in Corollary 1, its Julia set has positive measure. Then there are many maps really satisfying the conditions of Theorem 1.

Theorem 2: *Let $f(z) : \mathbb{C} \to \mathbb{C}$ be a finite type entire function. Then $f(z)$ is recurrent on its Julia set.*

Theorem 3: *Suppose $f(z) = mz + P(e^z) + Q(e^{-z}) : \mathbb{C} \to \mathbb{C}$, where $P(z)$ and $Q(z)$ are polynomials of degree $d, m \in \mathbb{Z}$, then $f(z)$ is not ergodic on its Julia set.*

Theorem 4: *Suppose $F(z) = z^m e^{P(z)+Q(\frac{1}{z})} : \mathbb{C}^* \to \mathbb{C}^*$, where $P(z)$ and $Q(z)$ are polynomials with the same degree, then $F(z)$ is not ergodic on its Julia set.*

[6], [8] and *[12]* only showed results for e^z . Mr. Qiu Weiyuan showed a result for λe^z, where $\lambda \in C_\rho \subset M = \{\lambda \in \mathbb{C} : J(\lambda e^z) = \mathbb{C}\}$, by the same method of Lyubich[8]. All of these are the particular cases of Theorem 2 by corollaries below.

Corollary 2: *Suppose that $E_\lambda(z) = \lambda e^z$, then for any $\lambda \in M, E_\lambda$ is recurrent on $\mathbb{C}$.*

Corollary 3: *For any $\gamma \neq 0, \nu \neq 0$, let $f(z) = \gamma e^z + \nu e^{-z}$, then $f(z)$ is recurrent on $J(f)$.*

$\{f(z) = \gamma e^z + \nu e^{-z} : \gamma \neq 0, \nu \neq 0\}$ represents the same conformal conjugacy class as the sine family $\{\lambda \sin z : \lambda \neq 0\}$. We showed that $\lambda \sin z$ is recurrent but not ergodic on its Julia set.

1   The recurrence of some maps.

Theorem 1: *Assume $f : \mathbb{C}^* \to \mathbb{C}^*$ is a finite type holomorphic map. Then f is recurrent on its Julia set.*

Proof: If the theorem is false , f is not recurrent on $J(f)$. Then there exists a cross section $X \subset J(f)$ such that mes $(X) > 0$. Suppose that $\text{mes}(X) = C > 0$, we can find a disk $\triangle$ on $\mathbb{C}$ such that $\text{mes}(\triangle \cap X) > 0$. There is a conformal mapping which maps $\triangle$ to the unit disk. We might as well suppose that $\triangle$ is just the unit disk below.

Dividing $\triangle$ into two sectors $\triangle_{S_1}$ and $\triangle_{\overline{S}_1}$ averagely, there exists at least one sector such that the intersection of it and X has positive measure. If $\mathrm{mes}(\triangle_{S_1} \cap X) = c_1 > 0$ and $\mathrm{mes}(\triangle_{\overline{S}_1} \cap X) = 0$, we divide $\triangle_{S_1}$ into two sectors $\triangle_{S_2}$ and $\triangle_{\overline{S}_2}$ averagely. If $\mathrm{mes}(\triangle_{\overline{S}_2} \cap X) = 0$ and $\mathrm{mes}(\triangle_{S_2} \cap X) = c_1 > 0$, we divide $\triangle_{S_2}$ until there exists a n_0 such that $\mathrm{mes}(\triangle_{S_{n_0}} \cap X) > 0$, and $\mathrm{mes}(\triangle_{\overline{S}_{n_0}} \cap X) > 0$.

Now we repeat above arguments for $\triangle_{S_{n_0}}$ and $\triangle_{\overline{S}_{n_0}}$ respectively, there are 2^2 sectors in $\triangle$ and the intersection of X and any one of them has positive measure. We repeat above arguments for all these sectors respectively. Then for any $k > 0$, there exist 2^k sectors in $\triangle$ such that the intersection of X and any one of them has positive measure.

For any one of the 2^k sectors, denoted by T, there is a point $\alpha \in \partial T$ (i.e. the boudary of T on $\partial \triangle$), we might as well suppose $\alpha \in [0, 2\pi]$, such that for any $\varepsilon > 0, \mathrm{mes}(O_\alpha(\varepsilon) \cap X) > 0$, where

$$O_\alpha(\varepsilon) = \{z \in \triangle : \arg z \in [\alpha - \varepsilon, \alpha + \varepsilon]\}.$$

Taking $m > 2q + 2$, where q is the number of the singular points of f. For k large enough, there are 2^k sectors in X such that the intersection of X and any one of them has positive measure. We take m sectors $T_1, \cdots, T_m$ from them such that the intersection of $\overline{T}_j$ and $\overline{T}_i$ is empty for any $i \neq j, i, j = 1, \cdots, m$, where $\overline{T}_i$ is the closure of T_i. Then there is a $\alpha_j \in \partial T_j$ such that for any $\varepsilon > 0$,

$$\mathrm{mes}(O_{\alpha_j}(\varepsilon) \cap X) > 0,$$

Suppose that $[\alpha_j - \delta, \alpha_j + \delta] \subset \partial T_j$, where we take δ small enough and it will be definited finally by arguments below, then

$$\mathrm{mes}(O_{\alpha_j}(\delta) \cap X) > 0.$$

The next stage we define a family of q.c. homeomorphisms $\Phi_t(z) = \Phi(t, z) : W_0 \times \overline{\mathbb{C}} \to \overline{\mathbb{C}}$, where

$$W_0 = \{t = (t_1, \cdots, t_m) \in \mathbb{R}^m : |t_j| \leq \delta, 1 \leq j \leq m\}.$$

First, putting

$$\psi_j(\theta) = \begin{cases} \delta^2 \exp[\dfrac{\delta^2}{(\theta - \alpha_j)^2 - \delta^2}], & |\theta - \alpha_j| \leq \delta, \\ 0, & |\theta - \alpha_j| > \delta, \end{cases}$$

then there exists a M such that $|\psi_j'(\theta)| \leq M\delta$. Let

$$\varphi_j(z) = re^{i[\theta + \psi_j(\theta)]}, z = re^{i\theta} \in \triangle,$$

then we have

$$\mu_j(z) = \frac{\partial_{\overline{z}} \varphi_j(z)}{\partial_z \varphi_j(z)} = e^{2i\theta} \cdot \frac{-\psi_j'(\theta)}{2 + \psi_j'(\theta)}.$$

Let
$$\mu_0(t,z) = \sum_{j=1}^{m} t_j \mu_j(z), \quad \text{for any} \quad t \in W_0 \quad \text{and} \quad z \in \Delta,$$

we get
$$\|\mu_0(t,z)\|_\infty \le m\delta \cdot \max_{j,z} |\mu_j(z)|$$
$$\le m\delta \cdot \frac{M\delta}{2 - M\delta}.$$

We take δ such that $M\delta < 1$ and $m\delta < 1$, then
$$\|\mu_0(t,z)\|_\infty \le k < 1.$$

Secondly denoting $X_0 = \Delta \cap X, X_n = f^n(X_0)$, then X_n is a cross section of f for any n. So that $f : X_n \to X_{n+1}$ is injective. We can define $\mu_n(t,z)$ in X_n successively for $n = 1, 2, \cdots$ by

$$\mu_{n+1}(t,z) = \mu_n(t, f^{-1}(z)) \cdot \frac{\overline{(f^{-1})'(z)}}{(f^{-1})'(z)}.$$

Suppose that $X_{n,m}$ satisfies $f^m(X_{n,m}) = X_n$, where $X_{n,0} = X_n$. We define $\mu_{n,m}(t,z)$ in $X_{n,m}$ successively for $m = 0, 1, 2, \cdots$ by

$$\mu_{n,m}(t,z) = \mu_{n,m-1}(t, f(z)) \cdot \frac{\overline{f'(z)}}{f'(z)},$$

where $\mu_{n,0}(t,z) = \mu_n(t,z)$.

It follows that

$$\mu(t,z) = \begin{cases} \mu_{n,m}(t,z), & z \in X_{n,m}, \\ 0, & \text{otherwise,} \end{cases}$$

for $z \in \mathbb{C}$ and $t \in W_o$, moreover $\mu(t,z)$ is f–invariant, i.e.

$$\mu(t,z) = \mu(t, f(z)) \cdot \frac{\overline{f'(z)}}{f'(z)}.$$

According to the definitions of $\mu_0(t,z), \mu_{n,m}(t,z)$ and $\mu(t,z)$, it is easily verified that $\mu(t,z)$ satisfies conditions of the Measurable Riemann Mapping Theorem with a Parameter. This gives that there exists a family of q.c. homeomorphisms $\Phi_t(z) = \Phi(t,z) : W_0 \times \overline{\mathbb{C}} \to \overline{\mathbb{C}}$ fixing $0, B, \infty$, where B is not a singular point of f.

Denote $f_t = \Phi_t \circ f \circ \Phi_t^{-1}$. Applying Theorem 4.1 in [7], f_t is a holomorphic self–map on $\mathbb{C}^*$ for any $t \in W_0$. Let $V = \{\beta_1, \cdots, \beta_q\}$ be the set of the singular points of f, then $\{\Phi_t(\beta_1), \cdots, \Phi_t(\beta_q)\}$ is the set of the singular points of f_t. We assume an $A \in \mathbb{C}^*$ such that $f(A) = B$. Recalling that $m > 2q + 2$ and applying the Implicit Function Theorem, then there exists a $s \in W_0$ and in a neighborhood of s there exists an arc $\gamma : [0,1] \to W_0, u \mapsto \gamma(u) = t$ and $\gamma(0) = s$, such that for any $t \in \gamma$,

$$\Phi_t(\beta_j) = \Phi_s(\beta_j), \quad \text{where} \quad j = 1, \cdots, q,$$

$$\Phi_t(A) = \Phi_s(A).$$

Applying Lemma 2.2 and Lemma 2.3 in [4], we have

$$(\Phi_s^{-1} \circ \Phi_t) \circ f \circ (\Phi_s^{-1} \circ \Phi_t)^{-1} = f,$$

and

$$\Phi_s^{-1} \circ \Phi_t|_{J(f)} = Id,$$

for any $t \in \gamma$. Then

$$\Phi_s^{-1} \circ \Phi_t : J(f) \to J(f) \quad \text{and} \quad F(f) \to F(f).$$

For any $t, \Phi_t(z)$ is analytic on $\mathbb{C} - \bigcup_{n,m} X_{n,m}$ by the definition of $\Phi_t(z)$. Then $\Phi_s^{-1} \circ \Phi_t|_{F(f)}$ is analytic since $F(f) \subset \mathbb{C} - \bigcup_{n,m} X_{n,m}$. Let D be a stable region of $F(f)$. When D is bounded, applying the Maximum Principle, the maximum of $|\Phi_s^{-1} \circ \Phi_t(z) - z|$ is assumed on ∂D, i.e. the boundary of D. Since $\partial D \subset J(f)$ and $\Phi_s^{-1} \circ \Phi_t|_{J(f)} = Id$, then

$$\Phi_s^{-1} \circ \Phi_t|_D = Id.$$

When D is unbounded, $M(D)$ is bounded, where $M(z) = \frac{1}{z}$. We repeat the above argument for $M \circ \Phi_s^{-1} \circ \Phi_t \circ M^{-1}(z)$, where $\partial M(D) = M(\partial D) \cup \{0\}$ and $\Phi_t(\infty) = \infty$ for any $t \in \gamma$. Thus

$$M \circ \Phi_s^{-1} \circ \Phi_t \circ M^{-1}(z)|_{M(D)} = Id.$$

It follows that

$$\Phi_s^{-1} \circ \Phi_t|_D = Id.$$

To sum up, $\Phi_s^{-1} \circ \Phi_t|_{F(f)} = Id$, i.e. $\Phi_s(z) = \Phi_t(z)$ for any $t \in \gamma$ and any $z \in F(f)$. Thus $\Phi_s(z) = \Phi_t(z)$ for any $t \in \gamma$ and any $z \in \mathbb{C}$. This leads to

$$\mu(s, z) = \mu(t, z), \quad \text{for any} \quad z \in \mathbb{C} \quad \text{and} \quad t \in \gamma.$$

Then we get

$$\mu_0(t, z) = \mu_0(s, z), \quad \text{for any} \quad t \in \gamma \quad \text{and any} \quad z \in \triangle \cap X,$$

i.e.

$$\sum_{j=1}^{m} t_j \mu_j(z) = \sum_{j=1}^{m} s_j \mu_j(z),$$

where

$$\mu_j(z) = e^{2i\theta} \cdot \frac{-\psi_j'(\theta)}{2 + \psi_j'(\theta)}.$$

Thus

$$(*) \qquad \sum_{j=1}^{m} (t_j - s_j) \cdot \frac{\psi_j'(\theta)}{2 + \psi_j'(\theta)} = 0.$$

We choose m points $z_1, \cdots, z_m$ such that $z_j = r_j e^{i\theta_j}$ is in $T_j \cap X$ and $|\theta_j - \alpha_j| \leq \delta$, for any $j = 1, \cdots, m$. Then

$$a_{jj} = \frac{\psi'_j(\theta_j)}{2 + \psi'_j(\theta_j)} \neq 0, j = 1, \cdots, m$$

and

$$a_{il} = \frac{\psi'_i(\theta_l)}{2 + \psi'_i(\theta_l)} = 0, \quad \text{for any } i \neq l, i, l = 1, \cdots, m.$$

It follows that the solution of the equations set $(*)$ is

$$t_j - s_j = 0, j = 1, \cdots, m, \quad \text{for any} \quad t \in \gamma.$$

This is a contradiction. The hypothesis at the beginning of the proof is false, i.e. there does not exist a cross section X of f on $J(f)$, such that mes $(X) > 0$. Then $f(z)$ is recurrent on $J(f)$, establishing the theorem. ∎

Theorem 2: *Suppose that $f(z) : \mathbb{C} \to \mathbb{C}$ is a finite typpe entire function, then $f(z)$ is recurrent on its Julia set.*

We can prove Theorem 2 only by substituting Lemma 5 ([6]), Lemmas 6 and 7 ([9]) for Theorem 4.1 ([7]), and Lemmas 2.2 and 2.3 ([4]) in the proof of Theorem 1.

2   The ergodicity of some maps

In this section, we discuss the ergodicity of $f(z) = mz + P(e^z) + Q(e^{-z})$: $\mathbb{C} \to \mathbb{C}$ and $F(z) = z^m e^{P(z) + Q(\frac{1}{z})} : \mathbb{C}^* \to \mathbb{C}^*$, where $P(z)$ and $Q(z)$ are polynomials with the same degree $d, m \in \mathbb{Z}$. We shall generalize Theorem 1.4 in [11] .

Let $g(x) = (\frac{e^x}{e})^d : [1, +\infty) \mapsto [1, +\infty)$, we define a semiconjugacy for all z in $\mathbb{C}$:

$$\phi(z) = \inf\{x \geq 1 : \lim_{n \to +\infty} \frac{|\mathrm{Re} f^n(z)|}{g^n(x)} = 0\}.$$

It is easily verfied that $\phi \circ f(z) = g \circ \phi(z)$.

As $|\mathrm{Re}z| \to +\infty$, we have $|\mathrm{Re} f^n(z)| \sim E^n(|\mathrm{Re}z|)$ successively for $n = 1, 2, \cdots$, where $E(x) = e^{dx}$. Thus $\phi(z) \leq |\mathrm{Re}z| + \varepsilon(|\mathrm{Re}z|)$, and $\varepsilon \to 0$ as $|\mathrm{Re}z| \to +\infty$.

Lemma 1: *For fixed $\varepsilon > 0$, let*

$$E = \{z \in \mathbb{C} : |\phi(z) - |\mathrm{Re}z|| < \varepsilon\}.$$

Then density $(E, B(z, 1)) \to 1$ as $|\mathrm{Re}z| \to +\infty$, i.e. when $|\mathrm{Re}z|$ is large enough, $\phi(z)$ is very close to $|\mathrm{Re}z|$.

Proof: Taking $x > |\mathrm{Re}z|$ large enough and $h_k = 2g^k(x - \varepsilon)$. As in the proof of Theorem 1 in [5], we denote

$$R(h_k) = \{z \in \mathbb{C} : |\mathrm{Re}z| > h_k\},$$

$$E_0 = \{B(z, 1)\} = \{F_0\},$$

$$E_1 = \{F_1 : F_1 \subset F_0 \in E_0, f(F_1) \in \mathrm{pack} f(F_0)\},$$

then we define inductively for $k = 2, 3, \cdots$,

$$E_k = \{F_k : \text{there exists a} \quad F_{k-1} \quad \text{such that} \quad F_k \subset F_{k-1} \in E_{k-1},$$

$$\text{and} \quad f^k(F_k) \in \mathrm{pack} f^k(F_{k-1})\},$$

and we define for any k

$$\overline{E}_k = \bigcup_{F_k \in E_k} F_k,$$

$$E' = \bigcap_{k=0}^{\infty} \overline{E}_k = \{z \in \mathbb{C} : |\mathrm{Re} f^k(z)| \geq h_k \quad \text{for any} \quad k\},$$

where pack $f^k(F_{k-1})$ is the set of all the disjoint small disks in $R(h_k) \cap f^k(F_{k-1})$. For any $F \in E_k$, we have

$$\mathrm{density}(\overline{E}_{k+1}, F) \geq \frac{1}{M^2} \mathrm{density}(f^{k+1}(\overline{E}_{k+1}), f^{k+1}(F))$$

$$= \frac{1}{M^2} \cdot \frac{\mathrm{Vol}\ (f(B) \cap R(h_{k+1}))}{\mathrm{Vol}(f(B))},$$

$$\text{where} \quad f^k(F) = B \in \mathrm{pack}\ f^k(F_{k-1}), \quad \text{and} \quad F_{k-1} \in E_{k-1}$$

$$\geq \frac{1}{M^2} \cdot \frac{O(e^{2dh_k}) - 2h_{k+1} \cdot O(e^{dh_k})}{O(e^{2dh_k})}$$

$$= 1 - O(\frac{h_{k+1}}{e^{dh_k}}) = \triangle_k,$$

where M is a constant not depending on n. Thus

$$\mathrm{density}(E', B(z, 1)) \geq \prod_{k=0}^{\infty} \triangle_k$$

$$= \prod_{k=0}^{\infty} (1 - O(\frac{h_{k+1}}{e^{dh_k}})),$$

where

$$\frac{h_{k+1}}{e^{dh_k}} = \frac{2g^{k+1}(x - \varepsilon)}{e^{2dg^k(x-\varepsilon)}} = \frac{2}{e^d} \cdot \frac{e^{dg^k(x-\varepsilon)}}{e^{2dg^k(x-\varepsilon)}}$$

$$= \frac{2}{e^d} \cdot \frac{1}{e^{dg^k(x-\varepsilon)}}.$$

This gives

$$1 \geq \text{density } (E', B(z,1)) \geq \prod_{k=0}^{\infty} (1 - O(\frac{h_{k+1}}{e^{dh_k}}))$$

$$= 1 - O(\sum_{k=1}^{\infty} \frac{h_{k+1}}{e^{dh_k}}) + \cdots$$

$$\to 1, \quad \text{as} \quad x \to +\infty.$$

For any ω in E', where E' is contained in $B(z,1)$, there exist a ξ such that

$$0 = \lim_{n\to\infty} \frac{|\text{Re}f^n(\omega)|}{g^n(\xi)} \geq 2 \lim_{n\to\infty} \frac{g^n(x-\varepsilon)}{g^n(\xi)}.$$

It follows that $\phi(\omega) \geq x - \varepsilon > |\text{Re}\omega| - \varepsilon$ as we take x large enough. On the other hand, we have $\phi(z) \leq |\text{Re}z| + \varepsilon$ for any z such that $|\text{Re}z|$ is large enough. So that we can say $\omega \in E$, i.e. $E' \subset E$. Thus

$$1 \geq \text{density } (E, B(z,1))$$

$$\geq \text{density}(E', B(z,1)) \to 1, \text{as} \quad |\text{Re}z| \to +\infty.$$

Lemma 2: *For any interval $I \subset [1, +\infty), \phi^{-1}(I)$ has positive measure.*

Proof: Applying Lemma 1 , $\{z \in \mathbb{C} : \phi(z) \in [x-1, x+1]\}$ contains all of the points z such that $|\text{Re}z|$ is quite close to x where x is large enough. So that it has positive measure. For any interval $I \subset [1, +\infty), g^n(I)$ contains $[x-1, x+1]$ as n is large enough. Moreover,

$$\phi(f^n(\phi^{-1}(I))) = g^n(I) \supset [x-1, x+1],$$

this gives that $f^n(\phi^{-1}(I))$ has positive measure, and

$$f^n(\phi^{-1}(I)) \supset \{z : ||\text{Re}z| - x| < \varepsilon\},$$

where ε is small enough. According to the continuity of f^n, the inverse image of an open set is an open set. Thus $\phi^{-1}(I)$ contains a small open set, i.e. $\phi^{-1}(I)$ has positive measure, establishing the lemma.

Theorem 3: *Let $f(z) = mz + P(e^z) + Q(e^{-z}) : \mathbb{C} \to \mathbb{C}$, where $P(z)$ and $Q(z)$ are polynomials with the same degree $d, m \in \mathbb{Z}$. Then $f(z)$ is not ergodic on its Julia set.*

Proof: We first recall that any z such that $\phi(z) > 1$ tends to infinity and consequently lies in $J(f)$ by Theorem 1 in [3] . For a fixed $x > 1$, we can take two disjoint intervals I and J such that $\bar{I} \subset (g^{-1}(x), x)$ and $\bar{J} \subset (g^{-1}(x), x)$. Thus $\phi^{-1}(I)$ and $\phi^{-1}(J)$ have positive measure.

Denote the set made up of the large orbit of I (resp. J) under g by $O_g(I)$ (resp. $O_g(J)$). Let $O_f(\phi^{-1}(I))$ (resp . $O_f(\phi^{-1}(J))$) be the set made up of the large orbit of $\phi^{-1}(I)$ under f. Then both $O_f(\phi^{-1}(I))$ and $O_f(\phi^{-1}(J))$ have positive measure, and $O_g(I) \cap O_g(J)$ is empty.

If there exists a point $z \in O_f(\phi^{-1}(I)) \cap O_f(\phi^{-1}(J))$, then for some $k, n, m \in \mathbb{N}, \alpha \in I$ and $\beta \in J$,

$$f^k(z) = f^n(\phi^{-1}(\alpha)) = f^m(\phi^{-1}(\beta)),$$

and that satisfies

$$\phi(f^n(\phi^{-1}(\alpha))) = \phi(f^m(\phi^{-1}(\beta))).$$

According to $\phi \circ f^n(z) = g^n \circ \phi(z)$, we have

$$g^n(\phi(\phi^{-1}(\alpha)) = g^m(\phi(\phi^{-1}(\beta))),$$

that is

$$g^n(\alpha) = g^m(\beta).$$

This contradicts that $O_g(I) \cap O_g(J)$ is empty. So that $O_f(\phi^{-1}(I)) \cap O_f(\phi^{-1}(J))$ is empty .

Thus there exist two disjoint sets made up of large orbits of f in $J(f)$, and which have positive measure. This gives that $f(z)$ is not ergodic on $J(f)$, establishing the theorem. ∎

Theorem 4: *Assume that $f(z) = mz + P(e^z) + Q(e^{-z}) : \mathbb{C} \to \mathbb{C}$ and $F(z) = z^m e^{P(z)+Q(\frac{1}{z})} : \mathbb{C}^* \to \mathbb{C}^*$, where $P(z)$ and $Q(z)$ are polynomials with the same degree, $m \in \mathbb{Z}$. Then $F(z)$ is not ergodic on its Julia set.*

Proof: First notice that $f(z)$ and $F(z)$ satisfy the commutative diagram

$$\begin{array}{ccc} \mathbb{C} & \xrightarrow{f} & \mathbb{C} \\ \pi(z) = e^z \downarrow & & \downarrow \pi(z) = e^z \\ \mathbb{C}^* & \xrightarrow{F} & \mathbb{C}^* \end{array}$$

and $\pi(J(f)) = \pi(J(F)), \pi^{-1}(J(F)) = J(f)$ by the results in [4].

From the proof of Theorem 3, there are two disjoint intervals I and J in $(1, +\infty)$ such that $O_f(\phi^{-1}(I))$ and $O_f(\phi^{-1}(J))$ are two disjoint positive measure sets made up of large orbits on $J(f)$. Denote $A = \phi^{-1}(I)$ and $B = \phi^{-1}(J)$, then $\pi(A)$ and $\pi(B)$ are contained in $J(F)$ and have positive measure.

For any two points z and w satisfying $w = z + 2k\pi i$ for a $k \in \mathbb{Z}$, we calculate successively for $n = 1, 2, \cdots$

$$f^n(z + 2k\pi i) = f^n(z) + 2km^n\pi i.$$

That is, for any n, $\mathrm{Re} f^n(z + 2k\pi i) = \mathrm{Re} f^n(z)$. This leads to

$$\phi(z + 2k\pi i) = \phi(z), \text{i.e. } \phi(\omega) = \phi(z).$$

Let $O_F(\pi(A))$ (resp . $O_F(\pi(B))$) be the set made up of the large orbit of $\pi(A)$ (resp. $\pi(B)$) under F, then $O_F(\pi(A))$ (resp. $O_F(\pi(B))$) has positive measure. If $O_F(\pi(A)) \cap O_F(\pi(B))$ is not empty, there exist two points $z_1 \in A$ and $z_2 \in B$, and $n, m \in \mathbb{N}$, such that

$$F^n(\pi(z_1)) = F^m(\pi(z_2)).$$

According to the above commutative diagram, we have

$$\pi \circ f^n(z_1) = \pi \circ f^m(z_2).$$

Then there exists a $k \in \mathbb{Z}$ such that

$$f^n(z_1) = f^m(z_2) + 2k\pi i.$$

Thus we have

$$\phi(f^n(z_1)) = \phi(f^m(z_2)),$$

and that is

$$g^n(\phi(z_1)) = g^m(\phi(z_2)),$$

where $z_1 \in \phi^{-1}(I)$ and $z_2 \in \phi^{-1}(J)$, i.e. $\phi(z_1) \in I$ and $\phi(z_2) \in J$. This contradicts that $O_g(I) \cap O_g(J)$ is empty. Then $O_F(\pi(A)) \cap O_F(\pi(B))$ is empty.

We find two disjoint positive measure sets $O_F(\pi(A))$ and $O_F(\pi(B))$ in $J(F)$ made up of large orbits of F. This means that $F(z)$ is not ergodic on $J(F)$. ∎

Acknowledgements. I would like to thank Professor Lü Yinian and Professor Li Zhong for their patient help and encouragement, and I also wish to thank Peng Guiai, Tsui Guizhen and Yang Guoxiao for helpful conversations.

References

[1]   L. Ahlfors, Lectures on Quasiconformal Mappings, Van Nostrand, 1966.

[2]   L. Ahlfors and L. Bers, Riemann's mapping theorem for variable metrics, Ann. of Math., **72** (1960), 385–404.

[3]   A. E. Eremenko and M. Yu. Lyubich, Dynamical properties of some classes of entire function, preprint.

[4]   Liping Fang, Complex analytic dynamics on $\mathbb{C}^*$ (in Chinese), Acta Mathematica Sinica, **34** (1991) , no. 5, 611–621.

[5]   Liping Fang, Area of Julia sets of holomorphic self–maps on $\mathbb{C}^*$, *Acta Mathematica Sinica*, New Series, 9 (1993), no.2 160-165.

[6] E.Ghys, L. Goldberg and D.Sullivan, On the measurable dynamics of $z \mapsto \exp(z)$, Ergod. Th. and Dynam. Sys., **5** (1985), 329–335.

[7] L. Keen, Dynamics of holomorphic maps of $\mathbb{C}^*$, in Holomorphic Functions and Moduli, Vol. I, ed. D. Drasin. Springer, Berlin – NewYork, 1988.

[8] M. Yu. Lyubich, The measurable dynamics of the exponential map, Siberian Journ. Math., B. **28** (1987), no. 5, 111–127.

[9] Yinian Lü, On the proof of Sullivan's eventural periodicity theorem, Acta Math. Sinca, New Series, **5** (1989), no. 4. 355–364.

[10] O. Lehto and K. I. Virtanen, Quasiconformal Mappings in the plane, Springer–Verlag, 1973.

[11] C. McMullen, Area and Hausdorff dimension of Julia sets of entire functions , Trans. A.M.S., **300** (1987), no.1., 329–342.

[12] M. Rees, The exponential map is not recurrent, Math. Z., **191** (1986), no. 4, 593–598.

[13] D.Sullivan, Quasiconforal homeomorphisms and dynamics I: Solution of the Fatou–Julia problem on wandering domains, Ann. of Math., **122** (1985) , no. 3, 402–418.

[14] D. Sullivan, Quasiconformal homeomorphisms and dynamics II, Acta Math., **155** (1985), 243–260.

Key words: cross section, recurrent, ergodic

1980 Mathematics Subject Classification 30D05, 58F13

The real part of small entire functions

P.C.Fenton
University of Otago
New Zealand

Given an entire function f, let us write, as usual, $M(r)$ for the maximum modulus of f, and $A(r)$ and $B(r)$ for the minimum and maximum of the real part of f. Of course we always have

$$-M(r) \le A(r) \le B(r) \le M(r),$$

but in fact the outer inequalities are, for most values of r, almost equalities.

Theorem A: (Wiman,1916) $A(r) \sim -M(r)$ and $B(r) \sim M(r)$ as $r \to \infty$ outside a set of finite logarithmic measure.

More recently Hayman showed that these asymptotic relations can be sharpened if a larger exceptional set is admitted.

Theorem B: (Hayman, 1974) *Suppose that f is an entire function and let*

$$p = \limsup_{r \to \infty} \frac{\log \log M(r)}{\log \log r}, \quad 2 \le p \le \infty.$$

Let $\alpha(p) = 1 - p^{-1}$. *Then, given* $K > 0$,

$$B(r) > M(r)\left\{1 - \frac{K\pi^2\alpha(p)}{2\log M(r)}\right\}, \quad -A(r) > M(r)\left\{1 - \frac{K\pi^2\alpha(p)}{2\log M(r)}\right\}, \quad (1)$$

for all r outside a set of lower logarithmic density $\le K^{-1}$.

The constant $\pi^2\alpha(p)/2$ is sharp, as Hayman himself showed, but the estimate for the exceptional set is nor right: it is in the nature of results of this sort that an assumption concerning the upper growth of f should be reflected in a conclusion about the upper size of the exceptional set. The estimate for the exceptional set is the focus of all that follows.

There is a version of Theorem B for case $1 < p < 2$, which we shall have occasion to refer to again; the inequalities then are better and the estimate for the exceptional set takes the right form.

The proof of Theorem B is in the two parts. (The case $p = \infty$ is set aside here, although it involves no essential difficulty. Thus f is small in the sence that it has order zero.) First, using arguments of Wiman-Valiron type, it is shown that, given $p^* > p$,

$$B(r) > M(r)\left\{1 - (\frac{1}{2}\pi^2 + o(1))\frac{b(r)}{a(r)^2} + O(a(r)^{-3/(2\alpha(p^*))})\right\} \quad (2)$$

Proceedings of the International Conference on Complex
Analysis at the Nankai Institute of Mathematics, 1992 pp. 90-94
©INTERNATIONAL PRESS 1994

for all r outside a set of zero logarithmic density. Here $a(r)$ and $b(r)$ are the first and second logarithmic derivatives of $\log M(r)$. There is a similar inequality for $-A(r)$ but for brevity let us restrict our considerations to $B(r)$. Further progress depends on obtaining simultaneous estimates for $b(r)/a(r)^2$ -which should be about $\alpha(p)/\log M(r)$ -and $a(r)^{-3/(2\alpha(p^*))}$, which should be $o(\log M(r))^{-1}$. This is effected by a growth lemma.

Lemma A: (Hayman, 1974) *Let $\varphi(x)$ be a positive, increasing and convex function for $x > X$, and suppose that*

$$\liminf_{x\to\infty}\frac{\log\varphi(x)}{\log x} \le p \le \limsup_{x\to\infty}\frac{\log\varphi(x)}{\log x}, \quad p > 1. \tag{3}$$

Given $K > 1$ and $a < \alpha(p)$, then

$$\frac{\varphi(x)\varphi''(x)}{\varphi'(x)^2} \le K\alpha(p) \quad and \quad \varphi'(x) > \varphi(x)^a \tag{4}$$

outside a set of lower density $\le K^{-1}$.

These estimates, taking $\varphi(x) = \log M(e^x)$,do the trick but it is here that the difficulty with the exceptional set arises; for the lemma does not discriminate, in its conclusion about the exceptional set, between the different possible hypotheses concerning p. The lemma has the right form, however, when p is the lower limit, and using this case and a version of the Wiman-Valiron method applicable to entire functions of finite lower growth it is possible to prove:

Lemma C: (Fenton, 1979) *If*

$$p = \liminf_{r\to\infty}\frac{\log\log M(r)}{\log\log r}, \quad p > 1, \tag{5}$$

then, given $K > 1$, (1) holds outside a set of lower logarithmic density $\le K^{-1}$.

One would expect the companion theorem to hold, with the obvious change from *lower* to *upper* throughout. This may be so but further considerations of the growth lemma suggests it may not be. For, given $p > 1$ it is possible to construct $\varphi(x)$ for which p is the upper limit in (3) and which is linear for certain arbitrarily long stretches of x.It can thus be arranged that the second inequality in (3) fails on a set of upper density 1, and it is therefore impossible to improve the lemma in the desired way.

The inequalities in (3) are used for distinct purposes. The vital one is the first of them, the other simply enabling us to ignore the error term in (2). Can the error term be estimated by some other means and Lemma A sharpened by eliminating the requirement that the second inequality holds ?

Let us take up this question of the error term in (2).

Using a refinement of methods developed elsewhere *[2, 3]*, it is possible to prove:

Lemma 1: *Suppose that f is an entire function, and let $p^* > 1$ and $R > 1$ be given. Then, for all r in $[1, R]$ outside a subset of logarithmic measure $O(a(R)^{1/(p^*-1)})$,*

(i) if $p^* \geq 2$ we have (2);
(ii) if $1 < p^* < 2$ we have

$$B(r) > M(r)\{1 - o(logM(r))^{-1}\}. \tag{6}$$

The usefulness of Lemma 1 lies in the fact that no prior knowledge of the growth of f is assumed. On the other hand it is vacuous if $a(R)$ is so large that the exceptional set might possibly occupy all of $[1, R]$. The inequalities of (i) and (ii) are those established by Hayman, and the technicalities involved in proving them are no different. Both depend on certain fundamental inequalities for the terms of the Taylor series for f.

Rather than appealing to Hayman,s growth lemma we use Lemma 1 recursively to estimate the error term in (2). That done, we shall turn to a truncated form of Lemma A, to do with the first inequality of(3).

For an entire function f satisfying the hypothesis of Theorem A we have, as is easily shown,

$$\frac{\log M(r)}{\log r} \leq a(r) \leq (\log r)^{p-1+o(1)}. \tag{7}$$

Let us say that the error term in (2) is em satisfactory for a given $p^* > 1$ and $\epsilon > 0$ -for it is then negligible - if

$$a(r)^{-3/(2\alpha(p^*))} < \epsilon / \log M(r). \tag{8}$$

Given $\epsilon > 0$, first choose $p^* > p$ arbitrarily, say $p^* = p+1$, and let R_1 be the largest value in $[1, R]$ at which the error term is not satisfactory. We have,then, (2) with a satisfactory error term on $[R_1, R]$ outside a subset of logarithmic measure $O(a(R)^{1/(p^*-1)}) = o(\log R)$, using (7). Further, since (8) fails at R_1, we deduce, making use of (7), that

$$a(R_1) \leq \epsilon^{-2}(\log R_1)^{2-\delta}, \tag{9}$$

where $\delta > 0$ depends on p^*.

Now apply Lemma 1 again, but this time on $[1, R_1]$ with $p^* = 3$. We let R_2 be the largest value in $[1, R_1]$ at which the error term is not satisfactory, and deduce, arguing as before, that (2) holds with a satisfactory error term on $[R_2, R_1]$ outside a subset of logarithmic measure $O(a(R_1)^{1/(p^*-1)}) = o(\log R_1) = o(\log R)$, using (9). Moreover

$$a(R_2) \leq \epsilon^{-4/5}(\log R_2)^{4/5}, \tag{10}$$

At this point we refer to the second part of Lemma 1 , with $p^* = 1.9$ say, and conclude that the inequality of Lemma 1 (ii) holds on $[1, R_2]$ outside a subset of logarithmic measure $O(a(R_2)^{1/(p^*-1)}) = o(\log R_2) = o(\log R)$, using (10).

The upshot of this is that we have, for each r in $[1, R]$ outside a subset of logarithmic measure $o(\log R)$, either (2) with a satisfactory error term, or (6). Thus

outside a set of zero logarithmic density, either (2)

holds with a satisfactory error term, or (6) holds. $\tag{11}$

It remains to estimate $b(r)/a(r)^2$ in (2). This is done by way of

Lemma 2: *(Fenton, 1993) Let $\varphi(x)$ be an increasing, convex function for $x > X$, and let*

$$p = \limsup_{x \to \infty} \frac{\log \varphi(x)}{\log x}, \quad 1 \le p \le \infty.$$

Then, given $K > 1$,

$$\frac{\varphi(x)\varphi''(x)}{\varphi'(x)^2} \le K\alpha(p)$$

outside a set of upper logarithmic density no more than K^{-1}.

Applying Lemma 2 with $\varphi(x) = \log M(e^X)$, we obtain

$$\frac{b(r)}{a(r)^2} \le \frac{K\alpha(p)}{\log M(r)}$$

outside a set of upper $\log - \log$ density no more than K^{-1}. (Note $\log - \log$ here, not log. It can be shown that, unless $p = \infty$, *upper logarithmic density* cannot be replaced by *upper density* in Lemma 2.) Combining this with (11) and using the fact that a set of zero logarithmic density necessarily has zero $\log - \log$ density *[1]*, we get

$$\text{either } B(r) > M(r) \left\{ 1 - \frac{K\pi^2\alpha(p) + \epsilon}{2\log M(r)} \right\}$$

$$\text{or } B(r) > M(r) \left\{ 1 - \frac{o(1)}{\log M(r)} \right\}$$

-and note that the second is stronger than the first - outside a set of upper $\log - \log$ density no more than K^{-1}.Using a standard argument the ϵ can be removed and we thus have, taking account of Theorem C,

Theorem 1: *Let*

$$p = \liminf_{r \to \infty} \frac{\log\log M(r)}{\log\log r}, \quad P = \limsup_{r \to \infty} \frac{\log\log M(r)}{\log\log r}.$$

Given $K > 1$,

$$B(r) > M(r) \left\{ 1 - \frac{K\pi^2\alpha(p)}{2\log M(r)} \right\}$$

outside a set of lower logarithmic density no more than K^{-1};and

$$B(r) > M(r) \left\{ 1 - \frac{K\pi^2\alpha(p)}{2\log M(r)} \right\}$$

outside a set of upper $\log - \log$ density no more than K^{-1}.

Doubtless the asymmetry is an aberration. It is a consequence of the relations between the various densities that Theorem 1 would be strengthened if it were rendered symmetric by making either half the analogue of the other.

References

[1]   P.D.Barry, The minimum modulus of small entire and subharmonic functions, *Proc. London Math. Soc.* 12(1962), 445-495.

[2]   P.C.Fenton, Some results of Wiman-Valiron type for integral functions of finite lower order, *Ann.Math.*, 103(1976), 237-252.

[3]   P.C.Fenton, Wiman-Valiron theory for entire functions of finite lower growth, *Trans. Amer. Math.Soc.*, 252(1979), 221-232.

[4]   P.C.Fenton, The growth of for convex functions, to appear in *Illinois J.Math.*(1993).

[5]   W.K.Hayman, The local growth of power series: a survey of the Wiman-Valiron method, *Canad. Math. Bull.*, 17(1974), 317-358.

Uniform Approximation: Holomorphic, Harmonic, Subharmonic[1]

P.M. Gauthier

Département de mathématiques et de statistique et Centre de recherches mathématiques
Université de Montréal, CP 6128-A, Montréal, H3C 3J7, Canada
(Dedicated to the memory of Professor Shen Xie-chang)

1 Introduction

Approximation theory can be approached from two viewpoints: the qualitative or the quantitative viewpoint. Qualitative approximation is more theoretical. It is concerned with density theorems, that is, with the mere *possibility* of approximation. Quantitative approximation takes a more practical stance. It is concerned with actually finding efficient algorithms for *performing* the approximations.

There is another way of classifying approximation theory from two viewpoints: smoothing-type approximation and extension-type approximation. The first type of approximation seeks to approximate a given function by a nicer function defined on (more or less) the same domain whereas extension-type approximation seeks to approximate a given function by a function (of the same sort but) defined on a larger domain.

The present notes deal only with qualitative extension-type approximation. For an overview of the state of the subject in the early 1980's, the reader may consult, for example, *[6]*, *[11]* and *[19]*. However, there have been many subsequent developments. We shall discuss a (very) few of these as well as a selection of classical results.

The present notes follow quite closely the lectures which I gave at Nankai Institute. I wish to thank the Nankai Institute and Professor Yang Lo for organizing this special year on complex analysis and for allowing me the privilege of participating.

2 Iversen's Maximum Principle

Theorem 1: [Maximum Principle] If Ω is a <u>bounded</u> open subset of the finite complex plane $\mathbf{C}$, then

$$sup_\Omega s = sup_{\partial\Omega} s,$$

for all s subharmonic in Ω.

---

[1] This research was supported by NSERC(Canada) and FCAR(Québec.)

Proceedings of the International Conference on Complex
Analysis at the Nankai Institute of Mathematics, 1992 pp. 95-113
©INTERNATIONAL PRESS 1994

The right side of the above equality requires some explanation, since the function s may not be defined on the boundary of Ω. We adopt the following convention:

$$sup_{\partial\Omega}s = sup_{y\in\partial\Omega}\{\overline{lim}_{x\to y}s(x)\}.$$

The following trivial examples show that the Maximum Principle is not in general true if Ω is not bounded.

Example 1 $\Omega = (|z| > 1)$, $s(z) = |z|$.

Example 2 $\Omega = (Re z > 0)$, $s(z) = |e^z|$.

It may come as a surprise that the Maximum Principle, nevertheless, does hold for some unbounded sets. Moreover, we shall give a simple characterization of such sets.

We have stated the maximum principle for subharmonic functions, which implies the maximum principle for harmonic functions as well as for holomorphic functions (since the absolute value of a holomorphic function is subharmonic). Harmonic functions are classically defined on open subsets of $\mathbf{R}^n$, whereas the natural domain of definition for a holomorphic function is a Riemann surface. Riemannian manifolds generalize both kinds of domains. We shall say that $\tilde{\Omega}$ is a second countable compactification of a Riemannian manifold Ω if $\tilde{\Omega}$ is a compact Hausdorff space satisfiying the second axiom of countability and containing (a homeomorphic copy of) Ω as an open (not necessarily dense) subset. We denote by $\tilde{\partial}\Omega = \tilde{\Omega} \setminus \Omega$ the ideal boundary of Ω in $\tilde{\Omega}$. A subset E of the ideal boundary $\tilde{\partial}\Omega = \tilde{\Omega}\setminus\Omega$ is said to be *accessible* (from Ω) if there exists a continuous path $\sigma : [0, +\infty) \to \Omega$ which is eventually in each neighbourhood of E. That is, for each neighbourhood V of E in $\tilde{\Omega}$, there is a t_V such that $\sigma(t) \in V$ for each $t > t_V$. An accessible set need not contain an accessible point. An upper semicontinuous function $s : \Omega \to [-\infty, +\infty)$ is said to be *subharmonic* if for any relatively compact open set V in Ω, any harmonic function which dominates s on the boundary of ∂V dominates s on V.

The following result, obtained jointly with Chen Huaihui *[7]*, characterizes those subsets of the boundary which can be disregarded in the maximum principle. Earlier versions are due independently to Sahakian *[30]* and Gauthier-Grothmann-Hengartner *[16]*, but especially (and I thank Alex Eremenko *[8]* for bringing this to my attention) to Iversen *[25]*(p. 24).

Theorem 2: *[Generalized Maximum Principle] Let E be a closed subset of $\tilde{\partial}\Omega$. A necessary and sufficient condition in order that*

$$sup_{\Omega}s = sup_{\tilde{\partial}\Omega\setminus E}s,$$

for all s subharmonic on Ω, is that E be not accessible.

There are two instances which are of particular interest. Firstly, there is the case where Ω is a (not necessarily bounded) open subset of $\mathbf{R}^n$ and $\tilde{\Omega} = \bar{\Omega}\cup\{\infty\}$. Let us say that the maximum principle holds on such an open set Ω if the conclusion of Theorem 1 holds for Ω. We have the following generalization of Theorem 1.

Corollary 1: *The maximum principle holds on an open subset Ω of $\mathbf{R}^n$ if and only if ∞ is not accessible from Ω.*

Note that in Examples 1 and 2 above, ∞ is accessible, whereas if Ω is bounded, then ∞ is not accessible. A less trivial example is the following.

Example 3 *Let*

$$\Omega = \{(x,y) \in \mathbf{R}^2 : 0 < x < 1, \ 0 < y < x^{-1}|sin(x^{-1})| + 1\}.$$

Then Ω is an unbounded domain from which ∞ is not accessible.

Remark. When the maximum principle for unbounded domains is mentioned, the first thought that usually comes to mind is the Phragmén-Lindelöf Theorem which for nice domains may be paraphrased as follows:
(boundary estimate) + (global estimate) $\Rightarrow$ (better global estimate).
We wish to emphasize that we assume *no* global estimates. Thus, we do not assume that s is bounded; rather, we *infer* that s is bounded - and by the same bound as on the boundary!
The sufficiency in Corollary 1 follows from the contrapositive of the following.

Theorem 3: *[Fuglede-Iversen [9]] If s is a subharmonic function on an open subset Ω of $\mathbf{R}^n$ and*

$$sup_\Omega s > sup_{\partial\Omega} s,$$

then there is a a continuous path $\sigma : [0, +\infty) \to \Omega$ along which s tends to $sup_\Omega s$.

A second instance of Theorem 2 which is of particular interest is when Ω is a bounded subset of $\mathbf{R}^n$ and $\tilde{\Omega} = \bar{\Omega}$.

Corollary 2: *Let E be a closed subset of the boundary of a bounded open subset Ω of $\mathbf{R}^n$. Then, a necessary and sufficient condition in order that*

$$sup_\Omega s = sup_{\partial\Omega \setminus E} s,$$

for all s subharmonic on Ω, is that E be not accessible.

Example 4 *Let Ω be the square $\{0 < x < 1, \ 0 < y < 1\}$ with the following segments removed:*

$$x = 1/(2^j), \ 0 < y \leq 2/3, \ and \ x = 1/(2^j + 1), \ 1/3 \leq y < 1; j = 1, 2, \ldots.$$

Then, the set $E_1 = \{x = 0, 0 \leq y \leq 1\}$ is accessible whereas the set $E_2 = \{x = 0, 0 \leq y \leq 1/2\}$ is not.

Under the additional assumption that s is bounded (a Phragmén-Lindelöf type assumption), the sufficiency in Corollary 2 is an immediate consequence of the two-constants theorem of Nevanlinna and the fact that any non-accessible set is of harmonic measure zero. We present an amusing proof of the latter. Recall that the harmonic measure at a point $x \in \Omega$ of a Borel set $E \subset \partial\Omega$ is the probability that Brownian motion starting at x first leave Ω through E. Since Brownian motion is continuous, if E is not accessible, it is *impossible*, hence *improbable*, that Brownian motion leave through E! Again, we repeat that this proof works only under the additional assumption that s is bounded.

For a proof of the sufficiency in Theorem 2, we refer the reader to *[7]*; the necessity will be established in subsequent sections as an application of approximations and extensions.

3   Runge Theorems

The year 1885 was a very special year for the qualitative theory of approximation, for that was the year that both Weierstrass and Runge proved their famous approximation theorems. To state these and related results, we introduce some notation. If X is a topological space, we denote by $C(X)$ the space of continuous complex valued functions on X, and if X is a subset of a complex manifold Y, we denote by $H(X)$ the space of holomorphic functions on (open sets containing) X.

Theorem 4: *[Weierstrass] Let I be a closed interval on the real axis. Then, for each $f \in C(I)$ and for each $\epsilon > 0$, there is a polynomial p such that*

$$|f - p| < \epsilon.$$

Theorem 5: *[Runge] Let K be a compact subset of the complex plane $\mathbf{C}$ such that $\mathbf{C} \setminus K$ is connected. Then, for each $f \in H(K)$ and for each $\epsilon > 0$, there is a polynomial p such that*
$$|f - p| < \epsilon.$$

Both of these theorems have been generalized. The generalizations of the Weierstrass Theorem due to Mergelyan and Arakelyan are better known than the generalization of the Runge Theorem due to Roth. For lack of time, we shall, therefore, present only the generalization of Roth. Although the theorems of Mergelyan and Arakelyan are more powerful than that of Roth, we shall see that Roth's theorem is nevertheless sufficiently strong to yield several surprising applications. We begin by reformulating Runge's Theorem.

Theorem 6: *[Runge] Let $W \subset \mathbf{C}$ be open. Then, each function in $H(W)$ can be approximated by polynomials uniformly on compact subsets of W if and only if $\mathbf{C} \setminus W$ is connected.*

Definition: Let W be an open subset of a Riemann surface Ω. The pair (W, Ω) is said to be a *Runge* pair (for compact sets) if for each $f \in H(W)$, for each compact $F \subset W$ and for each $\epsilon > 0$, there is a $g \in H(\Omega)$ such that $|f - g| < \epsilon$ on F.

Notation: Let us denote by $\Omega^* = \Omega \cup \{*\}$ the Alexandrov one-point compactification of Ω, where $*$ denotes the ideal Alexandrov point at infinity.

Riemann surface theory can be subdivided into two areas of study: the study of compact surfaces and the study of open (non compact) surfaces. Each of these has its principal theorem. For compact surfaces it is the Riemann-Roch Theorem while for open surfaces it is the following theorem of Behnke and Stein *[5]* which is in fact the natural extension of Runge's Theorem to Riemann surfaces. It may also be said that, to a great extent, this extension of Runge's Theorem, which Behnke and Stein published in 1949, inspired the remarkable flourishment of several complex variables which took place in the Séminaire Henri Cartan in the succeeding decade.

Theorem 7: *[Behnke-Stein] Let W be an open subset of an open Riemann surface Ω. The following are equivalent:*

a) (W, Ω) is a Runge pair for compact sets;

b) $\Omega^* \setminus W$ is connected.

The previous theorem generalizes Runge's Theorem in that, in place of approximation by functions holomorphic on a plane domain Ω, it deals with approximation by functions holomorphic on a Riemann surface Ω. In 1938, Alice Roth *[28]* had already generalized the Runge Theorem, but in a different direction. She proved an extension of Theorem 5 in which she considered approximation on *unbounded* closed subsets of the plane **C** rather than on compact subsets. Then, in 1973 *[29]* (Note the time span!), she generalized her own result by approximating on closed subsets of an arbitrary plane open set Ω, where "closed" means "closed in the relative topology of Ω". Here, then, is the Runge Theorem of Roth.

Theorem 8: *[Roth] Let F be a closed subset of a plane open set Ω such that $\Omega^* \setminus F$ is both connected and locally connected. Then, for each $f \in H(F)$ and for each $\epsilon > 0$, there is a function $g \in H(\Omega)$ such that*

$$|f - g| < \epsilon.$$

To give a better appreciation of the scope of Alice Roth's achievement, we point out two difficulties in passing from approximation on compact sets to approximation on closed sets. First of all, approximation theory often makes use of functional analysis. The space $C(F)$ is a topological space (with the topology of uniform convergence) and it is a vector space. However, if F is unbounded, $C(F)$ is not a topological vector space. Indeed, scalar multiplication is not continuous: If λ_n are non-zero scalars with $\lambda_n \to 0$, and f is unbounded, then $\lambda_n \cdot f \not\to 0$. A second difficulty is that proofs in complex approximation often use integral representations. On unbounded sets, integration is of course possible, but more subtle.

With the hope of making the reader more comfortable with the topological conditions in the above Runge-type theorems, we present some examples in which we take Ω to be the complex plane $\mathbf{C}$.

Example 5 $\mathbf{C}^* \setminus F$ *is connected if and only if* $\mathbf{C} \setminus F$ *has no bounded components. Thus, if* $F = \mathbf{R}$, *then* $\mathbf{C}^* \setminus \mathbf{R}$ *is connected. (Note that* $\mathbf{C} \setminus \mathbf{R}$ *is not!)*

Example 6   *If* F *is the unit circle* $(|z| = 1)$, *then* $\mathbf{C}^* \setminus F$ *is not connected.*

Example 7   *Let*

$$F = [0, +\infty) \cup \{z = x + iy : x = (1/y)|sin(1/y)|, 0 < y \leq 1\}.$$

This set is known as Arakelyan's glove. It has the property that both $\mathbf{C} \setminus F$ *and* $\mathbf{C}^* \setminus F$ *are connected but* $\mathbf{C}^* \setminus F$ *is not locally connected.*

Definition: A family $\mathcal{A}$ of sets in Ω is said to have *no long islands* if for each compact subset K of Ω there is a compact set Q in Ω such that any member of $\mathcal{A}$ which meets K is necessarily contained in Q.

Lemma 1: *The following conditions are equivalent:*
1. $\Omega^* \setminus F$ *is locally connected;*
2. $\Omega^* \setminus F$ *is locally connected at* $*$;
3. for each compact set K *in* Ω, *the family of bounded components of* $\Omega \setminus (F \cup K)$ *has no long islands.*

If $\Omega = \mathbf{C}$, *these conditions are also equivalent to the following:*
4. for each $r > 0$, *there exists* $r' > r$ *such that any two points* a, b *in* $\overline{\mathbf{C}} \setminus F$ *and outside of the disc* $D_{r'}$, *centered at* 0 *and of radius* r', *can be joined by a path in* $\overline{\mathbf{C}} \setminus F$ *and outside of* D_r.

Approximation on unbounded sets has many applications (to pure mathematics). We now present several of these. The first application is a "counterexample" to one of the best known theorems of function theory.

Application 1: *["Counterexample" to Liouville Theorem] There exists a nonconstant entire function which is bounded on each line.*

Proof: Let
$$F_1 = (y \geq 1/x) \cup (x \leq 0) \cup (y \leq 0)$$
and choose a point z_0 not in F_1. Set $F = F_1 \cup z_0$. Then $\mathbf{C}^* \setminus F$ is connected and locally connected. Define f to be 0 at the point z_0 and 1 on the set F_1. Then $f \in H(F)$ and so by Roth's Theorem there is a $g \in H(\mathbf{C})$ such that $|f - g| < 1/2$ on F. Since each line is, except for a compact portion, contained in F, it follows that g is bounded on each line. The function g cannot be constant since it approximates two different constants (0 and 1) too well. In fact we can do better (or worse, depending on how perverse we are). Set $g_0 = (g(z) - g(z_0))/(z - z_0)$. Then g_0 is also a nonconstant entire function and g_0 is not only bounded, but even tends to 0 on each line!

In order to give the next application we recall the beautiful approximation theorem which Whtney published in 1934. Let $C^\omega(\mathbf{R})$ denote the set of analytic functions on $\mathbf{R}$.

Theorem 9: *[Whitney] For each $f \in C(\mathbf{R})$ and for each $\epsilon > 0$, there is a function $g \in C^\omega(\mathbf{R})$ such that*

$$|f - g| < \epsilon.$$

The classical Dirichlet problem for the half-plane was solved by R. Nevanlinna in 1925. As an application of approximation theorems on unbounded sets, we now present an extremely short and elegant solution to this problem which was introduced by W. Kaplan in 1955. Let $U = \{z = x + iy : y > 0\}$ denote the open upper half of the complex plane $\mathbf{C}$ and let $\mathbf{R}$ denote the real axis in $\mathbf{C}$. Also, for $\varphi \in C(\mathbf{R})$, we denote by P_φ the Poisson integral of φ for the half-plane U.

Application 2: *[Dirichlet problem for half-plane] Given $\varphi \in C(\mathbf{R})$ find $u \in C(\overline{U})$ such that u is harmonic on U and $u = \varphi$ on $\mathbf{R}$.*

Solution: By Whitney's Theorem there is a $g \in C^\omega(\mathbf{R})$ such that $|\varphi - g| < 1$. Extend g to $\tilde{g} \in H(\mathbf{R})$. By Roth's Theorem, there is an $f \in H(\mathbf{C})$ such that $|f - \tilde{g}| < 1$. The function

$$u = \mathrm{Re}f + P_{\varphi - \mathrm{Re}f}$$

is then a solution of the given Dirichlet problem. One might ask why we did not simply take the Poisson integral of the function φ itself. But recall that the function φ is an *arbitrary* continuous function on $\mathbf{R}$ and so its Poisson integral may not even converge. However, the Poisson integral of any *bounded* continuous function does exist.

Application 3: *[Domains of holomorphy] Every domain $\Omega \subset \mathbf{C}$ is a domain of holomorphy. That is, there exists $g \in H(\Omega)$ such that g does not extend holomorphically to a (strictly) larger domain.*

Proof: Let $\{z_n\}$ and $\{z_n'\}$ be disjoint sequences in Ω having no limit point in Ω and having every point of $\partial\Omega$ as limit point. Set

$$F = \bigcup_{n=1}^{\infty} \{z_n, z_n'\}.$$

Then, $\Omega^* \setminus F$ is connected and locally connected. Also, the function

$$f = \begin{cases} 0 \text{ at each } z_n, \\ 1 \text{ at each } z_n', \end{cases}$$

is in $H(F)$. Thus by Roth's Theorem, there is a $g \in H(\Omega)$ such that $|f - g| < 1/3$ on F. Since the function g is discontinuous at each point of $\partial\Omega$, it cannot extend holomorphically to any domain containing Ω.

Application 4: [Mittag-Leffler Theorem] Let D be a plane domain, $\{z_n\}$ a sequence in D without limit points in D, and for each n, let s_n be a function holomorphic in a punctured neighbourhood of z_n. Then, there exists a function

$$g \in H(D \setminus \bigcup_{n=1}^{\infty} \{z_n\})$$

such that, for all n, $g - s_n \in H\{z_n\}$.

Proof: For each n we may choose a disc D_n centered at z_n such that the function s_n is holomorphic on $\overline{D}_n \setminus \{z_n\}$ and such that the family $\{\overline{D}_n\}$ is disjoint and locally finite. Set

$$\Omega = D \setminus \bigcup_{n=1}^{\infty} \{z_n\}, \ F = \bigcup_{n=1}^{\infty} (\overline{D}_n \setminus \{z_n\}).$$

Then, $\Omega^* \setminus F$ is connected and locally connected. Moreover, the function f, defined on F by setting $f = s_n$ on $\overline{D}_n \setminus \{z_n\}$ for each n, is holomorphic on F. Thus, by Roth's Theorem, there is a function $g \in H(\Omega)$ such that $|f - g| < 1$. The function g has the required properties. Notice that, whereas in the usual formulation of the Mittag-Leffler Theorem the prescribed singularities are poles, in the present formulation they are *arbitrary* isolated singularities.

Rosay and Rudin *[27]* have given a short and elegant proof of Arakelyan's Theorem (which implies Roth's Theorem). Arakelyan also showed that the condition in Roth's Theorem is not only sufficient for approximation but also necessary.

The necessity remains true on open Riemann surfaces. That is, if F is a closed subset of an open Riemann surface Ω such that for each $f \in H(F)$ and for each $\epsilon > 0$, there is a function $g \in H(\Omega)$ such that $|f - g| < \epsilon$, then $\Omega^* \setminus F$ is both connected and locally connected. However, the sufficiency fails! That is, although the compact Runge Theorem holds on Riemann surfaces (Theorem 7), the closed Runge Theorem (Theorem 8) fails to extend. The construction of a counterexample *[19]* uses the existence of certain surfaces first constructed by Myrberg (see, e.g. *[31]*(Chapter I, Section 10).

Example 8   *There exists an open Riemann surface X having a closed parametric disc K whose complement $X \setminus K$ admits no non-constant bounded holomorphic functions.*

Proof: Let X_1 and X_2 be two copies of the complex plane $\mathbf{C}$ each having the following slits:

$$(1/(2k+1), 1/(2k)), \ k = 1, 2, \ldots.$$

The Riemann surface X is formed by joining X_1 and X_2 in the usual way along their slits. Let D be that portion of X which lies over the open unit disc in $\mathbf{C}$. Let K be a closed disc on X_1 which is disjoint from D and suppose $f \in H(X \setminus K)$. Let $p(z) = \tilde{z}$ be the idempotent automorphism of X which maps a point $z \in X$ to the corresponding point $\tilde{z}$ on the other sheet of X. Note that, at the branch points (which lie over the points $1/j$), $z = \tilde{z}$. Now set

$$\varphi(z) = [f(z) - f(\tilde{z})]^2, \ z \in D.$$

Then $\varphi \in H(D)$ and $\varphi(z) = \varphi(\tilde{z})$. Thus, φ is well defined on the *unslit* punctured unit disc $0 < |z| < 1$. If f is bounded, then the isolated singularity of φ at the origin is removable and so φ extends to a function holomorphic in the unit disc. Since $\varphi(1/j) = 0$, $j = 1, 2, \ldots$, it follows that $\varphi = 0$. From the definition of φ, this means that on D, the function f takes the same value on both sheets. Thus, $f|X_2$ extends holomorphically, to the slits of X_2. But X_2 with its slits closed up is just the plane $\mathbf{C}$ punctured at the origin. Since f is bounded, f extends to a bounded entire function, which by Liouville's Theorem is necessarily constant. Since $X \setminus K$ is connected, f is constant on all of $X \setminus K$.

Example 9 (*[19]*) *Let X and K be as in the above example and let a and b be distinct points of K°. Then, the analog of the closed Runge Theorem 8 fails on the Riemann surface $\Omega = X \setminus \{a\}$.*

Proof: Set $F = (X \setminus K^\circ) \cup \{b\}$. Then F is a closed subset of Ω and $\Omega^* \setminus F$ is both connected and locally connected. Setting

$$f = \begin{cases} 0 \text{ on } X \setminus K^\circ, \\ 1 \text{ at } b, \end{cases}$$

we have that $f \in H(F)$. Suppose, now, that $g \in H(\Omega)$ and $|f - g| < 1/2$ on F. Then, by example 8, the function $g|_{X \setminus K}$ is constant and hence g is constant. But this contradicts the fact that g approximates both 0 and 1 within $1/2$.

The problem of characterizing those closed subsets F of a Riemann surface Ω for which every holomorphic function on F can be approximated by functions holomorphic on Ω remains open. In subsequent sections, we will consider analogous questions for harmonic approximation.

4   Harmonic and Subharmonic Analogs

Throughout this section, Ω will denote an open Riemannian manifold (or more precisely, a "noncompact smooth orientable Riemannian manifold") and W will denote an open subset of Ω. We say that (W, Ω) is a subharmonic extension pair for *compact* sets if for every function u subharmonic on W, and for each compact subset E of W, there is a function $\tilde{u}$, subharmonic on Ω, such that $\tilde{u} = u$ on E. We say that (W, Ω) is a subharmonic (respectively harmonic) Runge pair for *compact* sets if for every function u suharmonic (respectively harmonic) on W, for each compact subset E of W and for each $\epsilon > 0$, there is a function u_ϵ, subharmonic (respectively harmonic) on Ω, such that $u - \epsilon \le u_\epsilon \le u + \epsilon$ on E. Similarly, we define the respective notions of subharmonic extension pair, and subharmonic Runge pair, for *closed* sets by replacing compact subsets E of W, in the above definitions, by subsets E of W which are closed (in Ω, not just in W). We define the notions of *continuous* subharmonic extension pairs and *continuous* subharmonic Runge pairs for compact sets by replacing "subharmonic" by "continuous subharmonic" in the above definitions. Finally, we say that (W, Ω) is a continuous subharmonic (respectively harmonic) Runge pair for closed sets if for every function u continuous subharmonic (respectively harmonic) on W, there is a sequence of functions u_n, continuous subharmonic (respectively harmonic) on Ω which converges pointwise to u on W and such that the convergence is uniform on each closed subset E of W.

Theorem 10: *The following are equivalent:*

   a) (W, Ω) is a harmonic Runge pair for compact sets;

   b) (W, Ω) is a harmonic Runge pair for closed sets;

   c) (W, Ω) is a (continuous) subharmonic Runge pair for compact sets;

   d) (W, Ω) is a (continuous) subharmonic Runge pair for closed sets;

   e) (W, Ω) is a (continuous) subharmonic extension pair for compact sets;

   f) (W, Ω) is a (continuous) subharmonic extension pair for closed sets;

   g) $\Omega^* \setminus W$ is connected.

The preceding theorem is due to various authors (see *[14]* for more information on this). Notice that the equivalence of a) and g) is the harmonic analog of the Behnke-Stein Theorem. Similarly, it can be shown that the holomorphic analog of g) $\to$ b) is equivalent to the statement in Roth's Theorem.

As in the complex case, the theory of harmonic and subharmonic approximations and extensions on closed sets allows interesting applications.

Application 5: *[Necessity in maximum principle] Suppose E is an accessible closed subset of $\tilde{\partial}\Omega$. Then, there exists a function s subharmonic on Ω such that*

$$sup_\Omega s > sup_{\tilde{\partial}\Omega \setminus E} s.$$

Proof: Let σ be a path in Ω which tends to E. We may assume that σ is simple and hence it is possible to construct open sets U and V such that

$$\sigma \subset U \subset \overline{U} \subset V \subset \overline{V} \subset \Omega,$$

where the closures are to be taken in Ω and, setting $W = U \cup (\Omega \setminus \overline{U})$, we have that $\Omega^* \setminus W$ is connected. If we define u to be 1 on U and 0 on $\Omega \setminus \overline{U}$, then, by the implication g) $\to$ f), there is a function s, subharmonic (and continuous) on all of Ω which is 1 on σ and 0 on the "neighbourhood" $\Omega \setminus V$ of $\tilde{\partial}\Omega \setminus E$. It follows that the function s has the required properties. In fact, by the implication g) $\to$ b) we even have a *harmonic* counterexample to the generalized maximum principle for this E. A detailed justification of the preceding topological assertions can be given following the methods of *[2]*.

5   Walsh Theorems

We have seen that, near the end of the 19th century, K. Runge considered (quite successfully) the problem of describing those sets K for which every holomorphic function on K can be approximated by polynomials. The problem, then, quite naturally arose of approximating functions, given on K, by functions holomorphic on K. Of course, any function which can be uniformly approximated on K by functions holomorphic on K is necessarily continuous on K and holomorphic in the interior of K, and early in the present century, J. L. Walsh *[15]* showed that in some very important cases, no more is required of the approximatees. The problem, then, quite naturally arose of characterizing those sets for which *every* function continuous on K and holomorphic on the interior of K can be approximated by functions holomorphic on K. In this section, we present the complete solution to this problem which was found, first for the analogous harmonic problem by M.V. Keldysh, and then in the complex case by A.G. Vitushkin, both of the Russian school. The solutions to these problems turn out to be local and can be described in terms of capacity and, in the harmonic case, the notion of thinness.

Recall that the *fine topology* on a Riemannian manifold Ω is the topology generated by the subharmonic functions on Ω.

Theorem 11:   *For a subset E and a point x_0 of $\mathbf{R}^n$, the following are equivalent:*
  a)   *E is thin at x_0;*

  b)   *$(\mathbf{R}^n \setminus E) \cup \{x_0\}$ is a fine neighbourhood of $\{x_0\}$;*

  c)   *Brownian motion starting from x_0 remains in $\mathbf{R}^n \setminus E$ for positive time.*

The equivalence of a) and b) persists on Riemannian manifolds. The notion of Brownian motion can also be introduced on Riemannian manifolds, but, in any case, we will not require the equivalence of the last condition to the first two.

Let us denote by $\mathcal{C}(E)$, $\mathcal{C}_a(E)$ and $\mathcal{C}_{ac}(E)$ respectively the capacity, analytic capacity and continuous analytic capacity of a set E. Without stating the definitions of the various capacities, we recall that they have the following important property of characterizing removable sets for their associated classes of functions. Namely, a set is removable for the class of bounded harmonic, bounded holomorphic or (uniformly) continuous holomorphic functions if and only if it is respectively of harmonic, analytic or continuous analytic capacity zero.

We denote by $h(F)$ the class of functions harmonic on (a neighbourhood of) a set F and (as before), by $H(F)$ the class of functions holomorphic on F. Let $\overline{h}(F)$ (respectively, $\overline{H}(F)$) denote the functions on F which are uniform limits of functions in $h(F)$ (respectively, $H(F)$). Further, we denote by $a(F)$ (respectively, $A(F)$) the classes of functions continuous on F and harmonic (respectively, holomorphic) on the interior of F. Clearly, we have $\overline{h}(F) \subset a(F)$ and $\overline{H}(F) \subset A(F)$.

Theorem 12: *[Keldysh, Deny, Labrèche] Let F be a closed subset of a domain Ω of Euclidean space $\mathbf{R}^n$. The following are equivalent:*
  a)   *$\overline{h}(F) = a(F)$;*
  b)   *for each $x \in F$, there is a closed ball $\overline{B}_x \subset \Omega$ such that*

$$\overline{h}(F \cap \overline{B}_x) = a(F \cap \overline{B}_x);$$

c) *for each open U in Ω,*

$$\mathcal{C}(U \setminus F) = \mathcal{C}(U \setminus F^0);$$

d) $\Omega \setminus F$ *and* $\Omega \setminus F^0$ *are thin at the same points of* F.

This theorem was first shown for F compact by Keldysh and later, independently, by Deny. The closed case is due to Labrèche. Recently [1] this result has been extended to Riemannian manifolds (omitting condition c)).

We now give two examples of compact sets K for which approximation fails, that is, $\overline{h}(K) \neq a(K)$, followed by an application involving a closed set F where approximation is possible.

Example 10 *There is a set $K \subset \mathbf{R}^2$ of the form*

$$K = \overline{D} \setminus \bigcup_{j=1}^{\infty} D_j$$

such that $\overline{h}(K) \neq a(K)$, where $\overline{D}$ is the unit disc and $\{D_j\}$ is a sequence of discs having disjoint closures in D.

Such an example can be constructed [20], for example, by choosing the sequence $\{D_j\}$ in the upper half-disc and clustering precisely to the whole interval $[0, 1]$ and nowhere else. It follows that $\mathbf{R}^2 \setminus K^0$ is not thin at 0 since it contains the interval $[0, 1]$. On the other hand, the discs $\{D_j\}$ can be chosen so small that, by the Wiener criterion, their union, which locally is $\mathbf{R}^2 \setminus K$, is thin at 0.

If a compact subset K of $\mathbf{R}^2$ has connected complement, then automatically the complement is thin at no point of the boundary of K and so $\overline{h}(K) = a(K)$. This situation does not persist in higher dimensions. In fact there is the following striking example.

Example 11 *There is a compact starlike set $K \subset \mathbf{R}^3$ for which $\overline{h}(K) \neq a(K)$.*

Such an example was constructed by Bagby and the author [4] in the form

$$K = \overline{B} \setminus \bigcup_{j=1}^{\infty} S_j,$$

where $\overline{B}$ is the unit ball and $\{S_j\}$ is a sequence of Lebesgue spines having disjoint closures in B and clustering precisely to all points of the disc $\overline{D} = \overline{B} \cap \{x_3 = 0\}$ and to no other points. The argument that K has the desired properties is similar to that in the preceding example.

Let us denote an arbitrary point x of Euclidean n-space $\mathbf{R}^n$, $n \geq 2$, by $x = (x', x_n)$, where $x' \in \mathbf{R}^{n-1}$ and $x_n \in \mathbf{R}$. Let U denote the open upper half-space $\{x : x_n > 0\}$. As an illustration of the power of the above theorem on harmonic approximation on unbounded sets, we will now give an elegant solution to the higher dimensional analog of the Dirichlet problem for the half-plane.

Application 6: *[Dirichlet problem for half-space] Given* $\varphi \in C(\mathbf{R}^{n-1})$, *find* $u \in C(\overline{U})$ *such that* u *is harmonic on* U *and* $u(x',0) = \varphi(x')$ *for* $x' \in \mathbf{R}^{n-1}$.

Solution: By Labrèche's Theorem there is a $g \in h(\mathbf{R}^{n-1})$ such that $|\varphi - g| < 1$. By Section 4, there is an $f \in h(\mathbf{R}^n)$ such that $|f - g| < 1$. The function

$$u = f + P_{\varphi - f}$$

is then a solution of the given Dirichlet problem.

In regard to the preceding application, the reader may want to look at Gardiner's paper *[13]*, where a stronger result is proved.

The preceding theorem has a complex analog proved first by Vitushkin for compact F and later extended to closed sets by Nersesyan and Roth (see *[19]*).

Theorem 13: *[Vitushkin, Nersesyan, Roth] Let* F *be a closed subset of a domain* Ω *of the complex plane* **C**. *The following are equivalent:*

 a) $\overline{H}(F) = A(F)$;

 b) *for each* $z \in F$, *there is a closed disc* $\overline{D}_z \subset \Omega$ *such that*

$$\overline{H}(F \cap \overline{D}_z) = A(F \cap \overline{D}_z);$$

 c) *for each open* $U \subset \Omega$,

$$\mathcal{C}_{ac}(U \setminus F) = \mathcal{C}_{ac}(U \setminus F^0).$$

A particular, and most interesting, instance of this theorem is obtained by taking F to be the real axis. This yields Whitney's approximation theorem (as we have stated it) on uniform approximation of continuous functions on **R** by analytic ones. (Actually, Whitney obtained *better*-than-uniform approximation, and in $\mathbf{R}^n$, not just in **R**.) Thus, we are able to "bypass" the use of Whitney's Theorem in "our" earlier solution of the Dirichlet problem for the half-plane.

6   Individual Theorems

The theorems of the preceding section give necessary and sufficient conditions on a closed set F in order that each function belonging to the *class* of functions continuous on F and harmonic (respectively, holomorphic) on the interior of F can be approximated by functions harmonic (respectively, holomorphic) on F. Such theorems are called *class* theorems. Suppose now we consider an arbitrary closed set F, which may or may not satisfy the conditions of the class theorem, and hence there may be some functions in $a(F)$ (respectively, $A(F)$) which cannot be approximated. A more general question than that addressed by the class theorems is to ask just which functions on such an arbitrary F *can* be approximated. Theorems which deal with this latter question are called *individual function* theorems. In the harmonic case this problem has been solved on compact sets by Debiard and Gaveau in terms of fine potential theory.

Definition: Let x be a point of a finely open set V and let $E \subset \partial V$. We define the (fine) *harmonic measure* $\omega_x^V(E)$ as the probability that Brownian motion starting at x first exit V through E.

Remarks:

1. If V is an ordinary open set, then $\omega_x^V(E)$ is ordinary harmonic measure.
2. $\omega_x(E)$ is carried by the fine boundary of V.

Definition: A finely continuous function u defined on a finely open set U is said to be *finely harmonic* if there is a base $\mathcal{B}$ for the induced fine topology on U such that for all $V \in \mathcal{B}$, the fine closure of V is contained in U and

$$u(x) = \int u d\omega_x^V, \text{ for all } x \in V.$$

We remark that on a usual open set U, the harmonic functions are the finely harmonic functions which are locally bounded.

We may now state the complete characterization of approximable functions known as the Debiard and Gaveau theorem. This was originally proved for compact sets, later extended to closed sets of Euclidean space by Ladouceur and the author and then extended to closed subsets of Riemannian manifolds by Bagby and Blanchet *[1]*.

Theorem 14: *[Debiard-Gaveau theorem for closed sets] Suppose F is a closed subset of an open Riemannian manifold, and let $u \in C(F)$. Then the following are equivalent.*

  a) $u \in \overline{h}(F)$.

  b) u is finely harmonic on the fine interior of F.

The above individual function theorem of Debiard and Gaveau is stronger than the class theorem of Keldysh. Let us verify this by deriving the equivalence of a) and d) in the Keldysh theorem from the Debiard and Gaveau theorem.

Suppose, then, that $\Omega \setminus F$ and $\Omega \setminus F^0$ are thin at the same points and let $u \in a(F)$. We claim that $u \in \overline{h}(F)$. By the Debiard and Gaveau theorem, it is sufficient to show that u is finely harmonic on the fine interior of F. Let x be a fine interior point of F. We may assume that x lies on the boundary of F, for otherwise u is by hypothesis harmonic on an open neighbourhood of x and *a fortiori* finely harmonic in a fine neighbourhood of x. To say that the boundary point x lies on the fine interior of F is to say that $\Omega \setminus F$ is thin at x, but then, by hypotheses, $\Omega \setminus F^0$ is also thin at x. Thus, F^0 is a fine deleted neighbourhood of x. Since u is continuous on F it is bounded and so by a removable singularity theorem for finely harmonic functions, u extends to be finely harmonic at x. But since u is already continuously defined at x, u is itself this fine harmonic extension. Thus, $u \in \overline{h}(F)$. This shows that d) $\to$ a) in the Keldysh theorem.

The following lemma, first proved by Keldysh for compact sets, is fundamental. A set is said to be *thick* at a point if it is not thin at that point. A function u defined on F, peaks at a point $x \in F$ if $u(x) > u(y)$ for all $y \in F \setminus \{x\}$.

Lemma 2: *[Keldysh Lemma] Let F be a closed subset of a Riemannian manifold Ω and $x \in F$. Then, $\Omega \setminus F^0$ is thick at x if and only if x is a peak point for the class $a(F)$.*

Proof: Suppose, first, that x is a peak point for the class $a(F)$ and let u be a function in $a(F)$ which peaks at x. If $\Omega \setminus F^0$ were thin at x, then F^0 would be a deleted fine neighbourhood of x and, since u is bounded near x, by the theorem on removable singularities for finely harmonic functions, u would extend to be finely harmonic at x. But u is already continuously defined at x and so u itself would be finely harmonic at x. Since u peaks at x, this would violate the mean value property for finely harmonic functions. Thus, $\Omega \setminus F^0$ is thick at x.

Suppose, conversely, that $\Omega \setminus F^0$ is thick at x. Fix a parametric ball B centered at x. By the Wiener criterion, the thickness of the set $\Omega \setminus F^0$ at x is equivalent to the divergence of a certain Wiener series $\sum \lambda_n C(A_n \setminus F^0)$, where the A_n are appropriate "annuli" centered at x and the λ_n are appropriate constants. We may assume that the series $\sum \lambda_{2n} C(A_{2n} \setminus F^0)$ also diverges. Now, by a famous result of Ancona, for any $\epsilon_n > 0$, there is a compact set K_n contained in $A_{2n} \setminus F^0$ such that

$$C(K_n) > C(A_{2n} \setminus F^0) - \epsilon_n$$

and K_n is thick at each of its points. Set

$$G = \Omega \setminus \bigcup_{n=1}^{\infty} K_n.$$

The open set G is regular for the Dirichlet problem. Hence, there is a function $u \in a(\overline{G})$ which peaks at the point $x \in \partial G$. Since $F^0 \subset G$ it follows that u can be extended to a function in $a(F)$ which peaks at x. ■

In the above proof of the Keldysh Lemma, we have used the concept of capacity and the Wiener criterion for thinness, on a parametric ball in a Riemannian manifold. Here our concept of thinness comes from regarding the parametric ball as a harmonic space ("harmonic functions" defined by the Riemannian metric). We know of no reference for the Wiener criterion in this generality, although the Wiener criterion is of course well known when harmonic functions are defined by the Euclidean metric. However, Hervé *[24]* has shown that these two concepts of thinness in the parametric ball are the same. (A similar problem and solution occurs in *[1]*(1, Remark 8.2 (e)).)

We now present a proof that a) $\to$ d) in the Keldysh theorem. Suppose, then, that $\Omega \setminus F$ and $\Omega \setminus F^0$ are *not* thin at the same points and let us construct a $u \in a(F)$ which is not in $\overline{h}(F)$. There exists, by hypothesis, some point $x \in \partial F$ such that $\Omega \setminus F$ is thin at x but $\Omega \setminus F^0$ is not. By the Keldysh lemma, x is a peak point for the class $a(F)$. Thus, there is a function $u \in a(F)$ which peaks at x. By the maximum principle for finely harmonic functions, this function cannot be finely harmonic at x. Thus, by the theorem of Debiard and Gaveau this function u is not in $\overline{h}(F)$ and this shows that a) $\to$ d) in the Keldysh theorem.

Let us now turn to the individual function problem for holomorphic functions. A function f is said to be *finely holomorphic* if f' exists (in the fine sense) and is finely continuous. Let us denote by $\tilde{A}(F)$ the set of functions continuous on F and finely holomorphic on the fine interior of F. Then, as in the harmonic situation, we have the following inclusions:

$$\overline{H}(F) \subset \tilde{A}(F) \subset A(F).$$

However, in contrast to the harmonic case, in which the Debiard and Gaveau theorem asserts that the harmonic analogs of the first two classes coincide, Fuglede *[10]* has shown that the inclusions in the holomorphic case can be strict. Thus, the complex analog of the Debiard and Gaveau theorem fails.

Nevertheless, another type of individual function theorem does hold for complex approximation, namely, the well known theorem of Vitushkin which, in terms of certain estimates involving continuous analytic capacity, gives a complete description of those functions on a compact set which can be approximated by holomorphic functions. This powerful result of Vitushkin (which we do not state) has been extended to closed sets by Hadjiiski *[23]*.

7 Holomorphic vs. Harmonic

The holomorphic and harmonic Runge theorems which we have given for approximation on compact subsets of an open set W can both be stated in the same way.

Theorem 15: *Let W be an open subset of an open Riemann surface (respectively, Riemannian manifold) Ω. Then, (W, Ω) is a holomorphic (respectively, harmonic) Runge pair for compact sets if and only if $\Omega^* \setminus W$ is connected.*

We have seen that the same condition characterizes harmonic Runge pairs for approximation on *closed* subsets of W. It can be shown that the complex analog of this last statement is also true provided that Ω is a plane domain. However, this is not the case for Ω an arbitrary Riemann surface as our counterexample to the Roth theorem on Riemann surfaces shows.

Runge pairs (W, Ω) have been defined above, only when W is open in Ω. Replacing the open set W by a closed set in the definition of a Runge pair for closed sets, we shall say that for a closed subset F of a Riemann surface (respectively, Riemannian manifold) Ω, the pair (F, Ω) is a holomorphic (respectively, harmonic) Runge pair if each function holomorphic (respectively, harmonic) on F is the uniform limit of functions holomorphic (respectively, harmonic) on all of Ω. With this terminology, the theorem of Roth (with the necessity coming from Arakelyan) takes the following form.

Theorem 16: *Let F be a relatively closed subset of a plane domain Ω. Then, (F, Ω) is a holomorphic Runge pair if and only if $\Omega^* \setminus F$ is connected and locally connected.*

A partial analog of this result was found by Goldstein, Ow and myself for the harmonic situation.

Theorem 17: *([17, 18]) Let F be a relatively closed subset of a domain Ω in $\mathbf{R}^n$ such that $\Omega^* \setminus F$ is connected and locally connected. Then, (F, Ω) is a harmonic Runge pair.*

This result was extended to Riemannian manifolds by Bagby and Blanchet *[1]*(Theorem 9.3)

The following elementary example shows that, in contrast to the holomorphic situation, the conditions in the last theorem, although sufficient, are no longer necessary in the harmonic situation.

Example 12   *Let $F = (|z| = 1)$ and $\Omega = \mathbf{C}$. Then (F, Ω) is a harmonic Runge pair but $\mathbf{C}^* \setminus F$ is not connected.*

Proof: Suppose $u \in h(|z| = 1)$ and $\epsilon > 0$. Let $\tilde{u}$ be the solution of the Dirichlet problem on $(|z| \leq 1$, with $\tilde{u}(e^{i\theta}) = u(e^{i\theta})$. For $\rho < 1$, set $u_\rho(z) \equiv \tilde{u}(\rho z)$. Then, $u_\rho \in h(|z| \leq 1)$ and $u_\rho \to u$. There exists a ρ such that $|u_\rho - u| < \epsilon/2$ on F and by the harmonic Runge theorem, there is a $v \in h(\mathbf{C})$ such that $|v - u_\rho| < \epsilon/2$ on F. Thus, $u \in \overline{h}(\mathbf{C})$.

A complete characterization of harmonic Runge pairs (F, Ω), when F is a closed subset of Ω has been obtained only recently by Gardiner. In order to state Gardiner's theorem we introduce some terminology. A subset of Ω is said to be Ω-*bounded* if its closure is compact in Ω. Let F be a subset of Ω. An Ω-*hole* of F is an Ω-bounded component of the complement $\Omega \setminus F$. We denote by $\hat{F}$ the union of F and its holes.

Theorem 18: *[Gardiner [12]] Let F be a relatively closed subset of a domain Ω in $\mathbf{R}^n$. The following are equivalent:*
a) (F, Ω) is a subharmonic Runge pair;
b) (F, Ω) is a harmonic Runge pair;
c) (F, Ω) satisfies
   i) $\Omega \setminus \hat{F}$ and $\Omega \setminus F$ are thin at the same points of F,
   ii) for each compact $K \subset \Omega$ there is a compact $Q \subset \Omega$ which contains every Ω-hole of $F \cup K$ whose closure intersects K.

Of course, if F is compact, condition ii) is superfluous. When $n - 2$, one can replace i) by the condition that $\partial \hat{F} = \partial F$. It follows that, if we denote the boundary of the unbounded complementary component of a bounded subset K of the plane by $\partial_\infty K$, we have the following result.

Corollary 3:   *Let K be a compact subset of the complex plane $\mathbf{C}$. A necessary and sufficient condition in order that every $u \in h(K)$ be unifomly approximable by harmonic polynomials is that $\partial K = \partial_\infty K$.*

It is remarkable that, prior to Gardiner's beautiful theorem, the answer to such a natural (and old) problem as the characterization of harmonic Runge pairs, for closed subsets F, was unknown, even when F is compact.

Throughout these lectures, I have considered approximation only in the uniform norm (which to me seems the most natural). However, it is also of importance to study approximation in other norms and, in fact, the most interesting remarks concerning the topic of the present section "holomorphic uniform versus harmonic uniform approximation" involve other norms. Firstly, it follows from work of Khavin *[26]* that the problem of harmonic uniform approximation (for classes) in $\mathbf{R}^2$ is equivalent to that of holomorphic L^2-approximation in $\mathbf{C} = \mathbf{R}^2$. Moreover there are indications (for example, *[21]*) that the problem of holomorphic uniform approximation (for classes) is related (equivalent seems to be too strong a claim) to that of harmonic C^1-approximation.

In closing I wish to thank Professors Thomas Bagby and Stephen Gardiner for several helpful comments which corrected some of the deficiencies in the present notes.

References

[1]   Bagby, T. and Blanchet, P.: Uniform harmonic approximation on Riemannian manifolds. J. d'Analyse Math. (to appear).

[2]   Bagby, T., Cornea, A. and Gauthier, P.M.: Harmonic approximation on arcs *Constr. Approx* 9 (1993) 501-507.

[3]   Bagby, T. and Gauthier, P.M.: Approximation by harmonic functions on closed subsets of Riemann surfaces. J. d'Analyse Math. 51 (1988), 259-284.

[4]   Bagby, T. and Gauthier, P.M.: Uniform approximation by global harmonic functions, in Approximation by Solutions of Partial Differential Equations, 15-26, ed. by B. Fuglede, M. Goldstein, W. Haussmann, W. K. Hayman, and L. Rogge, NATO ASI Series, Kluwer Academic Publishers, 1992.

[5]   Behnke, H. and Stein, K.: Entwicklungen analytischer Funktionen auf Riemannschen Flächen. Math. Ann. 120 (1948), 430-461.

[6]   Burckel, R.B.: An Introduction to Classical Complex Analysis I. Academic Press, 1979.

[7]   Chen Huaihui and Gauthier, P.M.: A maximum principle for subharmonic and plurisubharmonic functions. Canad. Math. Bull. 35 (1992), 34-39.

[8]   Eremenko, A.E.: (private communication).

[9]   Fuglede, B.: Asymptotic paths for subharmonic functions and polygonal connectedness of fine domains. Springer Lecture Notes 814 (1980), 97-115.

[10]   Fuglede, B.: Sur les fonctions finement holomorphes. Ann. Inst. Fourier Grenoble 31 (1981), 57-88.

[11]   Gaier, D.: Lectures on Complex Approximation. Birkhaüser, 1987.

[12]   Gardiner, S.J.: Superharmonic extension and harmonic approximation (preprint).

[13]   Gardiner, S.J.: The Dirichlet and Neumann problems for harmonic functions in half-spaces, J. London Math. Soc. (2) 24 (1981), 502-512.

[14]   Gauthier, P.M.: Subharmonic extensions and approximations, Canad. Math. Bull. 37 (1994), 46-53.

[15]   Gauthier, P.M.: J.L. Walsh and Qualitative Approximation (to appear).

[16]   Gauthier, P.M., Grothmann, R. and Hengartner, W.: Asymptotic maximum principles for subharmonic and plurisubharmonic functions. Canad. J. Math. 40 (1988), 477-486.

[17]   Gauthier, P.M., Goldstein, M, and Ow, W.H.: Uniform approximation on closed

sets by harmonic functions with logarithmic singularities. Trans. Amer. Math. Soc. 261 (1980), 169-183.

[18] Gauthier, P.M., Goldstein, M, and Ow, W.H.: Uniform approximation on closed sets by harmonic functions with Newtonian singularities. J. London Math. Soc. (2) 28 (1983), 71-82.

[19] Gauthier, P.M. and Hengartner, W.: Approximation Uniforme Qualitative sur des Ensembles Non Bornés, Les Presses de l' Université de Montréal, 1982.

[20] Gauthier, P.M., W. Hengartner and Labrèche, M.: Approximation harmonique, approximation holomorphe et topologie. Canad. J. Math. 34 (1982), 216-219.

[21] Gauthier, P.M. and Paramonov, P.V.: Approximation by harmonic functions in the C^1-norm and harmonic C^1-content of compact subsets in $\mathbf{R}^n$ (Russian), *Mat. Zam.* 53 (1993), 21-30.

[22] Gol'berg, A.A.: (private communication).

[23] Hadjiiski, V.H.: Vitushkin's type theorems for meromorphic approximation on unbounded sets, in Proc. Conf. "Complex Analysis and Applications '81-Varna", 229-238, Bulgarian Acad. Sci., Sofia, 1984.

[24] Hervé, R.-M.: Quelques propriétés des fonctions.... Ann. Inst. Fourier (Grenoble) 15 (1965), 215-224.

[25] Iversen, F.: Recherches sur les fonctions inverses des fonctions méromorphes. Thèse, Helsingfors, 1914.

[26] Khavin, V.P.: Approximation in the mean by analytic functions (Russian). Dokl. Akad. Nauk SSSR 178 (1968), 1025-1028. English translation: Soviet Math. Doklady 9 (1968), 245-248.

[27] Rosay, J.-P. and Rudin, W.: Arakelian's approximation theorem. Amer. Math. Monthly 96 (1989), 432-434.

[28] Roth, A.: Approximationseigenschaften und Strahlengrenzwerte meromorpher und ganzer Funktionen. Comment. Math. Helv. 11 (1938), 77-125.

[29] Roth, A.: Meromorphe Approximationen. Comment. Math. Helv. 48 (1973), 151-176.

[30] Sahakian, R.Sh.: On a generalization of the maximum principle, Izv. Akad. Nauk Arm. SSR Mat. 22 (1987), 94-101; translation in Soviet J. Contemporary Math. Anal. 22 (1987), 94-102.

[31] Sario, L. and Nakai, M.: Classification Theory of Riemann Surfaces, Springer-Verlag, 1970.

Chebyshev Polynomials and Discrete Groups[1]

F. W. Gehring
University of Michigan
Ann Arbor, MI, USA

G. J. Martin
University of Auckland
Auckland, New Zealand

1 Introduction

The Chebyshev polynomials are defined by the recursive formula $T_0(w) = 1$, $T_1(w) = w$ and

$$T_{n+1}(w) = 2wT_n(w) - T_{n-1}(w) \tag{1}$$

for $n \geq 1$. The family $\{T_n(w)\}_{n\geq 0}$ constitutes one of the most important classes of orthogonal polynomials from many points of view, especially those of approximation and interpolation. In this paper we show how certain shifted Chebyshev polynomials act in a multiplicative fashion on the space of two generator discrete subgroups of the group of all Möbius transformations of the Riemann sphere $\overline{\mathbf{C}}$.

We then provide estimates on the growth of these polynomials for a fixed $z \in \mathbf{C}$ as n varies. These estimates can be used to extend Jørgensen's inequality [9] as well as other inequalities for discrete groups. An important consequence of Jørgensen's inequality is that certain groups are isolated in the topology of algebraic convergence. Here we give a quantitative measure of this fact for two generator groups with an elliptic generator. We also obtain asymptotically sharp estimates concerning the way loxodromic elements can degenerate into parabolic elements in sequences of discrete groups. Finally we conclude this paper with a sharp analog of Jørgensen's inequality for nonloxodromic elements.

We define the family of shifted Chebyshev polynomials $\{P_n(w)\}_{n\geq 0}$ by the rule

$$P_n(w) = 2T_n(1 + w/2) - 2. \tag{2}$$

We record here some elementary facts about $\{P_n(w)\}_{n\geq 0}$. It follows from (1) that the leading coefficient of $T_n(w)$ is 2^{n-1} and so by (2), that $P_n(w)$ is monic. As $T_n(1) = 1$, $n \geq 0$, we see $P_n(0) = 0$. Next from the well known identity [3]

$$T_n(\cosh(z)) = \cosh(nz), \tag{3}$$

<hr>

[1] Research supported in part by grants from the U. S. National Science Foundation and the Nankai Institute of Mathematics (FWG) and by the Auckland University Research Council (GJM).

Proceedings of the International Conference on Complex
Analysis at the Nankai Institute of Mathematics, 1992 pp. 114-125
©INTERNATIONAL PRESS 1994

we deduce the following lemma.

Lemma 1.1: $P_n(4\sinh^2(z)) = 4\sinh^2(nz)$.

Proof:

$$\begin{aligned}
P_n(4\sinh^2(z)) &= 2T_n(1 + 2\sinh^2(z)) - 2 \\
&= 2T_n(\cosh(2z)) - 2 \\
&= 2(\cosh(2nz) - 1) \\
&= 4\sinh^2(nz).
\end{aligned}$$

It follows that

$$u \in [-4, 0] \quad \text{if and only if} \quad P_n(u) \in [-4, 0] \tag{4}$$

and that

$$\lim_{w \to 0} \frac{P_n(w)}{w} = n^2. \tag{5}$$

In our applications we will be concerned with points in the complex plane where one of the polynomials P_n is small in modulus. The following easy consequence of Lemma 1.1 shows that points in $[-4, 0]$ have this property.

Corollary 1.2: If $u \in [-4, 0]$, then

$$\inf_{n \geq 1} |P_n(u)| = 0.$$

Proof: Choose $y \in [0, \pi]$ so that $u = -4\sin^2(y)$. Then

$$\inf_{n \geq 1} |P_n(u)| = \inf_{n \geq 1} |P_n(-4\sin^2(y))| = \inf_{n \geq 1} |4\sin^2(ny)| = 0$$

since the values $ny \bmod \pi$ are dense in $[0, \pi]$ if y is an irrational multiple of π, and the value 0 is attained if y is a rational multiple of π.

It is clear that for $w \in \mathbf{C} \setminus [-4, 0]$, $P_n(w) \to \infty$ as $n \to \infty$. Thus in choosing an n for which $|P_n(w)|$ is small we cannot appeal to a limiting process but rather must rely on a diophantine analysis. Here then is the necessary lemma. It is due to Zagier; a detailed proof can be found in *[12]*. We sketch the proof for the convenience of the reader.

Lemma 1.3: If $z = x + \imath y$ where $0 < x < \pi\sqrt{3}$, then there is an integer $n \geq 1$ such that

$$\cosh(nx) - \cos(ny) \leq \cosh\left(\sqrt{\frac{4\pi x}{\sqrt{3}}}\right) - 1.$$

Proof: Let $u = y/2\pi$ and $v = x/2\pi$. Then $w = u + \imath v$ lies in the upper half–plane and there is an element of the modular group which maps w into a fundamental domain all of whose points have imaginary part at least $\sqrt{3}/2$ [1]. We therefore deduce the existence of integers n and m such that

$$(nu + m)^2 + n^2 v^2 \leq \frac{2v}{\sqrt{3}}, \quad n \geq 0. \tag{6}$$

As $0 < x < \pi\sqrt{3}$, $n \geq 1$ and we obtain

$$\cosh(nx) - \cos(ny) = \cosh(nx) - \cos(ny + 2m\pi)$$

$$= \sum_{r=1}^{\infty} \frac{1}{(2r)!}[(nx)^{2r} - (-1)^r(ny + 2m\pi)^{2r}]$$

$$\leq \sum_{r=1}^{\infty} \frac{1}{(2r)!}[(nx)^2 + (ny + 2m\pi)^2]^r$$

$$\leq \sum_{r=1}^{\infty} \frac{1}{(2r)!}[4\pi x/\sqrt{3}]^r$$

$$= \cosh\left(\sqrt{\frac{4\pi x}{\sqrt{3}}}\right) - 1.$$

Theorem 1.4: *If $z = x + \imath y$ where $0 \leq x < \pi\sqrt{3}/2$, then*

$$\min_{n \geq 1} |P_n(4\sinh^2(z))| \leq 4\sinh^2\left(\sqrt{\frac{2\pi x}{\sqrt{3}}}\right). \tag{7}$$

Proof: If $0 < x < \pi\sqrt{3}/2$, then

$$\min_{n \geq 1} |P_n(4\sinh^2(z))| = \min_{n \geq 1} |4\sinh^2(nz)|$$

$$= 4\min_{n \geq 1} |\sinh(nx)\cos(ny) + \imath\cosh(nx)\sin(ny)|^2$$

$$= 2\min_{n \geq 1} |2\cosh^2(nx) - 2\cos^2(ny)|$$

$$= 2\min_{n \geq 1} |\cosh(2nx) - \cos(2ny)|$$

$$\leq 2\cosh\left(\sqrt{\frac{8\pi x}{\sqrt{3}}}\right) - 2$$

$$= 4\sinh^2\left(\sqrt{\frac{2\pi x}{\sqrt{3}}}\right)$$

by Lemma 1.3 with $2x$ and $2y$ in place of x and y. Otherwise $x = 0$,

$$4\sinh^2(z) \in [-4, 0]$$

and

$$4 \min_{n \geq 1} |P_n(4\sinh^2(z))| = 0 = 4\sinh^2\left(\sqrt{\frac{2\pi x}{\sqrt{3}}}\right)$$

by Corollary 1.2.

Remark: If $y = 0$, then (7) implies that

$$4\sinh^2(x) \leq 4\sinh^2\left(\sqrt{\frac{2\pi x}{\sqrt{3}}}\right)$$

and hence that $0 \leq x \leq 2\pi/\sqrt{3}$. Thus the upper bound for x in Theorem 1.4 cannot be increased by more than the factor $4/3$.

We can rephrase the above result in terms of values of the polynomials P_n.

Corollary 1.5: *If $|w| \leq 228$, then*

$$\min_{n \geq 1} |P_n(w)| \leq 4\sinh^2\left(\sqrt{\frac{\pi}{\sqrt{3}}\operatorname{arccosh}(\frac{|w + 4| + |w|}{4})}\right). \tag{8}$$

Proof: Choose $z = x + \imath y$ so that $x \geq 0$ and $w = 4\sinh^2(z)$. Then

$$228 \geq |w| = 4|\cosh^2(x) - \cos^2(y)| \geq 4\sinh^2(x)$$

and hence $x < \pi\sqrt{3}/2$. Next

$$|w + 4| + |w| = 4\cosh(2x)$$

and so

$$x = \frac{1}{2}\operatorname{arccosh}(\frac{|w + 4| + |w|}{4}).$$

Then by Theorem 1.4,

$$\min_{n \geq 1} |P_n(w)| = \min_{n \geq 1} |P_n(4\sinh^2(z))| \leq 4\sinh^2\left(\sqrt{\frac{2\pi x}{\sqrt{3}}}\right)$$

$$= 4\sinh^2\left(\sqrt{\frac{\pi}{\sqrt{3}}\operatorname{arccosh}(\frac{|w + 4| + |w|}{4})}\right).$$

The salient feature of the bound in (8) is that

$$\min_{n \geq 1} |P_n(w)| \to 0$$

uniformly as $w \to [-4, 0]$.

2   Discrete Groups

Let Möb(2) denote the group of all linear fractional transformations of the Riemann sphere $\overline{\mathbf{C}}$. Möb(2) is isomorphic as a topological group to PSL(2,C). Thus to each linear fractional transformation

$$f(z) = \frac{az+b}{cz+d}, \quad ad - bc = 1, \tag{9}$$

we can associate the matrix

$$A = \pm \begin{pmatrix} a & b \\ c & d \end{pmatrix}. \tag{10}$$

We shall use this identification, between Möbius transformations and matrices, without further comment. For $f, g \in$ Möb(2) the trace, $tr(f) = a + d$, is defined up to sign and the functions

$$\beta(f) = tr^2(f) - 4 \quad \text{and} \quad \gamma(f,g) = tr[f,g] - 2, \tag{11}$$

are well defined, where $[f,g] = fgf^{-1}g^{-1}$ is the multiplicative commutator. The three complex numbers $(\gamma(f,g), \beta(f), \beta(g))$ determine the two generator subgroup $G = \langle f, g \rangle$ up to conjugacy when $\gamma(f,g) \neq 0$ *[4]*. We call this triple of numbers the *parameters* of G and denote it by

$$par(G) = (\gamma(f,g), \beta(f), \beta(g)). \tag{12}$$

A group G of Möbius transformations is *discrete* if the identity is isolated in G. We refer the reader to *[5]* for various characterizations of discreteness in Möbius groups and a more detailed discussion of the above.

Let f be a Möbius transformation. If $\beta(f) \neq 0$, then we can conjugate f so that

$$f(z) = \lambda^2 z, \quad \lambda^2 = e^{\tau + i\theta}, \quad |\lambda| \geq 1. \tag{13}$$

If $|\lambda| = 1$, the transformation f is said to be *elliptic*. Otherwise f is *loxodromic*. When $\beta(f) = 0$ we can conjugate f to the form

$$f(z) = z + 1 \tag{14}$$

and we say that f is *parabolic*.

Every Möbius transformation extends to the 3–dimensional upper half–space via the Poincaré extension *[1]*. If f is not parabolic, the hyperbolic line joining the fixed points of f is called the *axis* of f. The action of f in the upper half–space is to translate along the axis by a hyperbolic distance $\tau = \tau(f)$ and rotate about the axis an angle $\theta = \theta(f)$. The number $\tau(f)$ is called the *translation length* of f and $\theta(f)$ is called the *angle of rotation* of f. Both quantities are conjugacy invariants and hence the normal form for f discussed above leads to the useful formulas,

$$\beta(f) = 4\sinh^2(\tau(f)/2 + i\theta(f)/2) \tag{15}$$

and

$$\beta(f^n) = 4\sinh^2(n\tau(f)/2 + \imath n\theta(f)/2), \tag{16}$$

where f^n denotes the n–fold composition of f with itself. The following two formulas also follow directly from (15),

$$\cosh(\tau(f)) = \frac{|\beta(f) + 4| + |\beta(f)|}{4} \tag{17}$$

and

$$\cos(\theta(f)) = \frac{|\beta(f) + 4| - |\beta(f)|}{4}. \tag{18}$$

(Cf. Lemma 4.4 of [6].)

If f and g are represented by matrices A and B, then we let

$$\|f\| = \|A\| \quad \text{and} \quad \|f - f^{-1}\| = \|A - A^{-1}\|, \tag{19}$$

where $\|\cdot\|$ is the usual Hilbert–Schmidt norm of a matrix.

Lemma 2.1:

$$\|f - f^{-1}\|^2 = 4\|f\|^2 - 2|\beta(f) + 4| \leq 4\|f\|^2.$$

Proof: If f is represented by

$$A = \pm \begin{pmatrix} a & b \\ c & d \end{pmatrix}, \quad ad - bc = 1,$$

then

$$\begin{aligned}
\|f - f^{-1}\|^2 &= 2|a - d|^2 + 4|b|^2 + 4|c|^2 \\
&= 4(|a|^2 + |b|^2 + |c|^2 + |d|^2) - 2|a + d|^2 \\
&= 4\|f\|^2 - 2|tr(f)|^2.
\end{aligned}$$

We will require the following relations between the parameters and the norms of elements.

Lemma 2.2:

$$|\beta(f)| \leq 2\|f\|^2 \quad and \quad |\gamma(f,g)| \leq \|f\|^2\|g\|^2.$$

Proof: By [5](Theorem 2.7) and Lemma 2.1,

$$|\beta(f)| \leq \frac{1}{2}\|f - f^{-1}\|^2 \leq 2\|f\|^2$$

and

$$|\gamma(f,g)| \leq \frac{1}{16}\|f - f^{-1}\|^2\|g - g^{-1}\|^2 \leq \|f\|^2\|g\|^2.$$

The connection between discrete groups and Chebyshev polynomials that concerns us here is a consequence of the following result.

Theorem 2.3: *Suppose that f and g are Möbius transformations and that f is not parabolic. Then*

$$\beta(f^n) = P_n(\beta(f)) \tag{20}$$

and

$$\gamma(f^n, g) = P_n(\beta(f))\gamma(f, g)/\beta(f). \tag{21}$$

Proof: Equation (20) follows from (15), (16) and Lemma 1.1. For (21) choose matrix representatives A and B for f and g of the form

$$A = \pm \begin{pmatrix} \lambda & 0 \\ 0 & 1/\lambda \end{pmatrix}, \quad B = \pm \begin{pmatrix} a & b \\ c & d \end{pmatrix}, \quad ad - bc = 1. \tag{22}$$

Then

$$\beta(f) = (\lambda - 1/\lambda)^2, \quad \gamma(f, g) = -bc(\lambda - 1/\lambda)^2 = -bc\beta(f) \tag{23}$$

whence

$$\gamma(f^n, g) = -bc\beta(f^n) \tag{24}$$

and (21) follows from (20) and (24).   ■

Theorem 2.4: *Suppose that f is a Möbius transformation and that f is not parabolic. If $0 \le \tau(f) < \pi\sqrt{3}$, then*

$$\min_{n \ge 1} |\beta(f^n)| \le 4\sinh^2\left(\sqrt{\frac{\pi\tau(f)}{\sqrt{3}}}\right). \tag{25}$$

Proof: Let $\tau = \tau(f)$ and $\theta = \theta(f)$. Then

$$\beta(f^n) = P_n(4\sinh^2(\tau/2 + \imath\theta/2))$$

and the desired conclusion follows from Theorem 1.4 with $x = \tau/2$.   ■

We now recall the following form of Jørgensen's inequality *[9]*. See *[6]*(Lemma 2.19).

Theorem 2.5: *Let $\langle f, g \rangle$ be a discrete group of Möbius transformations with $\gamma(f, g) \ne 0$ and $\gamma(f, g) \ne \beta(f)$. Then*

$$|\gamma(f, g)| + |\beta(f)| \ge 1. \tag{26}$$

We then have the following variant of Theorem 2.5.

Theorem 2.6: *Let $\langle f, g \rangle$ be a discrete group of Möbius transformations with $\gamma(f, g) \ne 0$ and $\gamma(f, g) \ne \beta(f)$. If f is loxodromic, then*

$$|\gamma(f, g)| + |\beta(f)| \ge \frac{\sin^2(\theta(f)/2)}{\sinh^2\left(\sqrt{\pi\tau(f)/\sqrt{3}}\right)}. \tag{27}$$

Proof: The group $\langle f^n, g \rangle$ is discrete as it is a subgroup of $\langle f, g \rangle$. Then $\beta(f^n) \neq 0$, $\gamma(f^n, g) \neq 0$, $\gamma(f^n, g) \neq \beta(f^n)$ and

$$|\beta(f^n)| \left(|\gamma(f, g)| + |\beta(f)| \right) \geq |\beta(f)| \tag{28}$$

by Theorems 2.3 and 2.5. Next

$$|\beta(f)| \geq 4 \sin^2(\theta(f)/2).$$

Hence (27) follows from Theorem 2.4 when $0 < \tau(f) < \pi\sqrt{3}$ and from Theorem 2.5 when $\tau(f) \geq \pi\sqrt{3}$. $\blacksquare$

The following theorem states roughly that in a discrete group, a loxodromic element cannot have a short translation length without the corresponding rotation angle also being small. This result gives a quantitative measure for the theorems of *[9]* which imply that in convergent sequences of discrete nonelementary groups, loxodromic elements cannot converge to elliptic elements.

Theorem 2.7: *Let $\langle f, g \rangle$ be a discrete group of Möbius transformations with $\gamma(f, g) \neq 0$. If f is loxodromic and if g is not of order 2, then*

$$\|f\|^2 \|g\|^2 \geq \frac{\sin^2(\theta(f)/2)}{2 \sinh^2\left(\sqrt{\pi\tau(f)/\sqrt{3}}\right)}. \tag{29}$$

Notice that the bound on the right hand side of (29) is asymptotic to

$$\frac{\sqrt{3}}{8\pi} \frac{\theta(f)^2}{\tau(f)} \tag{30}$$

as $\tau(f) \to 0$ and $\theta(f) \to 0$.

Proof: First, $\gamma(f) \neq \beta(f)$ since otherwise g would be elliptic of order 2 by *[6]*(Lemma 2.31). Then since $\|g\|^2 \geq 2$,

$$2\|f\|^2\|g\|^2| \geq |\gamma(f, g)| + |\beta(f)|$$

by Lemma 2.2 and we obtain the desired result from Theorem 2.6. $\blacksquare$

Remark: The relation between $\tau(f)$ and $\theta(f)$ in (29) is asymptotically sharp as $\tau(f) \to 0$ and $\theta(f) \to 0$. To show this we recall an example of Brooks and Matelski in *[2]*. For $n \geq 1$ choose $\lambda = \lambda(n)$ and $\mu = \mu(n)$ so that

$$2\sinh(\lambda/2) = \sin(\pi/n) \quad \text{and} \quad \sinh(\mu/2)\sin(\pi/n) = 1, \tag{31}$$

and let f_n and g_n be the Möbius transformations represented by the matrices

$$A_n = \pm \begin{pmatrix} \cosh(\lambda/2n + \imath\pi/n) & \sinh(\lambda/2n + \imath\pi/n)/\sinh(\lambda/2) \\ \sinh(\lambda/2n + \imath\pi/n)\sinh(\lambda/2) & \cosh(\lambda/2n + \imath\pi/n) \end{pmatrix}$$

and

$$B_n = \pm \begin{pmatrix} \cosh(\lambda/2) & e^{-\mu} \\ e^{\mu}\sinh^2(\lambda/2) & \cosh(\lambda/2) \end{pmatrix}.$$

Then $\cosh(\mu/2)\sin(\pi/n) > 1$ and $\langle f_n, g_n \rangle$ is discrete [2](pp. 178-180). Next it is easy to check that

$$\|f_n\|^2 \to 6, \quad \|g_n\|^2 \to 3, \quad |\gamma(f_n, g_n)| \to 4, \quad |\beta(f_n)| \to 0,$$

and

$$\tau(f_n) = \lambda/n \sim \pi/n^2, \quad \theta(f_n) = 2\pi/n$$

as $n \to \infty$. Hence

$$\|f_n\|^2 \|g_n\|^2 \to 18, \quad \theta(f_n)^2/\tau(f_n) \to 4\pi$$

and we see that the order in (30) is sharp and that the constant $\sqrt{3}/8\pi$ cannot be replaced by any number larger than $9/2\pi$.

This example also shows that inequality (27) is asymptotically sharp up to the factor of $4/\sqrt{3}$ as $\tau(f) \to 0$ and $\theta(f) \to 0$.

3   An inequality for nonloxodromics

The version of Jørgensen's inequality given above shows that a loxodromic element f in a discrete group cannot be nearly elliptic without $\|f\|$ being corresponding large. We derive here another sharp analog of Jørgensen's inequality for the case where f is not a loxodromic. We begin with the following formula for the hyperbolic distance $\delta(f, g)$ between the axes of nonparabolic Möbius transformations.

Lemma 3.1: *Suppose that f and g are Möbius transformation and that f is nonparabolic. Then*

$$\cosh(\delta(f, gfg^{-1})) = \frac{|\gamma(f,g) - \beta(f)| + |\gamma(f,g)|}{|\beta(f)|}. \tag{32}$$

Proof: Let $\beta = \beta(f)$, $\gamma = \gamma(f, g)$ and $h = gfg^{-1}$. Then $\beta(h) = \beta$ and

$$\gamma(f, h) = \gamma(\gamma - \beta) \tag{33}$$

by [6](2.2). If $\gamma(f, h) \neq 0$, then f and h have no common fixed points,

$$\begin{aligned}
2\cosh^2(\delta(f,h)) &= \cosh(2\delta(f,h)) + 1 \\
&= \left| \frac{4\gamma(f,h)}{\beta(f)\beta(h)} + 1 \right| + \left| \frac{4\gamma(f,h)}{\beta(f)\beta(h)} \right| + 1 \\
&= \left| \frac{4\gamma(\gamma - \beta) + \beta^2}{\beta^2} \right| + \left| \frac{4\gamma(\gamma - \beta)}{\beta^2} \right| + 1 \\
&= 4\frac{|\gamma - \beta/2|^2 + |\gamma(\gamma - \beta| + |\beta/2|^2}{|\beta|^2} \\
&= 2\left(\frac{|\gamma - \beta| + |\gamma|}{|\beta|} \right)^2
\end{aligned}$$

by *[6]*(Lemma 4.4) and we obtain (32). If $\gamma(f,h) = 0$, then f and h have a common fixed point, $\delta(f,h) = 0$, $\gamma = 0$ or $\gamma = \beta$ by (33), and each side of (32) is equal to 1. ∎

Theorem 3.2: *Let $\langle f, g \rangle$ be a discrete group of Möbius transformations with $\gamma(f,g) \neq 0$ and $\gamma(f,g) \neq \beta(f)$. If f is an elliptic of order $n \geq 3$, then*

$$|\gamma(f,g) - \beta(f)| + |\gamma(f,g)| \geq c|\beta(f)|, \quad c = (1 + \sqrt{5})/3 \approx 1.078\ldots \qquad (34)$$

or $\langle f, gfg^{-1} \rangle$ is a finite elliptic group and $n = 3, 4$ or 5. The bound on the right hand side is sharp.

Proof: Since f and gfg^{-1} are both elliptics of order $n \geq 3$, $\delta(f, gfg^{-1}) = 0$ or

$$\delta(f, gfg^{-1}) \geq b(n) \qquad (35)$$

by *[6]*(Theorem 4.18). Here the numbers $b(n)$ are given by

$$\cosh(b(3)) = (1 + \sqrt{5})/3,$$
$$\cosh(b(4)) = (1 + \sqrt{3})/2,$$
$$\cosh(b(5)) = (5 + \sqrt{5})/5,$$
$$\cosh(b(6)) = 2$$

and

$$\cosh(b(n)) = \frac{\cos(2\pi/n)}{2\sin^2(\pi/n)}$$

for $n \geq 7$; moreover, the bound in (35) is sharp for each n. See *[6]*(Section 4) for details. Thus if $\delta(f, gfg^{-1}) \neq 0$, then

$$|\gamma(f,g) - \beta(f)| + |\gamma(f,g)| \geq \cosh(b(n))|\beta(f)| \qquad (36)$$

by (32) and (35) and we obtain (34). If $\delta(f, gfg^{-1}) = 0$, then the axes of f and gfg^{-1} meet in a common point of the upper half space and $\langle f, gfg^{-1} \rangle$ is a finite elliptic group with generators which have order $n \geq 3$ and no common fixed point. Thus $\langle f, gfg^{-1} \rangle$ is isomorphic to one of the groups A_4, S_4, A_5 and $n = 3, 4$ or 5. ∎

The above argument yields an inequality which is reminiscent of the Lie product form of Jørgensen's inequality in Theorem 2.5 where (26) is replaced by

$$|\gamma(f,g) - \beta(f)| + |\beta(f)| \geq 1.$$

See *[10]*(Proposition 3).

Theorem 3.3: *Let $\langle f, g \rangle$ be a discrete group of Möbius transformations with $\gamma(f,g) \neq 0$ and $\gamma(f,g) \neq \beta(f)$. If f is not loxodromic, then*

$$|\gamma(f,g) - \beta(f)| + |\gamma(f,g)| \geq 2\cos(2\pi/7) = 1.2469\ldots. \qquad (37)$$

The bound on the right hand side is sharp.

Proof: If f is parabolic, then $\beta(f) = 0$ and the Shimizu-Leutbecher inequality *[11]*(Proposition II.C.5) or Theorem 2.5 imply that

$$|\gamma(f,g) - \beta(f)| + |\gamma(f,g)| = 2|\gamma(f,g)| \geq 2.$$

Next if f is elliptic of order n, then

$$|\beta(f)| \geq 4\sin^2(\pi/n).$$

In this case the triangle inequality implies that

$$|\gamma(f,g) - \beta(f)| + |\gamma(f,g)| \geq 4\sin^2(\pi/n) \geq (5 - \sqrt{5})/5 = 1.3819\ldots$$

if $2 \leq n \leq 5$, while the above argument implies that $\delta(f, gfg^{-1}) \neq 0$ and hence, with (36), that

$$|\gamma(f,g) - \beta(f)| + |\gamma(f,g)| \geq 4\sin^2(\pi/n)\cosh(b(n)) \geq 2\cos(2\pi/7)$$

if $n \geq 6$. Inequality (37) holds with equality for the $(2,3,7)$ triangle group.   ∎

Remark: Theorem 3.2 also holds for the case where f is loxodromic, but with the constant $2\cos(2\pi/7)$ replaced by a smaller number. For example, if $\langle f,g \rangle$ is Fuchsian, then one can use Theorem 2.9 of *[8]* to prove that

$$|\gamma(f,g) - \beta(f)| + |\gamma(f,g)| \geq c \tag{38}$$

where

$$c = 2(\cos(2\pi/7) + \cos(\pi/7) - 1) = 1.0489\ldots.$$

The $(2,3,7)$ triangle group shows that this bound is sharp. For the general case where $\langle f,g \rangle$ is not Fuchsian, one can obtain (38) with

$$.8786\ldots \leq c \leq 1.0489\ldots$$

from Theorem 3.1 of *[7]*.

References

[1]   A. F. Beardon, *The Geometry of Discrete Groups*, Springer-Verlag 1983.

[2]   R. Brooks and J. P. Matelski, Collars in Kleinian groups, *Duke Math. J.* **49** (1982) 163-182.

[3]   P. J. Davis, *Interpolation and Approximation*, Blaisdell 1965.

[4]   F. W. Gehring and G. J. Martin, Stability and extremality in Jørgensen's inequality, *Complex Variables* **12** (1989) 277-282.

[5]   F. W. Gehring and G. J. Martin, Inequalities for Möbius transformations and discrete groups, *J. reine angew. Math.* **418** (1991) 31-76.

[6]   F. W. Gehring and G. J. Martin, Commutators, collars and the geometry of Möbius groups, *J. d'Analyse Math.* 63 (1994) 175-219

[7]   F. W. Gehring and G. J. Martin, Some universal constraints for discrete Möbius groups, *Paul Halmos: Celebrating 50 Years of Mathematics*, Springer-Verlag 1991, 205-220.

[8]   F. W. Gehring and G. J. Martin, The iterated commutator in a Fuchsian group, *Complex Variables* 21 (1993) 207-218

[9]   T. Jørgensen, On discrete groups of Möbius transformations, *Amer. J. Math.* **98** (1976) 739-749.

[10]   T. Jørgensen, Comments on a discreteness condition for subgroups of SL(2,C), *Can. J. Math.* **31** (1979) 87-92.

[11]   B. Maskit, *Kleinian Groups*, Springer-Verlag 1987.

[12]   R. Meyerhoff, A lower bound for the volume of hyperbolic 3–manifolds, *Can. J. Math.* **39** (1987) 1038-1056.

A Uniqueness Theorem For An Entire Function And Its Derivatives

Yongxing Gu and Jin Qiao
Department of Applied Mathematics, Chongqing University
Chongqing, Sichuan, 630044, P. R. China

Abstract: In this paper, we proved the following result: Suppose that f is a nonconstant entire function in the plane, $a(z), b(z)$ are two distinct small meromorphic functions, $a^{(k)} - b$ and $b^{(k)} - a$ are not simultaneously identically equal to zero. If f and $f^{(k)}$ share $a(z)$ and $b(z)$, CM, $(k \geq 1)$, then $f \equiv f^{(k)}$. We also showed that the condition that $a^{(k)} - b \equiv 0$ and $b^{(k)} - a \equiv 0$ can not hold simultaneously is necessary.

1 Introduction

We say that two meromorphic function f and g share the value a CM if $f(z) = a$ if and only if $g(z) = a$, where CM means to count multiplicities and we also say that f and g share a small function $a(z)$ CM if and only if the two functions $f - a$ and $g - a$ assume the same zeros with the same multiplicities.

In [3], L. A. Rubel and C. C. Yang proved the following

Theorem A: *If an entire function f and its first derivative f' share two finite values CM, then $f \equiv f'$.*

Recently, Jiahua Zheng and Shupei Wang [4] proved that

Theorem B: *If a nonconstant entire function f and its first derivative f' share two distinct small functions $a(z)(\not\equiv \infty), b(z)(\not\equiv \infty)$ CM, then $f \equiv f'$.*

In this paper we shall show that the conclusion of Theorem B still holds if f' is replaced by $f^{(k)}, k \in N$.

2 Notations and Lemmas

Under the assumptions of the theorem, we first introduce some notations, which is similar to that in *[1]*.

Proceedings of the International Conference on Complex
Analysis at the Nankai Institute of Mathematics, 1992 pp. 126-131
©INTERNATIONAL PRESS 1994

$C_0(z), C_1(z), \cdots$ denote the small functions which are not identically equal to zero; $P_k(W_1, W_2)$ denote the polynomials in two elements of degree less that k, and its coefficients are all small functions of f; $H_k(W_1, W_2)$ denote the homogeneous polynomials in two elements of degree k, and its coefficients are also small functions of f; $H_k^*(W_1, W_2)$ denote the polynomials of $H_k(W_1, W_2)$ which omit the terms W_1^k and W_2^k; $Q(e^h), Q_1(e^h), Q_2(e^h) \cdots$ denote the polynomials of the following form:

$$C_1(z)e^{\alpha_1 h} + C_2(z)e^{\alpha_2 h} + \cdots + C_n(z)e^{\alpha_n h}$$

where $\alpha_j (j = 1, 2, \cdots, n)$ are distinct rational values; E denote any set of finite linear measure on $0 < r < \infty$; we say a meromorphic function $a(z)$ is a small function of $f(z)$, if

$$T(r, a(z)) = o(T(r, f)) \qquad r \overline{\in} E$$

Each notations are not necessarily the same at each occurrence.

Assume that the conditions of the theorem are satisfied. Then we have

$$(1) \qquad \frac{f^{(k)} - a}{f - a} = e^{\phi_1(z)}$$

$$(2) \qquad \frac{f^{(k)} - b}{f - b} = e^{\phi_2(z)}$$

where ϕ_1, ϕ_2 are two entire functions.

Eliminating $f^{(k)}$ from (1) and (2) gives

$$(3) \qquad (e^{\phi_1} - e^{\phi_2})f = a(e^{\phi_1} - 1) - b(e^{\phi_2} - 1)$$

If $e^{\phi_2} - e^{\phi_1} \not\equiv 0$, then

$$(4) \qquad T(r, f) = O\{m(r, e^{\phi_1}) + m(r, e^{\phi_2})\}$$

Let

$$F = \{\alpha\phi_1 + \beta\phi_2 : \alpha, \beta \quad \text{two rational values and} \quad |\alpha| + |\beta| \neq 0\}$$

The following lemmas will be needed in the proof of our theorem:

Lemma 1: *[2]. Let $b_0(z), b_1(z), \cdots, b_n(z)$ be meromorphic functions and let $g_1(z), g_2(z), \cdots, g_n(z)$ be entire functions. Further suppose that*

$$\sum_{i=1}^{n} b_i(z)e^{g_i(z)} \equiv b_0(z),$$

$$T(r, b_i(z)) = o\left(\sum_{j=1}^{n} m(r, e^{g_j})\right) \qquad (r \notin E, \ i = 0, 1, \cdots, m).$$

Then we have an identity

$$\sum_{j=1}^{n} C_j b_j(z) e^{g_j(z)} \equiv 0$$

where the constants $C_j (j = 1, 2, \cdots, n)$ are not all zero.

Lemma 2: *[1]. Let $h(z)$ be an entire function satisfying*

$$Q(e^{h(z)}) \equiv 0,$$

where $Q(e^{h(z)})$ is a ploynomial of the previous denotation. Then

$$T(r, e^{h(z)}) = o(T(r, f)) \qquad (r \overline{\in} E)$$

Lemma 3: *If in addition to the assumptions of the theorem, we suppose that*

(5) $$e^{\phi_1} - e^{\phi_2} \not\equiv Constant.$$

Then

(6) $$(a(z) - b(z))(e^{\phi_1} - e^{\phi_2})^k e^{\phi_1 + \phi_2} = P_{k+1}(e^{\phi_1}, e^{\phi_2})$$

where

(7) $$\begin{aligned} P_{k+1}(e^{\phi_1}, e^{\phi_2}) = &(a^{(k)} - b)e^{(k+1)\phi_1} \\ &+ (-1)^{k+1}(b^{(k)} - a)e^{(k+1)\phi_2} \\ &+ H_{k+1}^*(e^{\phi_1}, e^{\phi_2}) + H_k(e^{\phi_1}, e^{\phi_2}) \end{aligned}$$

Proof: Differentiating the identity (3), we get

(8) $$(\phi_1' e^{\phi_1} - \phi_2' e^{\phi_2})f + (e^{\phi_1} - e^{\phi_2})f' = a\phi_1' e^{\phi_1} - b\phi_2' e^{\phi_2} + a'(e^{\phi_1} - 1) - b'(e^{\phi_2} - 1)$$

From (3) and (8), we have

(9) $$(e^{\phi_1} - e^{\phi_2})^2 f' = a' e^{2\phi_1} + b' e^{2\phi_2} + H_2^* + H_1$$

Differentiating this identity yields

(10) $$\begin{aligned} 2(\phi_1' e^{\phi_1} - \phi_2' e^{\phi_2})(e^{\phi_1} - e^{\phi_2})f' + (e^{\phi_1} - e^{\phi_2})^2 f'' = &\, a'' e^{2\phi_1} \\ &+ b'' e^{2\phi_2} + 2a' \phi_1' e^{2\phi_1} \\ &+ 2b' \phi_2' e^{2\phi_2} + H_2^* + H_1 \end{aligned}$$

Eliminating f' from (10) and (9) gives

$$(e^{\phi_1} - e^{\phi_2})^3 f'' = a'' e^{3\phi_1} - b'' e^{3\phi_2} + H_3^* + H_2$$

Now by successive calculation, we obtain

$$(11) \quad (e^{\phi_1} - e^{\phi_2})^{k+1} f^{(k)} = a^{(k)} e^{(k+1)\phi_1} + (-1)^{k+1} b^{(k)} e^{(k+1)\phi_2} + H_{k+1}^* + H_k$$

On the other hand, by (1) and (3) we get

$$(12) \quad (e^{\phi_1} - e^{\phi_2})^{k+1} f^{(k)} = (a-b)(e^{\phi_1} - e^{\phi_2})^k e^{\phi_1 + \phi_2} + (be^{\phi_1} - ae^{\phi_2})(e^{\phi_1} - e^{\phi_2})^k$$

(11) and (12) imply that the equality (6) holds. ∎

Lemma 4: *Assume that the conditions of Lemma 3 are satisfied. Then neither e^{ϕ_1} nor e^{ϕ_2} can be small function of f.*

Since the proof is similar to that of Lemma 4 in [1], we shall leave it out.

Lemma 5: *Assume that the conditions of Lemma 3 are satisfied. Then $e^{g(z)}$ can not be a small function of f, where $g(z)$ is an arbitrary function of F.*

Proof: Suppose that there exists a function

$$g(z) = \alpha_0 \phi_1 + \beta_0 \phi_2 \in F$$

such that $T(r, e^g) = o(T(r, f) \quad r(\overline{\in} E)$.

Without loss of generality we assume that $\beta_0 \neq 0$. Then

$$(13) \qquad \phi_2 = \frac{1}{\beta_0} g + \delta \phi_1$$

where $\delta = -\alpha_0/\beta_0$.

For our proof we shall consider four cases:

Case 1 $\delta \neq 0, -1, 1$;

Case 2 $\delta = 0$;

Case 3 $\delta = -1$;

Case 4 $\delta = 1$.

For simplicity, we only prove the Case 3. For the other three cases, one can refer to [1].

Now we have $\delta = -1$, substituting (13) into (6) gives

$$(a - b) \left(e^{\phi_1} - e^{\frac{1}{\beta_0} g} e^{-\phi_1} \right)^k e^{\frac{1}{\beta_0} g}$$

$$= (a^{(k)} - b) e^{(k+1)\phi_1} + (-1)^{k+1} (b^{(k)} - a) e^{\frac{k+1}{\beta_0} g} e^{-(k+1)\phi_1} + H_{k+1}^* + H_k$$

Multiplying the two sides of the equality by $e^{(k+1)\phi_1}$, we get

$$(14) \qquad (a - b) e^{\phi_1} \left(e^{2\phi_1} - e^{\frac{1}{\beta_0} g} \right)^k e^{\frac{1}{\beta_0} g} = (a^{(k)} - b) e^{(2k+2)\phi_1}$$
$$+ (-1)^{k+1} (b^{(k)} - a) e^{\frac{k+1}{\beta_0} g}$$
$$+ e^{(k+1)\phi_1} H_{k+1}^* + e^{(k+1)\phi_1} H_k$$

We may regard both sides of (14) as polynomials of e^{ϕ_1}. The degree of the left side is less that $2k+1$ and greater than 1. On the right hand, the degree of the term $e^{(k+1)\phi_1}H^*_{k+1}$ is less than $2k$ and more than 2, and for the term $e^{(k+1)\phi_1}H_k$, its degree is less than $2k+1$ and more than 1. In addition, $a^{(k)}-b$ and $b^{(k)}-a$ can not simultaneously identically equal to zero. Hence, we can rewrite (14) in the form

$$Q(e^{\phi_1}) \equiv 0$$

Combining Lemma 2 with Lemma 4 gives a contradiction.   ∎

Lemma 6: *[1] Under the assumption of the Lemma 3, $g_1 = \alpha_1\phi_1 + \beta_1\phi_2, g_2 = \alpha_2\phi_1 + \beta_2\phi_2$ are any two functions of F, then g_1 and g_2 are linearly indepenent if and only if*

$$\begin{vmatrix} \alpha_1 & \beta_1 \\ \alpha_2 & \beta_2 \end{vmatrix} \neq 0$$

Lemma 7: *If the conditions of Lemma 6 are satisfied, then*

$$(15) \qquad T(r,f) = O(m(r, e^{g_1}) + m(r, e^{g_2}))$$

Proof: Since $e^{\phi_1} - e^{\phi_2} \not\equiv$ constant, by (3) we know that (4) holds. On the other hand, g_1, g_2 are linearly independent, Lemma 6 gives

$$\begin{pmatrix} \phi_1 \\ \phi_2 \end{pmatrix} = \begin{pmatrix} \alpha_1 & \alpha_2 \\ \beta_1 & \beta_2 \end{pmatrix}^{-1} \begin{pmatrix} g_1 \\ g_2 \end{pmatrix}$$

Hence

$$m(r, e^{\phi_1}) + m(r, e^{\phi_2}) = O(m(r, e^{g_1}) + m(r, e^{g_2}))$$

From this and (4) we obtain (15).   ∎

Lemma 8:   *[1]. Under the assumption of the Lemma 3. Suppose that $g_i(i = 1, 2, \cdots, m) \in F$ and $g_1, g_2, \cdots, g_m$ are linearly independent. Then*

$$Q_1(e^{g_1}) + Q_2(e^{g_2}) + \cdots + Q_m(e^{g_m}) \not\equiv 0.$$

3   Proof of the Theorem

The proof of the theorem is similar to the of [1]. We apply the above Lemmas and discuss it in three cases
   (i) $e^{\phi_1} - e^{\phi_2} \equiv 0,$
   (ii) $e^{\phi_1} - e^{\phi_2} \equiv C$ $(C \neq 0),$
   (iii) $e^{\phi_1} - e^{\phi_2} \not\equiv$ Constant.
   For simplicity, we leave out the details.

Remark The condition that $a^{(k)} - b$ and $b^{(k)} - a$ can not be simultaneously identical zero is necessary. Otherwise, we have

$$\frac{(a-b)^k}{a-b} = -1.$$

So if $f \equiv f^{(k)}$, then we can deduce that

$$T(r, f) = O(T(r, a-b))$$

which contradicts with the condition of the theorem.

References

[1] Gu, Y. X, A uniqueness theorem for entire function and its differential polynomial (in Chinese, to appear)

[2] Hiromi, G. and Ozawa, M., On the existence of analytic mappings between two ultrahyperlliptic surfaces, Kodai Math. Sem. Rep., 17 (1965), 281-306.

[3] Rubel, L. A. and Yang, C. C., Values shared by an entire function and its derivatives, Lecture Notes in Math., Vol. 599, 101-103.

[4] Zheng, J. H. and Wang, S. P., On unicity properties of meromorphic functions and their derivatives, Advances in Math. (PRC), 3(1992), 334-341.

Towards Topological Classification Of Critically Finite Polynomials

Sen Hu
Centre de Mathematiques, Ecole Polytechnique
91128 Palaiseau, France, shu@orphee.polytechnique.fr

Yunping Jiang
Dept of Mathematics, Queens College/CUNY
Flushing, NY 11367, jiang@math.sunysb.edu or yunqc@qcvaxa.acc.qc.edu

Abstract: In this note we give a construction of branched cover maps of S^2 to itself which are topologically equivalent to critically finite polynomials in the sense of Thurston. The construction are on the invariant laminations. It applies to all degrees. This way we confirm the conjecture given by Goldberg and Milnor about the fixed point portrait of critically finite polynomials.

1   Background and definitions

Definition 1(critically finite rational map): Given a rational map $f : \mathbf{C} \to \mathbf{C}$, let P_f be the set of its critical points. Let $\Omega_f = \bigcup_{n \geq 1} f^{\circ n}(P_f)$ be the post-critical orbits of f. The map f is said to be critically finite if Ω_f is finite. In other words, all the critical points are either periodic or pre-periodic (forward iterate landing on a periodic orbit).

To describe critically finite rational maps is by means of constructing topological branched cover maps of the sphere. Similar definitions to Definition 1 can be given to branched cover maps. This means that we just replace the term, critical point, by the term, branched point. Now we can talk about critically finite branched cover maps f with the same notations, P_f for branched points and Ω_f for positive orbits of the branched points.

Definition 2(Thurston equivalence): Given two branched cover maps f and g from S^2 to itself with both Ω_f and Ω_g finite. The maps f and g are Thurston equivalent if there is an orientation preserving homeomorphism h from S^2 to itself such that $h \circ f = g \circ h$ on Ω_f and $h \circ f$ is isotopic to $g \circ h$ rel Ω_f, that is, isotopic is restricted to the space S^2/Ω_f. An equivalent class of branched cover maps is called a combinatorial equivalent class.

Given a critically finite branched cover map f of the sphere, we can construct an orbifold $\mathcal{O}_f$, or punctured sphere with marked points. All positive

Proceedings of the International Conference on Complex
Analysis at the Nankai Institute of Mathematics, 1992 pp. 132-143
©INTERNATIONAL PRESS 1994

critical orbits are removed as punctures. For the points in Ω_f, we will associate an integer with each of them. Consider the function $\nu : \Omega_f \to N$, we choose the smallest ν with the property that if $x, y \in \Omega_f, y \in f^{-1}(x)$, then $\nu(y)$ is a multiple of $\nu(x)\deg_y f$.

It was shown by Thurston that each combinatorial class contains at most one rational map up to conformal conjugacy. In this sense the critically finite rational maps are determined by their isotopic classes.

2   Invariant lamination

Now we confine ourselve to consider critically finite polynomials only. There is a natural way to describe the dynamics at least combinatorially, i.e. by laminations.

A polynomial f has an attracting critical fixed point ∞. It is known that if the orbits of critical points are bounded (true for critically finite case) then the Julia set of f is connected. For such a map we can take an open disk U of ∞ on which f is conjugate to $z \to z^d$ where d is the degree of f. Then take $V = \bigcup_{i \geq 0} f^{-i}(U)$, we see that V is still a disk because there is no critical point goes to ∞. It is known that $\partial V = $ Julia set of f. The conjugacy over V can be extended to ∂V as a continuous map h. Usually h is no longer a homeomorphism on the circle, i.e. there can be several preimages corresponding to one point in the Julia set. [The image of $\{re^{2\pi i\theta}|r \geq 0, \theta$ fixed $\}$ under the conjugacy is called external ray.] There is a remarkable property observed by Thurston *[8]* that the *convex hulls of preimages of the points in Julia set are disjoint*. If one connects all the preimages of every points in Julia set by line segments (leaves), it forms a lamination, this means that the closed set of leaves. Such a lamination is forward invariant under the map $z \to z^d$ as we can see from the conjugacy. It means the leaves will be mapped to leaves or point under the map f (forward invariant). The preimages of leaves will form leaves too. Given a forward invariant lamination one can saturate it to an invariant lamination if the forward invariant lamination satisfies the following condition:

$$\sum (\text{ degree of critical gap } -1) + \text{number of critical leaves} = d - 1$$

Here a critical leaf means a leaf mapped to a point by $z \to z^d$, a critical gap means a gap, a connected component of the complement of the leaves, mapped to other gaps with degree greater than one.

We see here that laminations serve as nice models to describe the dynamics of polynomials. What we will do in this note is to construct branched cover maps out of forward invariant laminations.

3   Two examples in the degree two case

a). The Douady rabbit given by a branched cover map

Consider the following set of numbers $\{\{0\}, \{1/7, 2/7, 4/7\}\}$ from the unit circle (θ is corresponding to $e^{2\pi i\theta}$) it forms a forward invariant lamination of degree two. We will construct a branched cover map from this lamination.

The three points $\{1/7, 2/7, 4/7\}$ form a triangle T_1. Their preimages $\{1/14, 9/14\}$ form another triangle T_2. Consider the complement of these two triangles T_1 and T_2 in the unit disk. It has five connected components, say U_0, U_1, U_2, U_3 and U_4 (see figure below). The circle is divided into 6 segments, let us denote them as B_{01} and B_{02}, B_1, B_2, B_3 and B_4. The set $\{B_{0j}\}_{j=0}^1 \cup \{B_i\}_{i=0}^4$ gives a Markov partition of the unit circle under the standard map $s(z) = z^2$. We see that B_1 is mapped to B_2, B_2 is mapped to $B_{01} \cup B_{02} \cup B_3 \cup B_4$, B_3 is mapped to B_2, B_4 is mapped to $B_{01} \cup B_{02} \cup B_3 \cup B_4$, B_{01} and B_{02} are mapped to B_1. We see that $\{B_{0j}\}_{j=0}^1 \cup \{B_i\}_{i=1}^4$ is a Markov partition on the unit circle under $s(z) = z^2$.

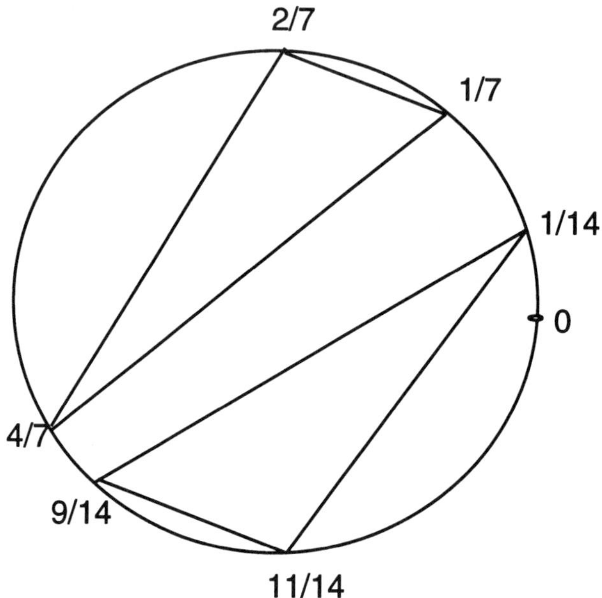

We put a branched point O in the component U_0 since s on the boundary of U_0 covers the boundary of U_1 twice. A map $f_{0,1}$ from U_0 to U_1 is a branched cover with only branched point O of degree two.

We can define the map over the disk as the continous extension of the map over the boundary. Since $B_{01} \mapsto B_1 \mapsto B_2 \mapsto B_{01}$ and $B_{02} \mapsto B_1 \mapsto B_2 \mapsto B_{01}$ are both admissible sequences, we can put into orbit of branched point according to this sequence. There is one ambiguity in the definition of the map f. We want to extend the map over each component, i.e. the disk with marked points of designated points of branched points. There are infinitely many isotopy classes of the map over disk with marked points if the number of marked points are more than one. However by considering the analytic structure the isotopy class will be limited to one. See section 5, part b).

Let us denote the copy of the unit disk we use to define the map f_0 as D_1. Take another copy D_2 of the unit disk and consider the map $s(z) = z^2$ of D_2.

Since the two maps, f_0 from D_1 to itself and s from D_2 to itself, are the same when they restrict on the boundaries of D_1 and D_2, we can glue the boundaries of D_1 and D_2 by the identity and collapse T_1 and T_2 into two points. The resulted space is a topological sphere S^2.

Since the branched points of f are periodic, by a theorem in Levy's thesis [5], there is no Thurston obstruction. So the map f is equivalent to a polynomial in the sense of Thurston. It is easy to check that it is equivalent to the Douady rabbit, $z \to z^2 + \omega(2-\omega)/4, \omega = e^{2\pi i/3}$.

Remark: What we will show in this paper is that the idea for the example above applies for all critically finite hyperbolic polynomials.

Remark: There is a close relationship between the invariant lamination and the Hubbard tree. The Hubbard tree is the dual graph of the pinched disk D^1/L. See for example the Douady rabbit.

b). The preperiodic case

In this case the lamination is similar except we add a critical leaf L_0 in the middle of leaves. The critical leaf split U_0 into two components U_{01}, U_{02}. We then define a map $k_{01,1}$ from U_{01} to U_1 and a map $k_{02,1}$ from U_{02} to U_1 are homeomorphisms which agree with $s(z) = z^2$ on the boundary $\partial U_0 = B_{01} \cup B_{02}$. Also we define a map j_0 from L_0 to a point. Again we can choose all the maps here as those we defined in a) such that they form a continuous map f_0 on the unit disk. We collapse the critical leaf L_0 to a point, which will be the branch point. One may denote this point as c and its image under j_0 as v. Since $\{B_{0j}\}_{j=0}^1 \cup \{B_i\}_{i=1}^4$ is a Markov partition on the unit circle under $s(z) = z^2$ and

$B_{01} \mapsto B_1 \mapsto B_2 \mapsto B_3 \mapsto B_2$ is an admissible sequence. we can find two point $c_1 \in B_{01}$ and $c_2 \in B_{02}$ such that their images are like $c_1 \to v \to v' \to v'' \to v'$ and $c_2 \to v \to v' \to v'' \to v'$ under s. The critical leaf now is just the line in U_0 connects c_1 and c_2.

Combining the arguments in a), we can get a branched cover f from S^2 to S^2.

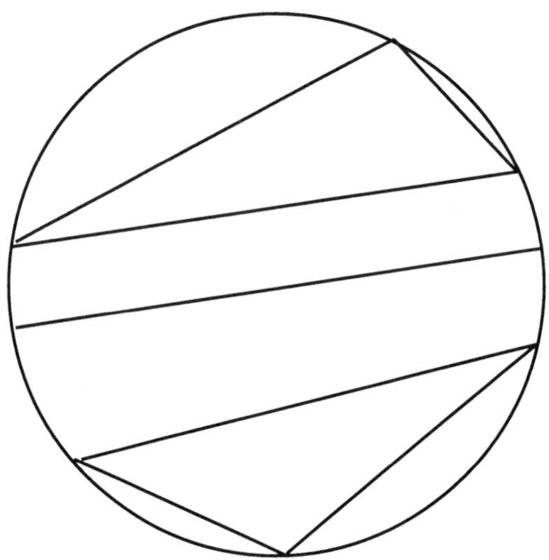

4   What happens for higher degree, fixed point portrait

Given a critically finite polynomial of degree d, it has several repelling fixed points all lying on the Julia set. Each fixed point has several external rays. The corresponding points in the unit circle of the end points of the external rays forms a forward invariant set under the map $z \to z^d$. Call such a set T, a rotation subset of S^1. Consider all the repelling fixed points, their corresponding rotatioal subsets $\{T_1, T_2, ..., T_k\}$ are called a fixed point portrait by L. Goldberg and J. Milnor [3]. They proved that the fixed point portrait has the following properties.

Theorem 4.1: ([3]) Let $\{T_1, T_2, ...T_k\}$ be the rotatioal subsets of a critically finite polynomial f of degree d, then it satisfies the following conditions:
- C_1: Each T_i is a rotation subset of S^1 in the sense that it is invariant under the map $z \to z^d$ and has a well defined rotation number p_i/q_i.
- C_2: The convex hulls of different T_i's are disjoint.
- C_3: The union of the rotatioal subsets of rotation number zero are $\{0, 1/(d-1), 2/(d-2), ..., (d-2)/(d-1)\}$.
- C_4: rotatioal subsets with non-zero rotation number are separated by rotatioal subset with zero rotation number.

Remark: From condition 2 we see that $\{T_1, T_2, ..., T_k\}$ forms a forward invariant lamination. The data given above, i.e. $\{T_1, T_2, ..., T_k\}$ satisfying the above conditions, is purely combinatorial. It was conjectured *[3]* that all such fixed point portraits can be realized by polynomials.

5 Realization of a fixed point portrait

In light of the above construction of the branched cover map, we give a construction of a branched cover map in all degree to realize Goldberg and Milnor's fixed point portrait. To show the map is equivalent to a polynomial we will use Thurston's topological characterization of critically finite rational maps.

a). Fixed point portrait and Markov partition of $s : z \to z^d$ of S^1.

Given a forward invariant lamination L, consider the preimage of L under $s :
z \to z^d$, we get another forward invariant lamination $L' = s^{-1}(L)$. To make sure that such a construction is possible we have to verify that the number of critical gaps and critical leaves satisfy the equation given in Section 2.

Proposition: Given a forward invariant lamination L arising from a fixed point portrait we have

$$\Sigma(\text{degree of critical gaps} - 1) + \text{number of critical leaves} = d - 1.$$

Proof:
We will construct a bijection of critical gaps to the set $\{0, 1/(d-1), 2/(d-2), ..., (d-2)/(d-1)\}$.

From the conditions C3 and C4 of Goldberg-Milnor's theorem we see that there are two kinds of components:

i). Gaps with only one leaf coming from non-zero rotation subset. Those gaps are not critical.

ii). Gaps with leaves coming from both non-zero rotation subset and zero rotation subset. We first consider the case with only two leaves. It looks like the following:

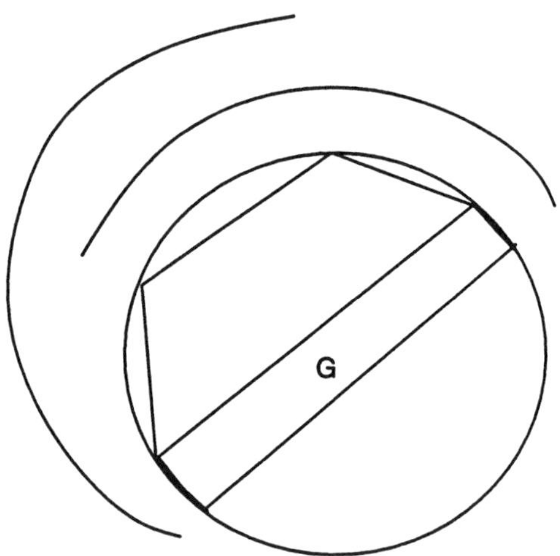

We look at the map of $G \cap S^1$ under s. Since one of the leaves is in the rotational subset and other leaf is invariant under s we get where is the other picture.

It is clearly a critical gap of degree two. If the other gap from the other side of the leaf of the zero rotational subset happen has only one non-zero leaf we got another critical gap of degree two. For this case we have the correspondence of the critical gaps to the zero rotational subset.

If the gap contains more than one leaves or point of zero rotaional subset we can subdivide it into several gaps by cutting along the leaves or points of zero rotational subset. Each piece will be a critical gap of degree two. Put them together we still have the correspondence of critical gaps to the leaves of zero rotational subset.   ∎

Now we interpret the forward invariant lamination as a Markov partition of $s : z \to z^d$ of S^1.

Definition: Markov partition of $s : z \to z^d$ of S^1 is a partition $\{U_i\}$ of S^1, such that

1.   Each U_i is a set of closed intervals contained in S^1;

2. $int(U_i) \cap int(U_j) = \phi$ if $i \neq j$;

3. $s(U_i)$ is a union of several intervals from $\{U_j\}$.

Given a forward invariant lamination L, consider its preimage $L^{'} = s^{-1}(L)$. Delete all the intersections of $L^{'}$ with S^1, we get a collection of open intervals $\{J_i\} \subset S^1$. Its closure is a partition of S^1. Two intervals J_i and J_j are said to be connected if the convex hull of J_i and J_j in D don't contain any leaf in $L^{'}$. It's an equivalence relation. Put all connected $J_i's$ together and let U_i be its closure. It is clear that $\{U_i\}$ forms a partition of S^1. Each U_i is called a component of the partition from $S^1/L^{'}$.

Proposition: *The partition $\{U_i\}$ constructed above is a Markov partition of $s : z \to z^d$ of S^1.*

Proof: It suffices to verify (3) in the definition.

By collapsing all convex hulls of L (or L') into points we got the quotient space S^1/L (or S^1/L') which are pinched disks, say V_i's. The boundary of those disks, ∂V_i, is a components of S^1/L. We call the disks the components of D^1/L.

We see that s maps each component of S^1/L' to S^1/L. Each component of S^1/L is union of several components of S^1/L'. So we see that the partition given by S^1/L' is a Markov partition of s. ∎

b). Construction of branched cover maps of S^2 to itself

To construct a branched cover map of S^2 to itself we need to put a branched point into components of D^1/L. From above we see that the map s over each gap V_i has well-defined degree. The gap V_i iscalled critical if degree of s over V_i is greater than one. It is clear that we should put branched point into V_i. However what we need is to locate the orbits of branched points. To do so we introduce another definition.

Definition: (Admissible sequence of the partition) A sequence $V_{i_1} \to V_{i_2} \to ... \to V_{i_k}$ is called an admissible sequence of the partition if we have the following:

1. Each V_{i_j} is a component of D^1/L',

2. $s(\partial V_{i_j}) \supset \partial V_{i_{j+1}}, j = 1, 2, ..., k$ and

3. $s(\partial V_{i_k}) \supset \partial V_{i_1}$.

Since s is expanding over S^1 we see that for each critical gap V, there exists an admissible sequence of the partition

$$V \to V' \to V'' \to ... \to V^{(k)}.$$

Put sequences of all critical gaps together we get an admissible sequence of gaps. With such a sequence we can construct a branched cover map of S^2 to itself. Actually what we need to do is to extend the map over S^1/L' to D^1/L'. Notice that L' is a s invariant lamination. If we take another copy of D^1 and map s we can glue s with the map above to get a map of S^2. The reason we can glue them together is they are agree over the boundary.

For each admissible sequence we put points corresponding to symbolic sequences. The points will be the points of orbits of branched points, denoted by $\mathcal{B}$. Now we only consider them as a finite set which address concides with the admissible sequences. We want to extend the map over S^1/L' to D^1/L' and sending the points above according to the address of the admissible sequences.

We now describe the extension over each component of D^1/L'. In doing so we keep in mind that what matters are only the isotopy class of the branched cover map. We can divide them into two cases.

1). If the component don't intersect $\mathcal{B}$, there will be a unique extension of the map over the boundary up to isotopy(Alexsanda's lemma).

2). If the component intersect with $\mathcal{B}$, we want to extend the map over the boundary and points in $\mathcal{B}$ according to their address. Since the set $\mathcal{B}$ comes from admissable sequence, we are able to find the image of $\mathcal{B} \cap G'$ in G. We can always extend the map over $\partial G' \cup (\mathcal{B} \cap G')$ to G'. Usually there are infinitely many isotopy classes for such extension. We only choose the trivial one from the consideration below.

Proposition: *Let M be a Riemann surface of punctured sphere with boundaries. Let $f : M \to M$ be an analytic homeomorphism. Then the isotopy classes of f is finite.*

Outline of the Proof: From the conformal structure of the Riemann surface there corresponds to a hyperbolic structure. The analytic automorphism is correspondent to isometry to the hyperbolic structure. The isotopy classes of isometry is finite.

From the proposition above we see that we should take the extension so that the isotopy class be trivial by the following consideration.

For each gap we construct a Riemann surface, or a subset of $\mathbf{C}$ with finitely many boundaries as the following.

i). First we consider the case that the gap is not critical. Draw small circles around each point in $\mathcal{B}$. Delete these small disks. Close the two external rays outside S^1. We get a Riemann surface S. See the picture below.

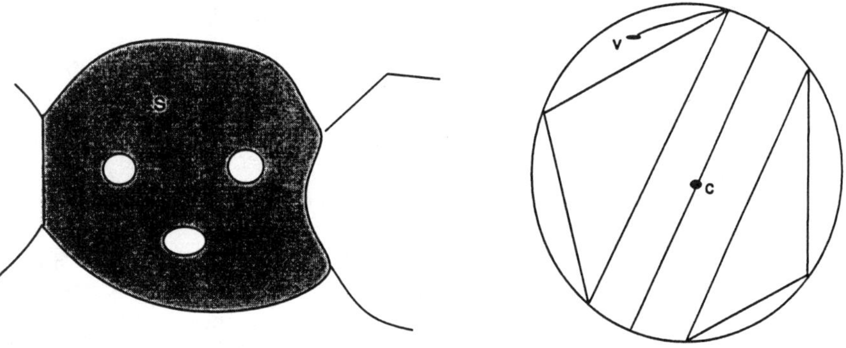

If the map is a polynomial map we see that the map is an analytic homeomorphism of this surface to another one of same topological type. From the proposition above we see the extension must be trivial.

ii). Now we consider how to extend the map over a critical gap. Actually we can subdivide the gap into several pieces by adding more leaves to decompose the map into several copies each of them is analytic homeomorphism. Then applying the argument above. For example, for degree two case, we draw one line in G to connect the branched value to ∂G, two lines in G' to connect the branched points to $\partial G'$. See the picture.

c). The maps are equivalent to polynomials.

To check that the branched cover maps defined above are equivalent to polynomials we use Thurston's topological characterization of rational maps. The theorem was stated and proved in [DH]. We will not state the theorem, instead we will use the consequence derived from the theorem.

Call a branched cover map of the sphere to itself a topologically polynomial map if the map has a fixed branched point (let's fixed it be the point ∞). There is a simple property for topologically polynomial map, i.e. a simple closed curve can be associated to a well-defined disk which is the disk bounded by the curve not containing ∞. This fact is crucial to have the following reduction shown in [F] that a critically finite topological polynomial map is equivalent to a critically finite polynomial if and only if there is no Levy cycle.

A Levy cycle for a critically finite branched cover map f is a collection of disjoint simple closed curves $\{\gamma_1, \gamma_2, ..., \gamma_s\}$ in S^2/Ω_f such that f maps γ_i homeomorphically to a curve which is isotopic to γ_{i+1}, all other components of $f^{-1}(\gamma_{i+1})$ are bounding some disks containing at most one point in Ω_f.

In our case all branched points of the branched cover maps constructed above are all periodic. It's quite easy to use the theorem above to show the main theorem in this paper.

Theorem: *All branched cover maps constructed above are equivalent to polynomials.*

Proof: Suppose on the contrary there exists a Levy cycle for a branched cover map. The map is a topological polynomial. A Levy cycle is associated to a set of disks which are mapped homeomorphically from one to another like the map over the curves. The points in Ω_f contained in those disks have to have a periodic point. By the assumption one of the points has to be branched point. Then the map over the disk cannot be a homeomorphism. Contradiction. ∎

Remark: Another approach to the realization of a fixed point portrait using Hubbard tree is in [P].

Remark: Our method here could be used to classify hyperbolic rational maps. Once we have a partition we can define kneading sequence, i.e. the itenerary of critical orbits of the polynomial. The kneading sequence is clearly a topological invariant. It's nearly complete in the sense that if we count the number of each periodic cycle of critical orbit then it becomes complete. For real multimode map there is another definition of kneading sequence. It's interesting to see if we can use the method in this paper to pick up the kneading sequences realized by polynomials.

d). An example in degree 3

Let's do the following example of $\{\{0, 1/2\}, \{1/8, 3/8\}\}$ in degree 3 to illustrate our construction above. There are three components of S^1/L, denoted by V_1, V_2, V_3. Take the preimages of $\{\{0, 1/2\}, \{1/8, 3/8\}\}$ under the map $z \to z^3$, and connect them to form non-intersecting leaves. We get 6 leaves, i.e. $\{0, 1/2\}$, $\{1/24, 11/24\}$, $\{1/8, 3/8\}$, $\{1/6, 1/3\}$, $\{2/3, 5/6\}$, $\{17/24, 19/24\}$. They divided the unit disk into 7 regions, namely, U_1 to U_7. Those are the components of S^1/L'.

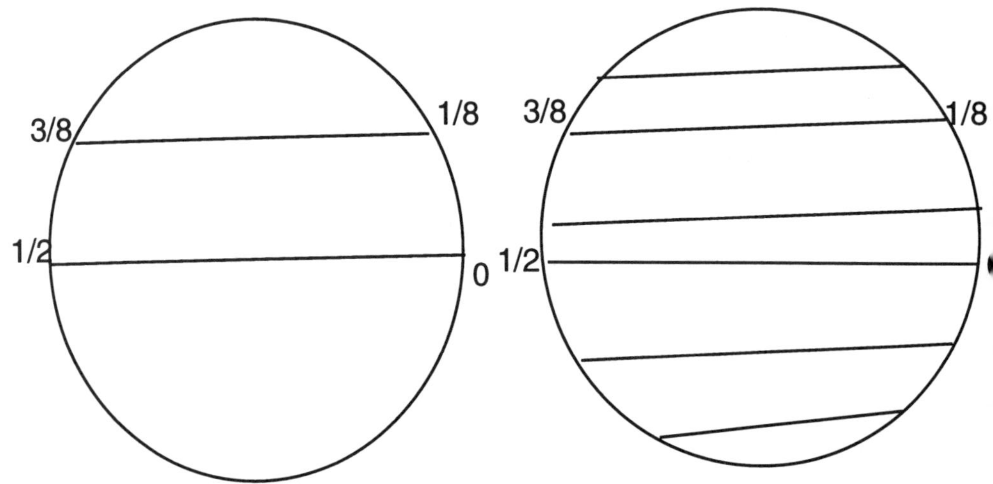

The map from S^1/L' to S^1/L is the following:

$$s(U_1) = V_3, s(U_2) = V_2, s(U_3) = 2V_1, s(U_4) = V_2, s(U_5) = 2V_3, s(U_6) = V_2, s(U_7) =$$

Here $2V_1$ means the component is covered twice by s. If one consider the maps over V_i's one can get the following:

$$s(V_1) = V_2 \cup V_3, s(V_2) = 2V_1 \cup V_2, s(V_3) = V_1 \cup V_2 \cup V_3$$

One can have several choices of the orbit of a branched point. The figure illustrate two examples.

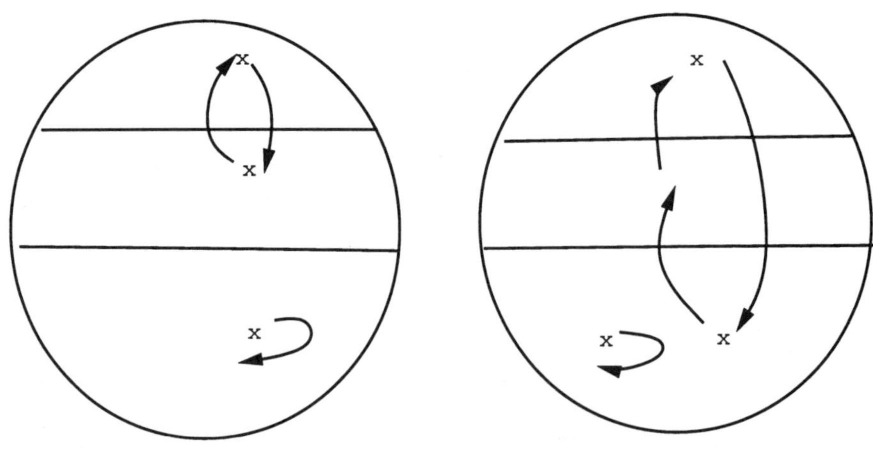

We can extend the map to the disk as described above and glue the map with another copy of s of D^1 to get a branched map of the sphere. As shown above the map is equivalent to a cubic polynomial.

Remark: There is a nice structure for the space of critically finite hyperbolic polynomials. For the periodic two cycle above we can zoom in a Mandelbrot set into it. By using this construction we can get all critically finite polynomials from simple models plus Mandelbrot set. The classification will be described in subsequent papers. The classification will be useful to compute the mapping class group of the hyperbolic component introduced by Sullivan.

Acknowledgement: We are indebted to W. Thurston to share his insight and his beautiful work. We would like to thank John Milnor and Tan Lei for very helpful discussions.

References

[1]   A. Douady and J. H. Hubbard, A Proof of Thurston's Topological Characterization of Rational Functions, Institute Mittag-Leffler Preprint (1984)

[2]   Y. Fisher, Thesis, Cornell University, 1989

[3]   L. R. Goldberg and J. Milnor, Fixed Points of Polynomial Maps, SUNY IMS Preprint, 1990

[4]   S. Hu and Y. Jiang, Toward classification of critically finite polynomials, IHES preprint, 1991.

[5]   S. Levy, Critically Finite Rational Maps, Princeton University, Ph.D. Thesis, 1985

[6]   A. Poirier, On the Realization of Fixed Point Portrait, SUNY/IMS preprint, 1991.

[7]   D. Sullivan, Quasiconformal Homeomorphism and Conformal Dynamics, III, preprint.

[8]   W. P. Thurston, On the Combinatorics and Dynamics of Iterated Rational Maps, Preprint, 1985

Key Words: Invariant lamination, fixed point portrait, kneading sequence. AMS 58F.

Measure properties of Julia sets of meromorphic functions

Janina Kotus

Institute of Mathematics, Technical University of Warsaw
Plac Politechniki 1, 00 661 Warszawa, Poland

1 Introduction

Let $f : \mathbf{C} \to \hat{\mathbf{C}}$ denote a transcendental meromorphic function. For $n \in \mathbf{N}$, f^n denotes the n-th iterate of f and $f^{-n} = (f^n)^{-1}$. The Fatou set $F(f)$ is the set of all the points $z \in \mathbf{C}$ such that (f^n), $n \in \mathbf{N}$, is defined, meromorphic and forms a normal family in some neighbourhood of z. The complement of $F(f)$ in $\hat{\mathbf{C}}$ is called the Julia set $J(f)$ of f.

Every transcendental meromorphic function belongs to one of the following cases: (i) f is entire (ii) f is a self-map of the punctured plane (*i.e.* $f = f_0$, where $f_0(z) = z_0 + (z - z_0)^{-k}\exp(g(z))$ with $k \in \mathbf{N}$ and a non-constant entire function g, (iii) f is neither entire nor a self-map of the punctured plane (*i.e.* f satisfies assumption A of [2, 3, 4, 5].

Let $S = \{z \in \mathbf{C} : \text{some branch of } f^{-1} \text{ has a singularity at } z\}$ and $P = \bigcup\limits_{n=0}^{\infty} f^n(S)$. Equivalently one may write $P = \{z \in \mathbf{C} : \text{for some } n \in \mathbf{N} \text{ some branch of } f^{-n} \text{ has a singularity at } z\}$. Denote the closure of P by $\bar{P}$. Let $E \subset \mathbf{C}$ be a compact set. If, for each $\mu > 0$, we put

$$H_\mu(E) = \liminf_{\varepsilon \to 0}\left(\sum_i (r_i)^\mu\right),$$

where the inf is taken over all possible covers of E with sets of diameter $r_i < \varepsilon$, then the Hausdorff dimension $\mathrm{HD}(E)$ of the set E is defined to be the unique value d satisfying

$$H_\mu(E) = \begin{cases} \infty & \text{for } \mu < d \\ 0 & \text{for } \mu > d. \end{cases}$$

Let m denote the planar Lebesgue measure. We say that a function f is ergodic on $\mathbf{C}$ with respect to m if for each measurable and f-invariant set $E \subset \mathbf{C}$ either $m(E) = 0$ or $m(\mathbf{C}\backslash E) = 0$.

In Section 2 and Section 3 we give the estimate of Hausdorff dimension of Julia sets, while in Section 4 we consider a class of functions which act ergodically on the Julia set being the whole plane.

---

[1] 1991 Mathematics Subject Classification:30D05, 58F08. Key words and Phrases: Meromorphic function, Julia set, Ergodicity, Hausdorf dimension

Proceedings of the International Conference on Complex
Analysis at the Nankai Institute of Mathematics, 1992 pp. 144-156

2 The Hausdorff dimension of Julia sets

It was shown by Garber[10] that for any rational function f, we have $0 < HD(J(f)) \leq 2$. The estimate of Brolin [6] [Theorem 12.2] that, for $c \geq 2 + 2^{\frac{1}{2}}$, $HD(J(f_c)) < \log 2 / \log 2q(c)$, where $f_c(z) = z^2 - c$ and $q(c) = \{c - \frac{1}{2} - (\frac{1}{4})^{\frac{1}{2}}\}^{\frac{1}{2}}$, shows that the lower bound of Garber's estimate is sharp. The upper bound of this estimate is also sharp as there are many rational function f such that $J(f) = \hat{\mathbf{C}}$ *e.g.* $f(z) = [(z - 2)/z]^2$. For a transcedental entire function f it is known that $HD(J(f)) \geq 1$. This follows from Baker's result [1] [Corollary to Thm. 1] which states that if f is a transcendental entire then $J(f)$ must contain non-degenerate continua. However it remains an open question whether exists a transcendental entire function whose Julia set has dimension exactly one. The upper bound for the Hausdorff dimension of Julia set for these function is two, since for *e.g.* $f(z) = e^z$ $J(f) = \hat{\mathbf{C}}$. Thus $1 \leq HD(J(F)) \leq 2$. For a transcendental meromorphic function f satisfying assumption A the estimate of $HD(J(f))$ is the same as for rational functions *i.e.* $0 < HD(J(f)) \leq 2$. The lower bound of this estimate is sharp (a result announced by Stallard) as well the upper bound since $f(z) = \pi i \tan z$ has a property $J(f) = \hat{\mathbf{C}}$.

Now, we give a lower bound for $HD(f)$ for meromorphic functions with infinitely many poles. Let $A = \{a_n : f(a_n) = \infty, n \in \mathbf{N}\}$ be the set of poles of f, b_n denotes the leading non-zero coefficient of principal part in the neighbourhood of the pole a_n and $D(a, r) = \{z : |z - a| < r\}$.

Theorem 1: *Let f be a meromorphic function satisfying*
  (i) A is infinite and $dist(A, \bar{P}) = d > 0$,
  (ii) there exist $\beta > 0, \xi \geq 0$ such that for $a_n \in A$ $|a_n| \sim n^\beta, |b_n| \sim n^{-\xi}$,
  (iii) there exist $m \in \mathbf{N}, \gamma < (\beta + \xi)/m$, and r_n such that $r_n \sim n^{-\gamma}$, and $|f'(z)| \sim m|b_n|/|z - a_n|^{m+1}$ for $z \in D(a_n, r_n)$.
  Then $HD(J(f)) \geq m/(\xi + (m + 1)\beta)$.

Above and in the sequel $|a_n| \sim n^\beta$ (analogously $|b_n| \sim n^{-\xi}$,etc.) means $C^{-1} \leq |a_n| n^{-\beta} \leq C$ for some constant $C > 0$ and all $n \geq n_0$. Roughly speaking the assertion (iii) enables us to replace f' by its principal part in r_n-neighbourhood of the pole a_n uniformly with respect to n.

Theorem 1 has been proved in [12]. We apply it to certain families of functions.

Let f be a meromorphic map on $\mathbf{C}$. For $\lambda \in \mathbf{C}$ let $f_\lambda(z) = \lambda f(z)$. Assume that further we consider only these λ for which $dist(A(f_\lambda), \bar{P}(f_\lambda)) > 0$. Recall that for meromorphic functions f with finitely many singular values (i.e. $card S(f) < \infty$) the Fatou-Sullivan classification of periodic component of $F(f)$ holds. This is a consequence of the general version of the Sullivan theorem proved in [5].

Example 1: Let $f_\lambda(z) = \lambda(\tan z)^m, m \in \mathbf{N}, \lambda \in \mathbf{C}$. Then $A(f_\lambda) = \{a_n = (n + 1/2)\pi : n \in \mathbf{Z}\}$ is the set of poles, and it is easy to see that $\beta = 1, \xi = 0$, so $HD(J(f_\lambda)) \geq m/(m + 1)$. In particular, for $m = 1$ $HD(J(f_\lambda)) \geq 1/2$.

Example 2: Let f_λ be a periodic function.

(i) If f_λ is simply periodic with finitely many poles in each strip of periodicity then $\beta = 1$ and $\xi = 0$. Thus $HD(J(f_\lambda)) \geq m/(m+1) \geq \frac{1}{2}$ and $HD(J(f_\lambda)) \to 1^-$ if $m \to \infty$.

(ii) Let f_λ be doubly periodic of order $m \in \mathbf{N}$ of the poles (e.g. an elliptic function), then $\xi = 0$. One may show that for these function $\beta = \frac{1}{2}$. Thus $HD(J(f_\lambda)) \geq 2m/(m+1) \geq 1$ and $HD(J(f_\lambda)) \to 2^-$ if $m \to \infty$.

The estimate given in Theorem 1 is sharp for some functions. To show this we must compare it with an upper bound for the Hausdorff dimension of Julia set for these functions.

Lemma 1: Let $f_\lambda(z) = \lambda \tan z$, $\lambda \in \mathbf{R}$ and $0 < |\lambda| \ll 1$. Then $HD(J(f_\lambda)) \leq 1/2 + |\lambda|^{1/2}/\pi + \mathcal{O}(|\lambda|)$.

In the proof we apply the following lemma.

Lemma 2: (cf. [8, p.32]) Let $D(z,r)$ denote a disc with center z and radius r. Then for every $0 < s < r$ and for every univalent map $g : D(z,r) \to \mathbf{C}$, the distortion of g in the disc $D(z,s)$ is bounded by the constant $L = L(s/r) = \{(r+s)/(r-s)\}^4$.

Proof of Lemma 1: For $|\lambda| < 1$ we have $f'_\lambda(0) = \lambda$, so 0 is an attracting fixed point. If additionally $\lambda > 0$, then there is a repelling fixed point $z_\lambda \in (0, \pi/2) \subset \mathbf{R}$, which belongs to the boundary of the attractive basin of 0. Analogously, for $-1 < \lambda < 0$ there is a repelling periodic point of order two with the same property. The functions f_λ have poles at the points of the set $A = \{a_n = (n+1/2)\pi : n \in \mathbf{N}\}$. Let $\delta_\lambda = |a_0 - z_\lambda|$, $I_n = (a_n - \delta_\lambda, a_n + \delta_\lambda) \subset \mathbf{R}$. Then $J(f_\lambda)$ is a Cantor subset of $\mathbf{R}$ and $J(f_\lambda) \subset \bigcup_{n=-\infty}^{\infty} I_n$, see [7]. We define the following family of sets. Let $B_0 = I_t$ for some $t \in \mathbf{Z}$, and

$\mathcal{B}_0 = \{B_0\}$,

$\mathcal{B}_1 = \{B_{1,n} : B_{1,n}$ is a component of $f^{-1}(I_n)$ for some $n \in \mathbf{Z}, B_{1,n} \subset B_0\}$,

...

$\mathcal{B}_k = \{B_{k,n} : B_{k,n}$ is a component of $f^{-k}(I_n)$ for some $n \in \mathbf{Z}, B_{k,n} \subset B_{k-1,l} \in \mathcal{B}_{k-1}\}$.

Thus $\mathcal{A}_k = \bigcup_{B_{k,n} \in \mathcal{B}_k} B_{k,n}$ is a covering of $J(f_\lambda)$ and $diam B_{k,n} \leq d_k \to 0$ if $k \to \infty$. To prove that $HD(J(f_\lambda)) \leq \alpha$ it is sufficient to show that, for each $k \in \mathbf{N}$,

$$\sum_{B_{k,n} \in \mathcal{B}_k} (diam B_{k,n})^\alpha \leq \sum_{B_{k-1,l} \in \mathcal{B}_{k-1}} (diam B_{k-1,l})^\alpha,$$

or equivalently that for each $k \in \mathbf{N}$ and each $B_{k-1,l} \in \mathcal{B}_{k-1}$

$$\sum_{B_{k,n} \in G(B_{k-1,l})} (diam B_{k,n})^\alpha \leq (diam B_{k-1,l})^\alpha,$$

where $G(B_{k-1,l}) = \{B_{k,n} \in \mathcal{B}_k : B_{k,n} \subset B_{k-1,l}\}$. By Lemma 2 the distortion $L(f^{-(k-1)}, I_l)$ is bounded by $L(\lambda) = \{(\frac{\pi}{2} + \delta_\lambda)/(\frac{\pi}{2} - \delta_\lambda)\}^4$, so

$$\frac{\sum_{B_{k,n} \in G(B_{k-1,l})} (diam\, B_{k,n})^\alpha}{(diam\, B_{k-1,l})^\alpha}$$

$$\leq L^\alpha(f^{k-1}, B_{k-1,l}) \frac{\sum_{B_{k,n} \in G(B_{k-1,l})} (diam\, f^{k-1}(B_{k,n}))^\alpha}{(diam\, f^{k-1}(B_{k-1,l}))^\alpha}$$

$$= L^\alpha(f^{-(k-1)}, I_l) \frac{\sum_{n=-\infty}^{+\infty} (diam\, f^{-1}(I_n) \cap I_l))^\alpha}{(diam\, I_l)^\alpha}$$

$$\leq 2L^\alpha(\lambda)(2\delta_\lambda)^{-\alpha} \sum_{n=1}^\infty (2\delta_\lambda \sup_{I_n} |(f^{-1})'|)^\alpha$$

$$= 2L^\alpha(\lambda) \sum_{n=1}^\infty (|\lambda|/\{\lambda^2 + [(n - \frac{1}{2})\pi - \delta_\lambda]^2\})^\alpha$$

$$\leq 2(|\lambda| L(\lambda))^\alpha [(\frac{\pi}{2} - \delta_\lambda)^{-2\alpha} + sum_{n=1} \infty (n\pi)^{-2\alpha}]$$

$$= 2(|\lambda| L(\lambda))^\alpha [(\frac{\pi}{2} - \delta_\lambda)^{-2\alpha} + \pi^{-2\alpha} \zeta(2\alpha)]$$

where ζ is the Riemann zeta function. If $\lambda > 1/2$ then $\zeta(2\alpha) < \infty$ and for $|\lambda|^\alpha \leq \pi^{2\alpha}(2\zeta(2\alpha))^{-1}$

$$\sum_{B_{k,n} \in G(B_{k-1,l})} (diam\, B_{k,n})^\alpha \leq (diam\, B_{k-1,l})^\alpha.$$

The inequality $|\lambda|^\alpha \leq \pi^{2\alpha}(2\zeta(2\alpha))^{-1}$ implies that $\alpha \leq 1/2 + |\lambda|^{1/2}/\pi + 2|\lambda|/\pi^2 = 1/2 + |\lambda|^{1/2}/\pi + \mathcal{O}(|\lambda|)$. $\blacksquare$

Lemma 1 together with Example 1 implies that the estimate in Theorem 1 is sharp for $f_\lambda(z) = \lambda \tan z$ and $\lambda \in \mathbf{R}, \lambda \to 0$. Note that assumptions of Theorem 1 are very general, so one can find different families of maps satisfying its assumptios with the same constants. For example for the families $f_\lambda(z) = 1/(\lambda + e^{-2z})$ with $\lambda \geq 0$ and $g_\lambda(z) = \lambda \tan z$ with $|\lambda| < 1$ we have $\beta = 1, \xi = 0, m = 1$, so $HD(J(f_\lambda)) \geq \frac{1}{2}$ and $HD(J(g_\lambda)) \geq \frac{1}{2}$. Now we show that using different methods which relay on investigation dynamics of the maps one can obtain a better estimate of the Hausdorff dimension of Julia sets for the family f_λ.

Theorem 2: *For $f_\lambda(z) = 1/(\lambda + e^{-2z})$, $\lambda > 0$, $J(f_\lambda)$ is a Cantor set. Moreover, the asymptotic estimate*

$$HD(J(f_\lambda)) \geq 1 - C(\log|\log \lambda|)^{-1}$$

holds for some $C > 0$ and $\lambda \to 0^+$.

Of course, the above inequality implies that our lower bound for $HD(J(f_\lambda))$ tends to 1 if λ tends to 0.

3   Proof of Theorem 2

The proof is based on the following result proved by McMullen *[14]*.

Lemma 3: *For each $k \in \mathbf{N}$, let $\mathcal{A}_k$ be a finite collection of disjoint compact subsets of $\mathbf{R}^n$, each of them has positive finite n-dimensional measure, and define $\mathcal{U}_k = \bigcup_{A_k \in \mathcal{A}_k} A_k$, $A = \bigcap_{k=1}^{\infty} \mathcal{U}_k$. Suppose also that, for each $A_k \in \mathcal{A}_k$, there exists $A_{k+1} \in \mathcal{A}_{k+1}$ and a unique $A_{k-1} \in \mathcal{A}_{k-1}$ such that $A_{k+1} \subset A_k \subset A_{k-1}$. If Δ_k, d_k satisfy, for each $A_k \in \mathcal{A}_k$, the conditions*

$$\frac{vol(\mathcal{U}_{k+1} \cap A_k)}{vol(A_k)} \geq \Delta_k,$$

and $diam A_k \leq d_k < 1, d_k \to 0$ as $k \to \infty$,
   then

$$HD(A) \geq n - \limsup_{k \to \infty} \sum_{j=1}^{k} \frac{|\log \Delta_j|}{\log d_k|}.$$

Let $f_\lambda(z) = 1/(\lambda + e^{-2z})$, $\lambda > 0$. We remind to the reader some of the essential properties of this family described in *[7]*. The function f_λ maps $\mathbf{R}$ diffeomorphically onto the interval with asymptotic values $0, 1/\lambda$ as the endpoints, so f_λ has an attracting fixed point $p_\lambda \in (0; 1/\lambda)$, and the entire real axis lies in the immediate basin of p_λ. In particular, both asymptotic values lie in the immediate basin of p_λ, and so there are discs about these points which lie in the basin. Taking preimages of these discs, it follows that there are half planes of the form $H_1 = \{z : \Re e z < \nu_1 = \nu_1(\lambda)\}$ and $H_2 = \{z : \Re e z > \nu_2 = \nu_2(\lambda)\}$ with $\nu_1 < p_\lambda < \nu_2$, which lie in the immediate basin of p_λ. Let $S_\mu = H_1 \cup H_2 \cup \{z : \nu_1 \leq \Re e z \leq \nu_2, |\Im m z - n\pi| < \mu, n \in \mathbf{Z}\}$, where $\mu = \mu(\lambda)$. Then $f_\lambda : S_\mu \longrightarrow S_\mu, so$ $S_\mu \subset F(f_\lambda)$. The complement of S_μ consists of infinitely many congruent rectangles R_m ,where $m \in \mathbf{Z}$ and R_m are indexed according to increasing imaginary parts. In each R_m f_λ has exactly one pole $s_m = -\frac{1}{2}\log \lambda + i(m+\frac{1}{2})\pi, m \in \mathbf{Z}$, and maps each R_m diffeomorficaly onto $\hat{\mathbf{C}} \backslash f_\lambda(S_\mu)$. Thus $J(f_\lambda) = \bigcap_{n \geq 0} f^{-n}(\hat{\mathbf{C}} \backslash S_\mu)$, so $J(f_\lambda)$ is a planar Cantor set, while $F(f_\lambda)$ is the attractive basin of p_λ.

Fix λ and assume that $0 < \lambda < \epsilon$ with ϵ small enough (we can take *e.g.* $\epsilon = 10^{10}$). As before we omit the index λ. Let s_K be a pole such that $-\frac{3}{2}\log \lambda \leq (K + \frac{1}{2})\pi \leq -\frac{3}{2}\log \lambda + \pi$, w_K be the preimage of s_K with $0 < \Im m w_K < \pi/2$, that is, $w_K = -\frac{1}{2}\log((1/s_K) - \lambda)$,

$$\Re e w_K \approx -\frac{1}{2}\log((\frac{5}{2}\log^2 \lambda + \frac{5}{2}\lambda \log^3 \lambda + \frac{25}{4}\lambda^2 \log \lambda)^{\frac{1}{2}}/(\frac{5}{2}\log^2 \lambda))$$

$$\approx -\frac{1}{2}\log((\frac{5}{2}\log^2 \lambda)^{\frac{1}{2}}) = \frac{1}{2}\log(-(2.5)^{\frac{1}{2}}\log \lambda),$$

$$\Im m w_K \approx -\frac{1}{2}\arctan(\frac{3}{2}\log \lambda(-\frac{1}{2}\log \lambda - \frac{5}{2}\log^2 \lambda)^{-1})$$

$$\approx -\frac{1}{2}\arctan 3 \approx 0.6245.$$

Above and in the sequel $G(\lambda) \approx H(\lambda)$ means $\lim G(\lambda)/H(\lambda) = 1$ if $\lambda \longrightarrow 0^+$.

Let

$$\xi = -\frac{1}{2}\log\lambda - \Re ew_K \approx -\frac{1}{2}\log\lambda - \frac{1}{2}\log(-(2.5)^{\frac{1}{2}}\log\lambda)$$

$$= \frac{1}{2}\log(-(2.5)^{\frac{1}{2}}\lambda\log\lambda),$$

$$\eta = \frac{1}{2} - \Im mw_K \approx 0.9463.$$

Define the sets:

$$B_m = \{z : \Re ew_K < \Re ez < -\frac{1}{2}\log\lambda, |\Im m(z - s_m)| < \eta\},$$

$$T = \bigcup_{m\in I} B_m, I = \mathbf{Z}\backslash\{-K,\dots,0,\dots,K-1\},$$

$$E = \{z : f^n(z) \in T \quad \text{for all} \quad n \in \mathbf{N}\}.$$

We want to show that $E \subset J(f)$ and estimate HD(E) from below.

Lemma 4: *If $f^j(z) \in T$, $j = 0,\dots,n-1$, then $|(f^n)'(z)| \geq C_1^n$, for a constant $C_1 > 1$.*

Proof:

Note that $|f'(z)| = 2|(\lambda + e^{-2z})e^z|^{-2}$. Taking the points $x_0 + iy$ in $B_m \subset T$, where x_0 is fixed, we have then $|f'(x_0 + iy)| \geq |f'(x_0 + i(\Im mw_K + m\pi))|$. Now, take $x + iy_0$ in B_m with $y_0 = \Im mw_K + m\pi$. Then $|f'(x + iy_0)| \geq |f'(\Re ew_K + iy_0)|$, and consequently $|f'(z)| \geq |f'(w_K + im\pi)|$ for all $z \in B_m$. As $|f'(w_K + im\pi)| = |f'(f^{-1}(s_K))| = 2|s_K(1 - \lambda s_K)| = (-\log\lambda)[10(1 + \lambda\log\lambda + \frac{5}{2}\lambda^2\log^2\lambda)]^{\frac{1}{2}} =: C_1, C_1 > 1$, then $|(f^n)'(z)| \geq C_1^n$ for $z \in B_m$, and by periodicity of f, for all $z \in T$.

Take the smallest $M \in \mathbf{N}$ satisfying $M \geq 10^{1/2}\pi^{-1}(-\log\lambda) + 1$. Note that $M > K$. Fix $N \in \mathbf{N}$ such that $N > M$. For a pole s_N, consider the preimage $f^{-1}(s_N)$ lying in B_0, and define $\eta_N = |\Im m(s_0 - f^{-1}(s_N))|$. We introduce auxiliary sets:

$$B_{m,N} = \left\{z : \Re ew_K < \Re ez < -\frac{1}{2}\log\lambda, \ \eta_N < |\Im m(z - s_m)| < \eta\right\},$$

$$T_N = \bigcup_{m\in I} B_{m,N}, \ E_N = \{z : f^n(z) \in T_N \ \text{for all } n \in \mathbf{N}\}.$$

Clearly $B_{m,N} \subset B_{m,N+1}, T_N \subset T_{N+1}, \ E_N \subset E_{N+1}$ for each $N \geq M$, $N \in \mathbf{N}$. Thus $\bigcup_{N>M} B_{m,N} = B_m, \bigcup_{N>M} T_N = T_m, \bigcup_{N>M} E_N = E$. To estimate from below the Hausdorff dimension of the set E, it is enough to find a lower bound for the dimension of each E_N.
■

Lemma 5: *For each $N \geq M, N \in \mathbf{N}$, the set E_N is contained in $J(f)$,so $E \subset J(f)$.*

Proof:
Suppose that $z_0 \in E_N \cap F(f)$.Then there is a $disc D = D(z_0, r) \subset F(f)$ and a subsequence of iterates (f^{n_k}) holomorphic on D which converges to a holomorphic function g. Hence $g'(z) \neq \infty$ in D. By Lemma 4 $g'(z_0) = \lim_{k \to \infty} (f^{n_k})'(z_0) = \infty$, so we arrive at contradiction. Thus $z_0 \in J(f)$. ∎

Note that $B_{m,N}$ has two components and $f(B_{m,N}) \cap f(B_{m,N}) \neq \emptyset$ for $m, n \in I_1 = \{n \in \mathbf{Z} : -(N+1) \leq n \leq -(K+1) \text{ or } K \leq n \leq N\}$. Take a specific set $B_{s,N}$, $s \in I_1$, and let $A_{N,1}$ be a component of $f^{-1}(B_{s,N})$ contained in T_N. Define the following collection of the sets inductively:

$$\mathcal{A}_{N,1} = \{A_{N,1}\},$$
$$\mathcal{A}_{N,2} = \{A_{N,2} : A_{N,2} \text{ is a component of } f^{-2}(B_{m,N}) \text{ for some } m \in I_1,$$
$$\text{and } A_{N,2} \subset A_{N,1}\},$$
$$\vdots \qquad \vdots \qquad \vdots$$
$$\mathcal{A}_{N,k} = \{A_{N,k} : A_{N,k} \text{ is a component of } f^{-k}(B_{m,N}) \text{ for some } m \in I_1,$$
$$\text{and } A_{N,k} \subset A_{N,k-1} \text{ for some } A_{N,k-1} \in \mathcal{A}_{N,k-1}\}.$$

Thus $\mathcal{A}_{N,k}$ consists of $(2(N - K + 1))^{k-1}$ sets $A_{N,k}, k \in \mathbf{N}$. Let $\mathcal{U}_{N,k} = \bigcup_{A_{N,k} \in \mathcal{A}_{N,k}} A_{N,k}$, and $A_N = \bigcap_{k=1}^{\infty} A_{N,k}$. Thus $A_N \subset E_N$ and so it is sufficient to find a lower bound for $HD(A_N)$.

Lemma 6: *For each $A_{N,k} \in \mathcal{A}_{N,k}, k \in \mathbf{N}$, $diam A_{N,k} \leq DC_1^{-k}$, where*

$$D = (\frac{1}{4}\log^2(-(2.5)^{\frac{1}{2}}\lambda \log \lambda) + 4\eta^2)^{\frac{1}{2}}$$

Proof:
By the definition of $A_{N,k}$ $f^k(A_{N,k})$ is a connected component of $B_{m,N}$ for some $m \in I_1$. As $B_{m,N} \in B_m$

$$diam f^k(A_{N,k}) \leq (\xi^2 + 4\eta^2)^{1/2} = (\frac{1}{4}\log^2(-(2.5)^{\frac{1}{2}}\lambda \log \lambda) + 4\eta^2)^{\frac{1}{2}} = D.$$

Moreover,$f^k(A_{N,k}) \subset T_N \subset T$, so by Lemma 4 $diam A_{N,k} \leq diam f^k(A_{N,k})/ \inf_{A_{N,k}} |(f^k)'| \leq$

DC_1^{-k}. ∎

The sets $A_{N,k}$ and constants $d_k = DC_1^{-k}$, $k \in \mathbf{N}$, satisfy the conditions of Lemma 3. To use this lemma we should find $\Delta_{N,k}$ such that for each $A_{N,k} \in \mathcal{A}_{N,k}$.

$$\frac{vol(\mathcal{U}_{N,k+1} \cap A_{N,k})}{vol(A_{N,k})} \geq \Delta_{N,k}. \tag{1}$$

Note that $\mathcal{U}_{N,k+1} \cap A_{N,k} = \bigcup_{A_{N,k+1} \in G(A_{N,k})} A_{N,k+1}$ where $G(A_{N,k}) = \{A_{N,k+1} :$
$A_{N,k+1} \in \mathcal{A}_{N,k+1} A_{N,k+1} \subset A_{N,k}\}$, so

$$\frac{vol(\mathcal{U}_{N,k+1} \cap A_{N,k})}{vol(A_{N,k})} \geq L(f^k, A_{N,k})^{-2} \sum_{A_{N,k+1} \in G(A_{N,k})} \frac{vol(f^k(A_{N,k+1}))}{vol(f^k(A_{N,k}))}. \tag{2}$$

Lemma 7: *For each* $A_{N,k} \in \mathcal{A}_{N,k}, k \in \mathbf{N}$*, the distortion* $L(f^k, A_{N,k})$ *is bounded above by a constant* $C_2 < 4$*.*

Proof:

The branches of f^{-k}, $k \in \mathbf{N}$, are univalent in T since $f^k(0)$ and $f^k(1/\lambda)$ are contained in $\mathbf{R}$. Thus f^k is a homeomorphism of $A_{N,k} \in \mathcal{A}_{N,k}$ onto $f^k(A_{N,k})$ and

$$L(f^k, A_{N,k}) = L(f^{-k}, f^k(A_{N,k})) = L(f^{-k}, B_{m,N}).$$

As $B_{m,N}$ is contained in a disc D of diameter $2s = diam B_m$, and $D \subset D(s_m, r)$ with $r = (K + 1/2)\pi$, it follows from Lemma 2 that

$$L(f^{-k}, B_{m,N}) \leq M(s/r) = ((r+s)/(r-s))^4$$
$$= \left(\frac{-\frac{3}{2} \log \lambda + \frac{1}{2} [\frac{1}{4} \log^2(-(2.5)^{\frac{1}{2}} \lambda \log \lambda) + 4\eta^2]^{\frac{1}{2}}}{-\frac{3}{2} \log \lambda - \frac{1}{2} [\frac{1}{4} \log^2(-(2.5)^{\frac{1}{2}} \lambda \log \lambda) + 4\eta^2]^{\frac{1}{2}}} \right)^4$$
$$\leq \left((1 + (0.0277 + \epsilon(\lambda))^{\frac{1}{2}})/(1 - (0.0277 + \epsilon(\lambda))^{\frac{1}{2}}) \right)^4,$$

where $\epsilon(\lambda)) = (\frac{1}{4} \log^2(-(2.5)^{\frac{1}{2}} \lambda \log \lambda) + 3.581)/(9 \log^2 \lambda)$. Thus $M(s/r) \leq (1.414)^4 \leq 3.9976 =: C_2 < 4$ if $\lambda < 10^{-10}$. ∎

Lemma 8: *There exist constants* $0 < \alpha_N, \beta_N < 1$*, and* $C_3, C_4 > 0$ *such that for each* $A_{N,k} \in \mathcal{A}_{N,k}, k \in \mathbf{N}$

$$C_3 - \alpha_N \leq \sum_{A_{N,k+1} \in G(A_{N,k})} vol(f^k(A_{N,k+1})) \leq C_4 - \beta_N.$$

Proof:

Let $W_m \subset B_m$ be defined by

$$W_m = \left\{ z : \mathfrak{Re} w_{K+2} < \mathfrak{Re} z < -\frac{1}{2} \log \lambda, \eta_N < |\mathfrak{Im}(z - s_m)| < \eta \right\},$$

$W_{m,N} = B_{m,N} \cap W_m$, where $w_{K+2} = f^{-1}(s_{K+2}), m \in I$. Then for each $A_{N,k} \in \mathcal{A}_{N,k}$

$$\sum_{A_{N,k+1} \in G(A_{N,k})} vol(f^k(A_{N,k+1})) \geq \frac{1}{2} vol(f^{-1}(T_N) \cap W_{m,N}), \tag{3}$$

where $f^k(A_{N,k})$ is a component of $B_{m,N}$ for some $m \in I$. Note that $f(B_m)$ intersects $B_{n,N}$ for $n \in I_2$ and covers $B_{n,N}$ for $n \in I_3$ where $I_2 = \{n \in \mathbf{Z} : -\frac{1}{2}(N-5) \leq n \leq -(K+1) \text{ or } K \leq n \leq \frac{1}{2}(N-3)\}, I_3 = I \backslash I_2$. Indeed, f maps $\{z : \Re e z = \Re e w_K\}$ onto the circle of radius $r_0 = r/(r^2 - \lambda^2)$ and centre $z_0 = -\lambda/(r^2 - \lambda^2)$, where $r = \exp(-2\Re e w_K)$. For $0 < \lambda < \epsilon, r_0 \approx -(2.5)^{\frac{1}{2}} \log \lambda, z_0 \approx -(2.5)\lambda \log^2 \lambda$, so $r_0/\pi < (N-1)/2$. The lines $\{z : \Im m z = (m + 1/2)\pi - \eta\}$ (resp. $\{z : \Im m z = (m + 1/2)\pi + \eta\}$) are mapped by f onto circles passing through $0, 1/\lambda$ and the pole s_K (resp. $s_{-(K+1)}$), while $f(\{z : \Re e z = -\frac{1}{2} \log \lambda\})$ is the line $\{z : \Re e z = 2/\lambda\}$.

As

$$vol(f^{-1}(B_m) \cap B_n) = \int_{B_m \cap f(B_n)} |(f^{-1})'(z)|^2$$

and $(f^{-1})'(z) = (2z(1 - \lambda z))^{-1}$, we have

$$\max_{B_m} |(f^{-1})'| \leq (4 \left[\frac{1}{4} \log^2(-(2.5)^{\frac{1}{2}} \log \lambda) + (m\pi)^2 \right] (1 + (\lambda m\pi)^2))^{-1},$$

$$\min_{B_m} |(f^{-1})'| \leq (4 \left[\frac{1}{4} \log^2 \lambda) + (m + 1/2 + \eta/\pi)^2 \pi^2 \right] (1 + (\lambda m\pi)^2))^{-1}.$$

In these estimates we take simply $|1 - \lambda z|^2 \approx 1 + (\lambda m\pi)^2$, what is sufficient approximation for our purposes.

Now, we estimate $vol(f^{-1}(\cup B_m) \cap W_n)$ for $m \in I_3$ and $n \in I$:

$$vol(f^{-1}(\cup B_m) \cap W_n) \geq 2 \sum_{m \geq (N-1)/2} vol(B_m)(\min_{B_m} |(f^{-1})'|^2)$$

$$\geq 2 \sum_{m \geq (N-1)/2} \frac{(-\frac{1}{2}\log(-(2.5)^{\frac{1}{2}}\lambda \log \lambda))2\nu}{4[\frac{1}{4}\log^2 \lambda) + (m + 1/2 + \eta/\pi)^2\pi^2](1 + (\lambda m\pi)^2)}$$

$$\geq \frac{1}{8}\pi(-\log(-(2.5)^{\frac{1}{2}}\lambda \log \lambda)) \int_a^\infty (\lambda^2\pi^4)^{-1}((x^2 + D_2^2)(x^2 + D_1^2))^{-1} dx,$$

where $2\eta > \pi/2$, $a = N/2 + \eta/\pi$, $D_1 = 1/(\lambda\pi)$, $D_2 = (-\log \lambda)/(2\pi)$,

$$= \frac{1}{8}\pi(-\log(-(2.5)^{\frac{1}{2}}\lambda \log \lambda)) \int_a^\infty \frac{(x^2 + D_2^2)^{-1} - (x^2 + D_1^2)^{-1}}{\lambda^2\pi^4(D_1^2 - D_2^2)} dx$$

$$= (8\pi)^{-1} \frac{-\log(-(2.5)^{\frac{1}{2}}\lambda \log \lambda)}{(1 - \frac{1}{4}\lambda^2 \log^2 \lambda)} \left[D_2^{-1} \arctan(\frac{x}{D_2}) - D_1^{-1} \arctan(\frac{x}{D_1}) \right]_a^\infty$$

$$= (8\pi)^{-1} \frac{(1 + G_1(\lambda))}{G_3(\lambda)} [\pi^2 - 2\pi \arctan((N\pi + 2\nu)/(-\log \lambda)) + G_2(\lambda)]$$

with $\quad G_1(\lambda) = -\log(-(2.5)^{\frac{1}{2}}\log\lambda)/(-\log\lambda),$

$$G_2(\lambda) = \pi\lambda\log\lambda(\pi/2 - \arctan(\lambda\pi N/2 + \eta\lambda)), \quad G_3(\lambda) = 1 - \frac{1}{4}\lambda^2\log^2\lambda;$$

$$\geq (8\pi)^{-1}(1 + G_1(\epsilon))$$
$$\times(\pi^2 - 2\pi\arctan(10^{\frac{1}{2}} + 2(\pi+1)(-\log^{-1}\epsilon)) + \pi^2\epsilon\log\epsilon/2)$$
$$\geq 0.0569 =: C_3'.$$

Thus for $n \in I$

$$vol\left(f^{-1}(\bigcup_{m\in I_3} B_m) \cap W_n\right) \geq C_3'. \tag{4}$$

Analogously we can find an upper bound for $vol\left(f^{-1}(\bigcup_{m\in I_3} B_m) \cap B_n\right)$, $n \in I$, namely

$$vol\left(f^{-1}(\bigcup_{m\in I_3} B_m) \cap B_n\right) \leq 2 \sum_{m\geq(N-1)/2} vol(B_m)(\max_{B_m}|(f^{-1})'|^2)$$

$$\leq 2 \sum_{m\geq(N-1)/2} \frac{(-\frac{1}{2}\log\lambda)\pi}{4[\frac{1}{4}\log^2(-(2.5)^{\frac{1}{2}}\log\lambda) + (m\pi)^2](1 + (\lambda m\pi)^2)}$$

$$\leq \frac{1}{4}\pi(-log\lambda)\int_{(N-3)/2}^{\infty} \frac{(x^2 + D_3^2)^{-1} - (x^2 + D_1^2)^{-1}}{\lambda^2\pi^4(D_1^2 - D_3^2)}dx,$$

where $D_3 = \log(-(2.5)^{\frac{1}{2}}\log\lambda)/(2\pi)$, and therefore

$$\leq (4\pi)^{-1}(-\log\lambda)[1 - \frac{1}{4}\lambda^2\log^2(-(2.5)^{\frac{1}{2}}\log\lambda)]^{-1}\left[D_3^{-1}\arctan(\frac{x}{D_3})\right]_{\frac{N-3}{2}}^{\infty}$$

$$\leq (2\pi)^{-1}(-\log\lambda)/[10^{\frac{1}{2}}(-\log\lambda)/\pi - 2] < 0.55/\pi =: C_4'.$$

Moreover, if $m \in I_4 = \{n \in \mathbf{Z} : -(N-5)/2 \leq n \leq -(K+2)$ or $(K+1) \leq n \leq (N-3)/2\}$, then using $vol(B_m\cap f(B_n)) \leq \pi(-\log\lambda)(-\frac{1}{2} - [2.5 - (k\pi/\log\lambda)^2]^{\frac{1}{2}})$ we obtain

$$vol\left(f^{-1}(\bigcup_{m\in I_4} B_m) \cap B_n\right)$$

$$\leq 2 \sum_{K+1}^{(N-3)/2} \frac{\pi(-\log\lambda)(-\frac{1}{2} - [2.5 - (k\pi/\log\lambda)^2]^{\frac{1}{2}}}{4[\frac{1}{4}\log^2(-(2.5)^{\frac{1}{2}}\log\lambda) + (m\pi)^2][1 + (\lambda m\pi)^2]}$$

$$\leq \frac{1}{2}\pi(-\log\lambda)\int_K^{(N-1)/2} \frac{(\frac{1}{2} - [2.5 - (x\pi/\log\lambda)^2]^{\frac{1}{2}}dx}{[\frac{1}{4}\log^2(-(2.5)^{\frac{1}{2}}\log\lambda) + (m\pi)^2][1 + (\lambda m\pi)^2]}$$

$$\leq \frac{1}{2}\pi(-\log\lambda)\int_K^{(N-1)/2}(-\frac{1}{2}-[2.5-(x\pi/\log\lambda)^2]^{\frac{1}{2}}(\pi x)^{-2}dx$$

$$\leq \frac{1}{2}\int_{\frac{1}{2}}^{(2.5)^{\frac{1}{2}}}(\frac{1}{2}-(2.5-y^2)^{\frac{1}{2}})y^{-2}dy < 0.0091 =: C"_4.$$

As $vol(f^{-1}(B_{-(K+1)}\cup B_K)\cap B_n) < 0.0001 =: C'''_4$ we have for $n\in I$

$$vol\left(f^{-1}(\bigcup_{m\in I}B_m)\cap B_n\right)\leq C'_4 + C"_4 + C'''_4 \leq 0.177 =: C_4. \tag{5}$$

Note that for $n\in I$

$$\lim_{N\to\infty}vol\left(f^{-1}(\bigcup_{m\in I}B_{m,N})\cap W_{n,N}\right) = vol(f^{-1}\left(\bigcup_{m\in I}B_m\right)\cap W_n), \tag{6}$$

so by (3)-(4) and (5)-(6) there are constants $C_3 = \frac{1}{2}C'_3$ and α_N, β_N such that

$$C_3 - \alpha_N \leq \sum_{A_{N,k+1}\in G(A_{N,k})} vol\left(f^k(A_{N,k+1})\right) \leq C_4 - \beta_N,$$

and $\alpha_N, \beta_N \to 0$ if $N\to\infty$ if $N\to\infty$. Lemma 7 and Lemma 8 together with (2) imply that $vol(U_{N,k+1}\cap A_{N,k})/vol(A_{N,k}) \geq (C_3 - \alpha_N)/(C_2^2 C_5)$ with $C_5 = -\log(-(2.5)^{\frac{1}{2}}\lambda\log\lambda) \geq vol(B_m) \geq vol(B_{m,N})$, that is, $\Delta_{N,k} = (C_3 - \alpha_N)/(C_2^2 C_5)$ by (1).

Now we can apply Lemma 3 to the sets $A_{N,k}$. Thus

$$HD(A_N) \geq 2 - \limsup_{k\to\infty}\frac{k|\log[C_3-\alpha_N)/(C_2^2 C_5)]|}{|\log(DC_1^{-k})|}$$
$$= 2 - \frac{|\log[C_3-\alpha_N)/(C_2^2 C_5)]|}{|\log(C_1)|}.$$

As $A_N\subset A_{N+1}$ and $\bigcup_{N\geq M}A_N = A$ we have

$$HD(A)\geq\lim_{N\to\infty}HD(A_N)\geq 2 - \frac{|\log(C_3/(C_2^2 C_5))|}{|\log(C_1)|}.$$

By the definition $A_N\subset E_N$ for each $N\geq M$ and $\bigcup_{N\geq M}E_N = E\subset J(f_\lambda)$, so

$$HD(J(f_\lambda)) \geq 2 - |\log(C_3/(C_2^2 C_5))|/|\log(C_1)|$$
$$= 2 - \frac{\log 0.0284 - \log 3.998^2 - \log(-\log(-(2.5)^{\frac{1}{2}}\lambda\log\lambda))|}{|\log(-\log\lambda) + \frac{1}{2}\log(10(1-\lambda\log\lambda + \frac{5}{2}\lambda^2\log^2\lambda))|}$$
$$\geq 2 - \frac{\log(|\log\lambda| - \log(1.581|\log\lambda|)) + 6.331}{\log(|\log\lambda|) + 1.152}$$
$$\geq 1 - C(\log|\log\lambda|)^{-1}$$

for some $C > 0$.

Our lower bound for $HD(J(f_\lambda))$ tends to 1 if $\lambda \to 0^+$, as follows from the above inequality valid for $0 < \lambda < 10^{-10}$. ∎

4   Ergodicity of Julia sets

Let z_0 be an isolated singularity of f^{-1}. Let g be a branch of f^{-1} that can be continued analytically (without bound) for $0 < |z - z_0| \leq \eta$. If we set $t = log(z - z_0)$, then $g(z_0 + e^t)$ can be continued without bound in the halfplane $H_t = \{t \in \mathbf{C} : \Re e t \leq \log \eta\}$ and because of the monodromy theorem the continuation is single-valued. If the branch g defined in this way is univalent in H_t, the point z_0 is called a *logarithmic singularity of* f^{-1}. We say that f is *eventually periodic* if for any $z \in S$ there exists $n = n(z) \geq 0$ such that $f^n(z)$ is a repelling periodic point.

For $p \in \mathbf{N}$ let $\mathcal{M}_p$ be the class of functions with card $S = p$, where f is transcendental entire or satisfies assumption A. Let $\mathcal{M} = \bigcup_{p=1}^{\infty} \mathcal{M}_p$.

The *maximum modulus* $M(r, f)$ of a meromorphic function f is defined as $M(r, f) = \max_{|z| \leq r}(|g_1(z)|^2 + |g_2(z)|^2)^{\frac{1}{2}}$ if f is a quotient of entire functions g_1, g_2 not having common factors. Function f is of *order* ρ if

$$\rho = \limsup_{r \to \infty}(\log \log M(r, f)/\log r)$$

(see *[11]*). In *[11]* we have proved the following theorem

Theorem 3: *Let $f \in \mathcal{M}$ be eventually periodic of finite order and at least one of branch of f^{-1} has a logarithmic singularity at some $z \in S$. Then $J(f) = \hat{\mathbf{C}}$ and f is ergodic on $J(f)$ with respect to the Lebesgue measure.*

The assumption of the theorem are essential. Exponential function $f(z) = e^z$ is of order one and has one singular value 0 which is a logarithmic singularity of f^{-1}.Its trajectory escapes to infinity, so $J(f) = \hat{\mathbf{C}}$ but f is not ergodic as it was shown by Lyubich*[13]*. McMullen proved in *[14]* that any member of the family $f_{a,b}(z) = sin(az + b), a \neq 0$, is not ergodic on its Julia sets. Observe that $f_{a,b} \in \mathcal{M}_2$, has order one but $f_{a,b}^{-1}$ has no finite logarithmic singularities. Next Fang*[9]* generalized this result to self-maps of $\mathbf{C}\backslash\{0\}$ with the form $f(z) = z^m \exp(P(z) + Q(1/z))$ and entire functions $f(z) = mz + P(e^z) + Q(e^{-z})$, where $m \in \mathbf{Z}, P, Q$ are polynomials of the same degree.

References

[1]   I.N.Baker.   The domains of normality of an entire function.   *Ann.   Acad. Sci.Fenn. Ser. A I. Math.*, 1:277-283, 1975.

[2]   I.N.Baker, J.Kotus, and Lü Yinian. Iterates of meromorphic functions I.*Ergodic Th. and Dynam. Sys.*, 11: 241-248, 1991.

[3]   I.N.Baker, J.Kotus, and Lü Yinian. Iterates of meromorphic functions II: Examples of wandering domains. *J.London Math.Soc.*, 2(42): 267-278, 1990.

[4]   I.N.Baker, J.Kotus, and Lü Yinian.Iterates of meromorphic functions III: Preperiodic domains.*Ergodic Th. and Dynam. Sys.*, 11:603-618,1991.

[5]   I.N.Baker, J.Kotus, and Lü Yinian.Iterates of meromorphic functions IV. to appear.

[6]   H.Brolin. Invariant sets under iteration of rational functions.*Arkiv för Mathematik*, 6:103-144, 1985.

[7]   R.L.Devaney and L.Keen. Dynamics of meromorphic maps with polynomial Schwarzian derivative.*Ann. Sci. Ecole Norm. Sup.*, 22:55-81,1989.

[8]   P.L.Duren.*Univalent Functions.* Springer, New York, 1983.

[9]   L.Fang. Complex Analytic Dynamics on $\mathbf{C}^*$. preprint.

[10] V.Garber. On the iteration of rational functions.*Math. Proc. Camb. Phil. Soc.*, 84:497-505, 1978.

[11] J.Kotus. On ergodicity of Julia sets of meromorphic functions. Preprint.

[12] J. Kotus. On the Hausdorf dimension of Julia sets of meromorphic functions II. Preprint.

[13] M.Yu.Lyubich. The measurable dynamics of the exponential map.*Siberian J. Matyh.*, B.28:111-127, 1987.

[14] C.McMullen. Area and Hausdorff dimension of Julia sets of entire functions.*Trans. A.M.S.*, 300:329-342,1987.

[15] T.Shimizu. On theory of meromorphic functions.*Japan J.Math.*, 6:119-171, 1929.

A unified treatment of some special classes of univalent functions

Wancang Ma[1] and David Minda[1,2]

*Department of Mathematical Sciences, University of Cincinnati
Cincinnati, Ohio 45221-0025, U.S.A.*

[1]Research partially supported by a National Science Foundation Grant

1   Introduction

Some special classes of univalent functions play an important role in geometric function theory because of their geometric properties. Many such classes have been introduced and studied; some become well-known, for example, the classes of convex, starlike, strongly convex and strongly starlike functions, to name just a few. It is fairly common that a function in one of these classes is characterized by either of the quantities $1 + zf''(z)/f'(z)$ or $zf'(z)/f(z)$ lying in a given region in the right half-plane; the region is often convex and symmetric with respect to the real axis. In the past, these families were introduced and studied one by one. It is interesting to note that we can obtain many of their analytic properties by a unified method. Moreover, we will not even assume the convexity of these regions in the right-half plane, but a much weaker condition, namely starlikeness with respect to 1.

To make our claim better understood, let us begin by introducing some notation. In this paper we always assume that φ is a holomorphic univalent function with positive real part in the unit disk $\mathbf{D} = \{z : |z| < 1\}$, $\varphi(\mathbf{D})$ is symmetric with respect to the real axis and starlike with respect to $\varphi(0) = 1$, and $\varphi'(0) > 0$. This forces $\varphi'(0) \le 2$.

Next, let $P(\varphi)$ be the family of holomorphic functions p in $\mathbf{D}$ such that $p(0) = 1$ and $p(\mathbf{D}) \subseteq \varphi(\mathbf{D})$, that is, $p \prec \varphi$. It is clear that each $P(\varphi)$ is a subfamily of the well-known class $P = P((1 + z)/(1 - z))$ of normalized functions in $\mathbf{D}$ with positive real part.

Define $C(\varphi)$ $(S^*(\varphi))$ to be the class of normalized univalent functions

$$f(z) = z + a_2 z^2 + a_3 z^3 + \cdots$$

in $\mathbf{D}$ such that $1 + zf''(z)/f'(z)$ $(zf'(z)/f(z)) \in P(\varphi)$. Note that $f \in C(\varphi)$ if and only if $zf' \in S^*(\varphi)$.

It is obvious that $C((1 + z)/(1 - z)) = C$ is the well-known class of normalized convex functions. If $0 < \alpha \le 1$, then $C((1 + z)^\alpha/(1 - z)^\alpha)$ is the class of

---

[2]Participation in the International Conference on Complex Analysis at the Nankai Institute of Mathematics, Nankai University, Tianjin, People's Republic of China, June 19-23, 1992 was funded by a travel grant from the Taft Faculty Committee of the University of Cincinnati.

Proceedings of the International Conference on Complex
Analysis at the Nankai Institute of Mathematics, 1992 pp. 157-169
©INTERNATIONAL PRESS 1994

strongly convex functions of order α. For $0 \le \beta < 1$, $C((1+(1-2\beta)z)/(1-z))$ is the class of convex functions of order β. When

$$\varphi(z) = 1 + \frac{2}{\pi^2}\left(\log\frac{1+\sqrt{z}}{1-\sqrt{z}}\right)^2, \tag{1}$$

$\varphi(\mathbf{D}) = \{w = u + iv : v^2 < 2u - 1\} = \{w : \operatorname{Re} w > |w - 1|\}$. It is proved in *[9]* and *[13]* that $C(\varphi)$ is the family UCV of normalized uniformly convex functions introduced recently by Goodman *[3]*. This class was defined geometrically. Precisely, $f \in UCV$ provided that for every circular arc γ contained in $\mathbf{D}$, with center also in $\mathbf{D}$, the image arc $f(\gamma)$ is convex.

Note that $S^*((1 + z)/(1 - z)) = S^*$, the well-known class of normalized starlike functions. Geometrically, $f \in S^*$ if and only if $f(\mathbf{D})$ is starlike with respect to the origin. The family $S^*((1 + z)^\alpha/(1 - z)^\alpha)$, $0 < \alpha \le 1$, consists of strongly starlike functions of order α; $f \in S^*((1 + z)^\alpha/(1 - z)^\alpha)$ if and only if for every $w \in f(\mathbf{D})$ a certain lens-shaped region with end points 0 and w lies in $f(\mathbf{D})$, which is a natural refinement of starlikeness *[8]*. For $0 \le \beta < 1$, $S^*((1+(1-2\beta)z)/(1-z))$ is the class S_β^* of starlike functions of order β. When φ is defined by formula (1), $S^*(\varphi)$ becomes the subclass of starlike functions corresponding to UCV; this class was studied by Rønning (*[13]*, *[14]*, who used the notation S_p) and is related to the class UST of normalized uniformly starlike functions introduced by Goodman *[4]*.

The classes $C(\varphi_{A,B})$ and $S^*(\varphi_{A,B})$ defined by $\varphi_{A,B}(z) = (1+Az)/(1+Bz)$, $-1 \le B < A \le 1$, have been studied by a number of authors. The paper *[6]* contains references to several papers in which these classes are studied.

As we have seen above, when φ varies, $C(\varphi)$ and $S^*(\varphi)$ generate a number of known subclasses of convex and starlike functions, respectively.

In Section 2, we present some functions in $C(\varphi)$ and $S^*(\varphi)$. Section 3 is devoted to subordination results; as corollaries we derive sharp distortion, growth, rotation and covering theorems for $C(\varphi)$ and $S^*(\varphi)$. Some coefficient problems are treated in Section 4; we obtain the sharp bound on the coefficient functional $|a_3 - \mu a_2^2|$, $-\infty < \mu < +\infty$, which implies the sharp upper bounds on the second and third coefficients. We also obtain the sharp order of growth for the coefficients in certain cases. In Section 5 we give some convolution properties.

By letting φ be the particular functions mentioned above, we can recover many basic properties of functions in various subclasses previously defined. Typically, we will not point out the numerous known results that can be obtained as special instances of our work.

2 Examples

In this section we introduce some functions in either $C(\varphi)$ or $S^*(\varphi)$, which turn out to be extremal for certain functionals. First, observe that $\varphi(z^{n-1})$ is contained in $P(\varphi)$ for $n = 2, 3, 4, \cdots$.

Define $k_{\varphi n}(n = 2, 3, 4, \cdots)$ by $k_{\varphi n}(0) = k'_{\varphi n}(0) - 1 = 0$ and

$$1 + \frac{z k''_{\varphi n}(z)}{k'_{\varphi n}(z)} = \varphi(z^{n-1}),$$

which is clearly in the class $C(\varphi)$. We write $k_{\varphi 2}$ simply as k_φ, which plays the role of the Koebe function for the family $C(\varphi)$.

Similarly, we define $h_{\varphi n}(n = 2, 3, 4, \cdots)$ by $h_{\varphi n}(0) = h'_{\varphi n}(0) - 1 = 0$ and

$$\frac{z h'_{\varphi n}(z)}{h_{\varphi n}(z)} = \varphi(z^{n-1}).$$

Then $h_{\varphi n}$ belongs to $S^*(\varphi)$ and $h_{\varphi 2}$, which we simply denote by h_φ, plays the role of the Koebe function in the family $S^*(\varphi)$. Note that $z k'_{\varphi n}(z) = h_{\varphi n}(z)$.

3   Subordination theorem and consequences

First, we derive some subordination results for $C(\varphi)$; as corollaries we obtain sharp distortion, growth, covering and rotation theorems.

Theorem 1: *Let $f \in C(\varphi)$. Then $z f''(z)/f'(z) \prec z k''_\varphi(z)/k'_\varphi(z)$ and $f'(z) \prec k'_\varphi(z)$.*

Proof: Let $p(z) = 1 + z f''(z)/f'(z)$. Then we know that $p \prec \varphi$. Thus, $p - 1 \prec \varphi - 1$, which is the same as $z f''(z)/f'(z) \prec z k''_\varphi(z)/k'_\varphi(z)$. Note that $(\varphi(z)-1)/\varphi'(0) \in S^*$ and $p \prec \varphi$ implies that $(p(z)-1)/\varphi'(0) \prec (\varphi(z)-1)/\varphi'(0)$. By using a result of Suffridge *[17]*, we may conclude that

$$\frac{1}{\varphi'(0)} \log f'(z) = \int_0^z \frac{p(\xi) - 1}{\xi \varphi'(0)} d\xi \prec \int_0^z \frac{\varphi(\xi) - 1}{\xi \varphi'(0)} d\xi = \frac{1}{\varphi'(0)} \log k'_\varphi(z).$$

Equivalently, $f' \prec k'_\varphi$. This completes the proof of Theorem 1.   ■

Corollary 1: *(Distortion Theorem for $C(\varphi)$). Assume $f \in C(\varphi)$ and $|z_0| = r < 1$. Then*

$$k'_\varphi(-r) \leq |f'(z_0)| \leq k'_\varphi(r).$$

Equality holds for some $z_0 \neq 0$ if and only if f is a rotation of k_φ.

Proof: Since $\varphi(\mathbf{D})$ is starlike with respect to 1, $\varphi(z) - 1$ is starlike with respect to 0. As $\psi(z) = -\log(1 - z) \in C$, we see that *[16]*

$$\log k'_\varphi(z) = \int_0^z \frac{\varphi(\xi) - 1}{\xi} d\xi = (\varphi - 1) * \psi(z)$$

is also (starlike) univalent in $\mathbf{D}$. Moreover, all coefficients of $\log k'_\varphi(z)$ are real because $\varphi(z)$ has real coefficients, which follows from the assumption that $\varphi(\mathbf{D})$

is symmetric with respect to the real axis. This implies that $\log k'_\varphi(x)$ maps $(-1, 1)$ one-to-one into the real axis. Since $\varphi'(0) > 0$ is equivalent to the fact that $\log k'_\varphi(x)$ has positive derivative at $x = 0$, we see that $\log k'_\varphi(x)$ is increasing on $(-1, 1)$. Because $\varphi - 1$ is typically-real in $\mathbf{D}$, we have that (see *[11]*(Theorem 2.14))

$$Re\left\{(1 - z^2)\frac{d}{dz}\log k'_\varphi(z)\right\} = Re\left\{(1 - z^2)\frac{\varphi(z) - 1}{z}\right\} > 0, \quad z \in \mathbf{D}.$$

Hence, $\log k'_\varphi(z)$ is convex in the direction of the imaginary axis.

As $\log f'(z) \prec \log k'_\varphi(z)$, we find from the subordination principle that for a point z_0 with $|z_0| = r$

$$\log k'_\varphi(-r) = \exp\{\log k'_\varphi(-r)\} = \exp\{\min_{|z|=r} Re\log k'_\varphi(z)\}$$
$$\leq \exp\{Re\log f'(z_0)\} = |f'(z_0)|$$
$$\leq \exp\{\max_{|z|=r} Re\log k'_\varphi(z)\} = \exp\{\log k'_\varphi(r)\} = k'_\varphi(r).$$

Here we have employed the fact that for a function convex in the direction of the imaginary axis and symmetric about the real axis, the image of $\{z : |r| < r\}$ is also convex in the direction of the imaginary axis; the corresponding result for the full class of functions convex in the direction of the imaginary axis is false *[5]*. Note that for some z_0 with $0 < |z_0| = r < 1$,

$$Re\{\log f'(z_0\} = \min_{|z|=r} Re\{\log k'_\varphi(z)\}$$

or

$$Re\{\log f'(z_0)\} = \max_{|z|=r} Re\{\log k'_\varphi(z)\}$$

if and only if $\log f'(z)$ is $\log k'_\varphi(e^{i\theta}z)$ for some $\theta \in \mathbf{R}$; this follows from the principle of subordination. Therefore, the assertion about the case of equality in the theorem holds. This completes our proof. $\blacksquare$

Corollary 2: *(Growth Theorem for $C(\varphi)$). Let $f \in C(\varphi)$ and $|z_0| = r < 1$. Then*

$$-k_\varphi(-r) \leq |f(z_0)| \leq k_\varphi(r).$$

Equality holds for some $z_0 \neq 0$ if and only if f is a rotation of k_φ.

Corollary 3: *(Covering Theorem for $C(\varphi)$). Suppose $f \in C(\varphi)$. Then either f is a rotation of k_φ or $f(\mathbf{D}) \supseteq \{w : |w| \leq -k_\varphi(-1)\}$. Here $-k_\varphi(-1)$ is understood to be the limit of $-k_\varphi(-r)$ as r tends to 1.*

Note that $-k_\varphi(-r)$ is increasing in $(0, 1)$ and bounded above by 1 (recall that each function $f \in C(\varphi)$ is normalized by $f(0) = f'(0) - 1 = 0$), so the limit of $-k_\varphi(-r)$ exists as r tends to 1.

The following rotation theorem follows from the subordination $f' \prec k'_\varphi$ given in Theorem 1.

Corollary 4: *(Rotation Theorem for $C(\varphi)$). Let $f \in C(\varphi)$ and $|z_0| = r < 1$. Then*

$$|arg\{f'(z_0)\}| \leq \max_{|z|=r} \, arg\{k'_\varphi(z)\}.$$

Equality holds for some $z_0 \neq 0$ if and only if f is a rotation of k_φ.

Next we state corresponding results for $S^*(\varphi)$. When a result follows from the simple correspondence between $S^*(\varphi)$ and $C(\varphi)$, we will omit its proof.

Theorem 1': *Let $f \in S^*(\varphi)$. Then $zf'(z)/f(z) \prec zh'_\varphi(z)/h_\varphi(z)$ and $f(z)/z \prec h_\varphi(z)/z$.*

Corollary 1': *(Growth Theorem for $S^*(\varphi)$). Assume $f \in S^*(\varphi)$ and $|z_0| = r < 1$. Then*

$$-h_\varphi(-r) \leq |f(z_0)| \leq h_\varphi(r).$$

Equality holds for some $z_0 \neq 0$ if and only if f is a rotation of h_φ.

By making use of standard techniques and Corollary 1', we obtain the following covering theorem for $S^*(\varphi)$. Note that $-h_\varphi(-r)$ is increasing in $(0,1)$ and bounded above by 1 (since each function $f \in S^*(\varphi)$ is normalized by $f(0) = f'(0) - 1 = 0$), so $\lim_{r \to 1} -h_\varphi(-r)$ exists.

Corollary 2': *(Covering Theorem for $S^*(\varphi)$). Suppose $f \in S^*(\varphi)$. Then either f is a rotation of h_φ or $f(\mathbf{D}) \supseteq \{w : |w| \leq -h_\varphi(-1)\}$. Here $-h_\varphi(-1)$ is defined to be $\lim_{r-1} -h_\varphi(-r)$.*

Corollary 3': *Let $f \in S^*(\varphi)$ and $|z_0| = r < 1$. Then.*

$$|arg\{f(z_0)/z_0\}| \leq \max_{|z|=r} \, arg\{h_\varphi(z)/z\}.$$

Equality holds for some $z_0 \neq 0$ if and only if f is a rotation of h_φ.

In order to give a distortion theorem for the class $S^*(\varphi)$ we will impose two extra conditions on φ, namely, $\min_{|z|=r} |\varphi(z)| = \varphi(-r)$ and $\max_{|z|=r} |\varphi(z)| = \varphi(r)$. These conditions are satisfied by those φ corresponding to many classes studied earlier by various authors, but do not hold for the fairly general functions φ we consider in this paper. For example, the function $\varphi(z) = 1 - i\alpha \log((1 + iz)/(1 - iz))$, $0 < \alpha < 2/\pi$ does not satisfy $\max_{|z|=r} |\varphi(z)| = \varphi(r)$ for all r, $0 < r < 1$, as $\varphi(r) \leq 1 + \pi\alpha/2$ while $\max_{|z|=r} |\varphi(z)| = O(-\log(1 - r))$ when $r \to 1$. For $0 < \lambda < 1$, the function

$$\varphi(z) = \sqrt{\lambda + (1 - \lambda) \left(\frac{1 + z}{1 - z}\right)^2}$$

does not satisfy $\min_{|z|=r} |\varphi(z)| = \varphi(-r)$ for all r, $0 < r < 1$, since $\min_{|z|=r} |\varphi(z)|$ tends to 0 as $r \to 1$ and $\varphi(-r) > \lambda$.

Theorem 2: *(Distortion Theorem for $S^*(\varphi)$ under some restrictions). Let $\min\limits_{|z|=r}|\varphi(z)| = \varphi(-r)$, $\max\limits_{|z|=r}|\varphi(z)| = \varphi(r)$, $f \in S^*(\varphi)$ and $|z_0| = r < 1$. Then*

$$h'_\varphi(-r) \le |f'(z_0)| \le h'_\varphi(r).$$

Equality holds for some $z_0 \ne 0$ if and only if f is a rotation of h_φ.

Proof: Let $q(z) = zf'(z)/f(z)$. Then $f \in S^*(\varphi)$ if and only if $q \prec \varphi$. By using the subordination principle, we see that

$$\varphi(-r) = \min_{|z|=r}|\varphi(z)| \le |q(z_0)| \le \max_{|z|=r}|\varphi(z)| = \varphi(r).$$

Thus, from Corollary 1', we have for $|z_0| = r$

$$h'_\varphi(-r) = \varphi(-r)h_\varphi(-r)/(-r) \le |q(z_0)f(z_0)/z_0| = |f'(z_0)| \le \varphi(r)h_\varphi(r)/r = h'_\varphi(r).$$

The assertion about equality follows from Corollary 1'. This completes the proof of Theorem 2. ◾

4   Coefficient bounds

In this section we discuss coefficient problems for $C(\varphi)$ and $S^*(\varphi)$. First, we prove a corresponding result for the class P.

Lemma 1: *Let $p_1(z) = 1 + c_1 z + c_2 z^2 + \cdots \in P$. Then*

$$|c_2 - vc_1^2| \le \begin{cases} -4v + 2 & \text{if } v \le 0, \\ 2 & \text{if } 0 \le v \le 1, \\ 4v - 2 & \text{if } v \ge 1. \end{cases}$$

When $v < 0$ or $v > 1$, equality holds if and only if $p_1(z)$ is $(1+z)/(1-z)$ or one of its rotations. If $0 < v < 1$, then equality holds if and only if $p_1(z)$ is $(1+z^2)/(1-z^2)$ or one if its rotations. Inequality becomes equality when $v = 0$ if and only if

$$p_1(z) = \left(\frac{1}{2} + \frac{1}{2}\lambda\right)\frac{1+z}{1-z} + \left(\frac{1}{2} - \frac{1}{2}\lambda\right)\frac{1-z}{1+z}, 0 \le \lambda \le 1,$$

or one of its rotations. While for $v = 1$, equality holds if and only if $p_1(z)$ is the reciprocal of one of the functions such that equality holds in the case of $v = 0$.

Remark. Although the above upper bound is sharp, it can be improved as follows when $0 < v < 1$:

$$|c_2 - vc_1^2| + v|c_1|^2 \le 2, \quad 0 < v \le 1/2$$

and

$$|c_2 - vc_1^2| + (1-v)|c_1|^2 \leq 2, \quad 1/2 \leq v < 1.$$

Proof of Lemma 1. If $v \geq 1$ then by using $|c_1^2 - c_2| \leq 2$ *[7]* and $|c_1| \leq 2$ we have

$$|c_2 - vc_1^2| \leq |c_1^2 - c_2| + (v-1)|c_1|^2 \leq 2 + 4(v-1) = 4v - 2.$$

If $v \leq 0$, then from $|c_1| \leq 2$ and $|c_2| \leq 2$ we obtain

$$|c_2 - vc_1^2| \leq (-v)|c_1|^2 + |c_2| \leq -4v + 2.$$

If $0 \leq v \leq 1/2$, then by using $\left|c_2 - \frac{1}{2}c_1^2\right| \leq 2 - \frac{1}{2}|c_1|^2$ (see *[10]*) we have

$$|c_2 - vc_1^2| + v|c_1|^2 \leq \left|c_2 - \frac{1}{2}c_1^2\right| + \left(\frac{1}{2} - v\right)|c_1|^2 + v|c_1|^2 \leq 2,$$

which is the stronger result stated in the remark above.

Similarly, if $1/2 \leq v \leq 1$, we get

$$|c_2 - vc_1^2| + v|c_1|^2 \leq \left|c_2 - \frac{1}{2}c_1^2\right| + \left(v - \frac{1}{2}\right)|c_1|^2 + (1-v)|c_1|^2 \leq 2.$$

When $v < 0$ or $v > 1$, equality holds if and only if $|c_1| = 2$, that is, $p_1(z)$ is equal to $(1+z)/(1-z)$ or one of its rotations (see *[11]*(p. 41)). If $0 < v < 1$, then equality holds if and only if $|c_1| = 0$ and $|c_2| = 2$, or equivalently, $p_1(z)$ is $(1+z^2)/(1-z^2)$ or one of its rotations *[11]*.

For inequality to become equality when $v = 0$ requires $|c_2| = 2$. Hence *[11]*,

$$p_1(z) = \left(\frac{1}{2} + \frac{1}{2}\lambda\right)\frac{1+z}{1-z} + \left(\frac{1}{2} - \frac{1}{2}\lambda\right)\frac{1-z}{1+z}, \quad 0 \leq \lambda \leq 1,$$

or one of its rotations.

Finally, when $v = 1$, equality means $|c_2 - c_1^2| = 2$, which holds if and only if $p_1(z)$ is the reciprocal of one of the functions such that equality holds in the case $v = 0$. The proof of the Lemma 1 is complete. ∎

Now we introduce the following functions in $C(\varphi)$. For $0 \leq \lambda \leq 1$, define f_λ and g_λ by $f_\lambda(0) = f_\lambda'(0) - 1 = g_\lambda(0) = g_\lambda'(0) - 1 = 0$,

$$1 + \frac{zf_\lambda''(z)}{f_\lambda'(z)} = \varphi\left(\frac{z(z+\lambda)}{1+\lambda z}\right)$$

and

$$1 + \frac{zg_\lambda''(z)}{g_\lambda'(z)} = \varphi\left(-\frac{z(z+\lambda)}{1+\lambda z}\right),$$

respectively. Then it is clear that both f_λ and g_λ belong to $C(\varphi)$. Also notice that $f_1 = k_\varphi$, $f_0 = k_{\varphi 3}$, $g_1(z) = -k_\varphi(-z)$ and $g_0(z) = -k_{\varphi 3}(-z)$.

Theorem 3: *Let $\varphi(z) = 1 + B_1 z + B_2 z^2 + \cdots$. If $f(z) = z + a_2 z^2 + a_3 z^3 + \cdots \in C(\varphi)$, then*

$$
\left| a_3 - \mu a_2^2 \right| \leq
\begin{cases}
\frac{1}{6}\left(B_2 - \frac{3}{2}\mu B_1^2 + B_1^2 \right) & \text{if } 3B_1^2 \mu \leq 2(B_2 + B_1^2 - B_1), \\[2mm]
\frac{1}{6} B_1 & \text{if } 2(B_2 + B_1^2 - B_1) \leq 3B_1^2 \mu \leq 2(B_2 + B_1^2 + B_1), \\[2mm]
\frac{1}{6}\left(-B_2 + \frac{3}{2}\mu B_1^2 - B_1^2 \right) & \text{if } 2(B_2 + B_1^2 + B_1) \leq 3B_1^2 \mu.
\end{cases}
$$

When $3B_1^2 \mu < 2(B_2 + B_1^2 - B_1)$ or $2(B_2 + B_1^2 + B_1) < 3B_1^2 \mu$, equality holds if and only if f is k_φ or one of its rotations. When $2(B_2 + B_1^2 - B_1) < 3B_1^2 \mu < 2(B_2 + B_1^2 + B_1)$, equality holds if and only if f is equal to $k_{\varphi 3}$ or one of its rotations. If $3B_1^2 \mu = 2(B_2 + B_1^2 - B_1)$, equality holds if and only if f is equal to f_λ or one of its rotations. When $3B_1^2 \mu = 2(B_2 + B_1^2 + B_1)$, inequality becomes equality if and only if f equals g_λ or one of its rotations.

Remark. When $2(B_2 + B_1^2 - B_1) < 3B_1^2 \mu < 2(B_2 + B_1^2 + B_1)$, the above inequality can be improved as follows:

$$
\left| a_3 - \mu a_2^2 \right| + \frac{2}{3B_1^2}\left(\frac{3}{2}\mu B_1^2 + B_1 - B_2 - B_1^2 \right) |a_2|^2 \leq \frac{1}{6} B_1
$$

if $2(B_2 + B_1^2 - B_1) < 3B_1^2 \mu \leq 2(B_2 + B_1^2)$ and

$$
\left| a_3 - \mu a_2^2 \right| + \frac{2}{3B_1^2}\left(-\frac{3}{2}\mu B_1^2 + B_1 + B_2 + B_1^2 \right) |a_2|^2 \leq \frac{1}{6} B_1
$$

if $2(B_2 + B_1^2) \leq 3B_1^2 \mu < 2(B_2 + B_1^2 + B_1)$.

Also, note that Theorem 3 yields the sharp upper bound on both the second and third coefficients of functions in $C(\varphi)$.

Proof of Theorem 3. Let $p(z) = 1 + z f''(z)/f'(z) = 1 + b_1 z + b_2 z^2 + \cdots$. Then $2a_2 = b_1$ and $6a_3 = b_2 + b_1^2$. Since φ is univalent and $p \prec \varphi$, we see that the function

$$
p_1(z) = \frac{1 + \varphi^{-1}(p(z))}{1 - \varphi^{-1}(p(z))} = 1 + c_1 z + c_2 z^2 + \cdots
$$

is holomorphic and has positive real part in $\mathbf{D}$, that is, $p_1 \in P$. Equivalently,

$$
p(z) = \varphi\left(\frac{p_1(z) - 1}{p_1(z) + 1} \right).
$$

From the power series expansion of φ, $\varphi(z) = 1 + B_1 z + B_2 z^2 + \cdots$, we can express b_n in terms of c_n. Precisely, $b_1 = \frac{1}{2} B_1 c_1$ and $b_2 = \frac{1}{2} B_1 \left(c_2 - \frac{1}{2} c_1^2 \right) + \frac{1}{4} B_2 c_1^2$. Therefore, we have

$$
a_2 = \frac{1}{4} B_2 c_1,
$$

$$
a_3 = \frac{1}{12} B_1 c_2 + \frac{1}{24}(B_2 + B_1^2 - B_1) c_1^2
$$

and

$$a_3 - \mu a_2^2 = \frac{1}{12} B_1 (c_2 - vc_1^2),$$

where

$$v = \frac{1}{2}\left(\frac{3}{2}\mu B_1 + 1 - \frac{B_2}{B_1} - B_1\right).$$

By using Lemma 1, we obtain the desired result. We omit further details. $\blacksquare$

We mention that the sharp upper bound on the coefficient functional $|a_3 - \mu a_2^2|$ over $S^*(\varphi)$ is an immediate consequence of Theorem 3 and the correspondence between $C(\varphi)$ and $S^*(\varphi)$. Also, by expressing the first two coefficients of inverse functions of functions in $C(\varphi)$ or $S^*(\varphi)$ in terms of coefficients of $p_1 \in P$, we can give the sharp upper bounds on the first two coefficients of inverses of functions in these two classes. We omit the details here.

Next, we investigate the sharp order of growth for the coefficients of functions in $C(\varphi)$ when $\varphi \in H^2$, the Hardy class of analytic functions in $\mathbf{D}$.

Theorem 4: *Let* $\varphi(z) = 1 + B_1 z + B_2 z^2 + \cdots \in H^2$. *For* $f(z) = z + a_2 z^2 + a_3 z^3 + \cdots \in C(\varphi)$, *we have the sharp order of growth* $|a_n| = O(1/n^2)$.

Proof: First, note that

$$k_{\varphi n}(z) = z + \frac{1}{n(n-1)} B_1 z^n + \cdots.$$

This implies the lower bound

$$\max\{|a_n| : f \in C(\varphi)\} \geq \frac{1}{n(n-1)} B_1.$$

Hence the order is best possible.

We now show that there exists a constant M such that $n^2(n-1)^2 |a_n|^2 \leq M^2$. Let $p(z) = 1 + zf''(z)/f'(z) = 1 + b_1 z + b_2 z^2 + \cdots$ and $k_\varphi(z) = z + A_2 z^2 + A_3 z^3 + \cdots$. Because $p \prec \varphi$ and $f' \prec k'_\varphi$, we have

$$\sum_{k=1}^{n-1} |b_k|^2 \leq \sum_{k=1}^{n-1} |B_k|^2$$

and

$$1 + \sum_{k=2}^{n-1} k^2 |a_k|^2 \leq 1 + \sum_{k=2}^{n-1} k^2 |A_k|^2,$$

by Rogosinski's result (*[12]*, see also *[2]*(p. 192)). From

$$n(n-1)a_n = \sum_{k=1}^{n-1} k a_k b_{n-k},$$

we get

$$n^2(n-1)^2|a_n|^2 \leq \left(\sum_{k=1}^{n-1}|b_k|^2\right)\left(1+\sum_{k=2}^{n-1}k^2|a_k|^2\right)$$
$$\leq \left(\sum_{k=2}^{n-1}|B_k|^2\right)\left(1+\sum_{k=2}^{n-1}k^2|A_k|^2\right).$$

Recall that $\varphi \in H^2$ if and only if

$$\sum_{k=1}^{\infty}|B_k|^2 < \infty.$$

Furthermore, observe that

$$1+\sum_{k=2}^{\infty}kA_k z^{k-1} = k'_\varphi(z) = \exp\left\{\sum_{k=1}^{\infty}k^{-1}B_k z^k\right\}.$$

From an inequality of Milin (see *[11]*(p. 81)) it follows that

$$1+\sum_{k=2}^{\infty}k^2|A_k|^2 \leq \exp\left\{\sum_{k=1}^{\infty}k^{-1}|B_k|^2\right\} < \infty.$$

The proof of Theorem 4 is complete if we set

$$M^2 = \left(\sum_{k=1}^{\infty}|B_k|^2\right)\left(1+\sum_{k=2}^{\infty}k^2|A_k|^2\right).$$

It is interesting to see that no matter what extra condition φ satisfies in addition to belonging to H^2, as long as φ is not constant in $\mathbf{D}$, the above order of growth cannot be improved. It is proved in *[9]* that the function defined by (1) belongs to H^2. For $0 < \alpha < 1/2$, the function $\varphi_\alpha(z) = \left(\frac{1+z}{1-z}\right)^\alpha$ belongs to H^2 (see *[1]*(p. 13)). Moreover, $\varphi_{A,B}(z) = (1+Az)(1+Bz)$, $-1 < B < A \leq 1$, is bounded and so belongs to H^2. Note that any function subordinate to a function in H^2 also belongs to H^2.

Corollary 4': *Let $\varphi \in H^2$ and $f(z) = z + a_2 z^2 + a_3 z^3 + \cdots \in S^*(\varphi)$. Then we have the sharp order of growth $|a_n| = O(1/n)$.*

5 Convolution properties

In this section, we investigate convolution properties of functions in $C(\varphi)$ and $S^*(\varphi)$ in the special situation that $\varphi(\mathbf{D})$ is convex. This includes most of the special cases considered earlier by other authors. In order to prove our results, we need the following result due to Ruscheweyh and Sheil-Small (see *[15]*(Theorem 2.4, pp. 54-55) and *[16]*).

Lemma: *Suppose either $g \in C$, $h \in S^*$ or else $g, h \in S^*_{1/2}$. Then for any analytic function G in $\mathbf{D}$,*

$$\frac{g * (hG)(z)}{g * h(z)} \in \overline{co}G(\mathbf{D}), \quad g \in \mathbf{D},$$

where $\overline{co}G(\mathbf{D})$ is the closed convex hall of $G(\mathbf{D})$.

Theorem 5: *Let $\varphi(\mathbf{D})$ be convex, $g \in C$ and $f \in C(\varphi)$ $(S^*(\varphi))$. Then $g * f \in C(\varphi)$ $(S^*(\varphi))$.*

Proof: We only prove the case in which $f \in C(\varphi)$; the other case can be proved similarly. Note that $h(z) = zf'(z) \in S^*$. Let

$$G(z) = 1 + \frac{zf''(z)}{f'(z)}.$$

Then $G(\mathbf{D}) \subset \varphi(\mathbf{D})$. It is easy to see that

$$g * h(z) = z(g * f)'(z)$$

and

$$g * (hG)(z) = z(g * f)'(z) + z^2(g * f)''(z).$$

Since $\varphi(\mathbf{D})$ is convex, we have from the Lemma above that

$$1 + z(g * f)''(z)/(g * f)'(z) = g * (hG)(z)/(g * h)(z)$$

lies in the closure of $\varphi(\mathbf{D})$. As $1 + z(g * f)''(z)/(g * f)'(z)$ is holomorphic in $\mathbf{D}$, we see that

$$1 + z(g * f)''(z)/(g * f)'(z) \in \varphi(\mathbf{D}).$$

This tells us that $g * f \in C(\varphi)$.

When $f \in S^*(\varphi)$, we set $h = f$ and $G(z) = zf'(z)/f(z)$ and show the desired result similarly. This completes our proof of Theorem 5. ∎

When $\varphi(z) = (1+z)/(1-z)$, Theorem 5 was proved in *[16]*. If $Re\{\varphi(z)\} > 1/2$ $(z \in \mathbf{D})$, we can replace the condition $g \in C$ by $g \in S^*_{1/2}$. For $g \in S^*_{1/2}$ and $f \in S^*(\varphi)$, where φ is defined by (1), the result was given in *[14]*.

Theorem 6: *Let $\varphi(\mathbf{D})$ be convex and $Re\{\varphi(z)\} > 1/2$ $(z \in \mathbf{D})$. If*

$$f(z) = \sum_{n=1}^{\infty} a_n z^n, \quad g(z) = \sum_{n=1}^{\infty} b_n z^n \in C(\varphi),$$

then

$$(zf') * g(z) = \sum_{n=1}^{\infty} n a_n b_n z^n \in C(\varphi).$$

Proof: Let $F_1(z) = zf'(z)$ and $G_1(z) = zg'(z)$. Then $F_1, G_1 \in S^*(\varphi)$. In particular, since $Re\{\varphi(z)\} > 1/2$, we see that $F_1, G_1 \in S^*_{1/2}$. Define

$$G(z) = 1 + \frac{zg''(z)}{g'(z)}.$$

This time we use the second condition of the above lemma and the rest of the proof is similar to the proof of Theorem 5. This completes the proof of Theorem 6. ∎

References

[1]   P. Duren, Theory of H^p Spaces, Academic Press, New York, 1970.

[2]   P. Duren, Univalent Functions, Grundlehren der Math. Wissenschaften, vol. 259, Springer-Verlag, New York, 1983.

[3]   A.W. Goodman, On uniformly convex functions, Ann. Polon. Math., 56(1991), 87-92.

[4]   A.W. Goodman, On uniformly starlike functions, J. Math. Anal. Appl., 155(1991), 364-370.

[5]   W. Hengartner and G. Schober, On schlicht mappings to domains convex in one direction, Comment. Math. Helv. 45(1970), 303-314.

[6]   M. Jahangiri, H. Silverman and E.M. Silvia, Classes of functions defined by subordination, New Trends in Geometric Function Theory and Applications, ed. by R. Parvatham and S. Ponnusamy, World Scientific, Singapore, 1991, pp. 34-41.

[7]   A.E. Livingston, The coefficients of multivalent close-to-convex functions, Proc. Amer. Math. Soc., 21(1969), 545-552.

[8]   W. Ma and D. Minda, An internal geometric characterization of strongly starlike functions, Ann. Univ. Mariae Curie-Sklodowska Sect. A, 45 (1991), 89-97.

[9]   W. Ma and D. Minda, Uniformly convex functions, Ann. Polon. Math., 57 (1992), 165-175.

[10]   W. Ma and D. Minda, Uniformly convex functions II, Ann. Polon. Math. 58 (1993), 275-285.

[11]   Ch. Pommerenke, Univalent Functions, Vandenhoeck and Ruprecht, Göttingen, 1975.

[12]   W.W. Rogosinski, On the coefficients of subordinate functions, Proc. London Math. Soc., 48(1943), 48-82.

[13]   F. Ronning, Uniformly convex functions and a corresponding class of starlike functions, Proc. Amer. Math. Soc., 118 (1993), 189-196.

[14]   F. Ronning, On starlike functions associated with parabolic regions, Ann. Univ. Mariae Curie-Sklodowska Sect. A, 45(1991), 117-122.

[15]   St. Ruscheweyh, Convolutions in Geometric Function Theory, Les Presses de l'Université de Montréal, Montréal, Canada, 1982.

[16] St. Ruscheweyh and T. Sheil-Small, Hadamard products of schlicht functions and the Pólya-Schoenberg conjecture, Comment. Math. Helv., 48(1973), 119-135.

[17] T.J. Suffridge, Some remarks on convex maps of the unit disk, Duke Math. J., 37(1970), 755-777.

Holomorphic Dynamics of a Topologically Complete Family

Weiyuan Qiu[1]

Department of Mathematics, Fudan University
Shanghai 200433, P. R. China

[1]Project supported by Chinese NNSF.

Abstract: Let $\mathcal{M} = \{f(z) = z^n \exp(P(z) + Q(1/z))|n \in \mathbf{Z}$, P and Q are polynomials of degree p and $q\}$. We discuss the J-stability and the structural stability of $f \in \mathcal{M}$ and prove that the set of structurally stable functions in $\mathcal{M}$ is an open dense subset of $\mathcal{M}$.

1 Introduction

This paper is concerned with the study on the holomorphic dynamics of the family of functions defined on the punctured plane $\mathbf{C}^* = \mathbf{C} \setminus \{0\}$

$$\mathcal{M} = \mathcal{M}_{n,p,q} = \{ \ f(z) = z^n \exp(P(z) + Q(1/z))|n \in \mathbf{Z},$$

$$P \text{ and } Q \text{ are polynomials of degree } p \text{ and } q\}.$$

L. Keen [6] proved that for fixed n, p, q, the family $\mathcal{M}$ is topologically complete in the sense that any holomorphic function on $\mathbf{C}^*$ which is topologically conjugate to a function in $\mathcal{M}$ is also conformally conjugate to a member of $\mathcal{M}$. It follows that we can consider the structure stability with in the family $\mathcal{M}$ and $\mathcal{M}$ can be parameterized by the coefficients of polynomials P and Q. We give the topology on $\mathcal{M}$ the induced topology of $\mathbf{C}^{p+q+1}$. This topology is equivalent to the compact-open topology. The dynamical decomposition of a topologically complete family of functions were discussed by many authors [3, 10, 9]. For example, the quadratic family is relative to the famous Mandelbrot set [3]. Mañé, Sad and Sullivan [10] and Lyubich [9] studied the stability of rational functions with degree $d \geq 2$. Their theory can be developed to study many other families of functions. Eremenko and Lyubich [4] studied the analogue problem for the family of entire functions in which every function has only finite many singularities and any two functions are topologically equivalent. The purpose of this paper is to study the stability of functions in the family $\mathcal{M}$. We prove that the set of structurally stable functions in $\mathcal{M}$ is an open dense subset of $\mathcal{M}$. The proof is based on the study on the asymptotic behaviour at essential singularities 0 and ∞ of $f \in \mathcal{M}$.

Relating to $f(z) = z^n \exp(P(z) + Q(1/z)) \in \mathcal{M}$, there is an entire function $F(z) = nz + P(e^z) + Q(e^{-z})$ which is semi-conjugate to f by the project map $\pi : \mathbf{C} \to \mathbf{C}^*, \pi(z) = e^z$. The result on stability are also true for $F(z)$.

Proceedings of the International Conference on Complex
Analysis at the Nankai Institute of Mathematics, 1992 pp. 170-179
©INTERNATIONAL PRESS 1994

2   Asymptotic Behaviour at Essential Singularities

Let n, p, q be integers and $p, q \geq 1$. Denote by $\mathcal{M} = \mathcal{M}_{n,p,q} = \{f : \mathbf{C}^* \to \mathbf{C}^* | f(z) = z^n \exp(P(z) + Q(1/z))$, where P and Q are polynomials of degree p and $q\}$. Then $\mathcal{M}$ can be parameterized by coefficients of polynomials P and Q. Denote by $P = P_{\mathbf{a}}(z) = a_p z^p + \cdots + a_1 z + a_0$ $(a_p \neq 0)$ and $Q = Q_{\mathbf{b}}(z) = b_q z^q + \cdots + b_1 z$ $(b_q \neq 0)$, where the indices $\mathbf{a} = (a_p, \cdots, a_1, a_0)$, $\mathbf{b} = (b_q, \cdots, b_1)$. Therefore, $f \in \mathcal{M}$ can be denoted by $f_{\mathbf{c}}$ with index $\mathbf{c} = (\mathbf{a}, \mathbf{b}) = (a_p, \cdots, a_0, b_q, \cdots, b_1)$

Let $f_{\mathbf{c}} \in \mathcal{M}$.

$$f_{\mathbf{c}}(z) = z^n \exp(P_{\mathbf{a}}(z) + Q_{\mathbf{b}}(1/z)), \qquad (1)$$

Then 0 and ∞ are essential singularities of f. We consider the asymptotic behaviour of f near the essential singularities. At first, $f_{\mathbf{c}}$ can be written as

$$f_{\mathbf{c}}(z) = z^n \exp(a_p z^p (1 + o(1))) \qquad \text{as } z \to \infty \qquad (2)$$

and

$$f_{\mathbf{c}}(z) = z^n \exp(b_q z^{-q} (1 + o(1))) \qquad \text{as } z \to 0 \qquad (3)$$

Let $D_1(\infty) = \{z \in \mathbf{C}^* | |z| > 1\}$ and $D_1(0) = \{z \in \mathbf{C}^* | |z| < 1\}$. Divide $D_1(\infty)$ into sectors $V_0, V_1, \ldots, V_{2p-1}$ by the rays $\{z \in D_1(\infty) | \mathrm{Re} a_p z^p = 0\}$:

$$V_l = \{z \in D_1(\infty) | \frac{l\pi}{p} < \arg z + \frac{\arg a_p}{p} < \frac{(l+1)\pi}{p}\} \qquad (4)$$

$l = 0, 1, \ldots, 2p-1$. From (2), $f_{\mathbf{c}}(z)$ tends to infinity along each ray in V_{2j} and $f_{\mathbf{c}}(z)$ tends to 0 along each ray in V_{2j+1}. Similarly, $D_1(0)$ can be divided into sectors $U_0, U_1, \ldots, U_{2q-1}$,

$$U_l = \{z \in D_1(0) | \frac{l\pi}{q} < \arg z + \frac{\arg b_q}{q} < \frac{(l+1)\pi}{q}\} \qquad (5)$$

$l = 0, 1, \ldots, 2q-1$. $f_{\mathbf{c}}(z)$ tends to 0 along rays in U_{2j} and tends to ∞ along rays in U_{2j+1}.

Recall that a point a is an asymptotic value of a function $f : \mathbf{C}^* \to \mathbf{C}^*$ if there exists a curve $\gamma(t)$ $(0 \leq t < \infty)$ which tends to ∞ or 0 as $t \to \infty$ such that $f(\gamma(t))$ tends to a as $t \to \infty$. Each $f_{\mathbf{c}} \in \mathcal{M}$ has exactly two asymptotic values 0 and ∞. All other singularities of $f_{\mathbf{c}}$ in $\mathbf{C}^*$ are algebraic singularities (critical values). And the number of critical points is not more than $p + q$, so is of critical values.

A domain D is called bounded if it is bounded away from both 0 and ∞. Otherwise it is called unbounded. If D is bounded away from ∞ but not from 0, call it unbounded at 0. If it is bounded away from 0 but not from ∞, call it unbounded at ∞.

Now define the exponential tracts of $f_{\mathbf{c}}$ as follows:

Let $N_R(\infty) = \{z \in \mathbf{C}^* | |z| > R\}$ and $N_r(0) = \{z \in \mathbf{C}^* | |z| < r\}$ be the neighborhoods of ∞ and 0 respectively. Take R sufficiently large and r sufficiently small such that both of $N_R(\infty)$ and $N_r(0)$ contain no critical values.

Consider the components of $f_{\mathbf{c}}^{-1}(N_R(\infty))$ and $f_{\mathbf{c}}^{-1}(N_r(0))$. Note that 0 and ∞ are asymptotic values and also essential singularities, each component T is an unbounded simply connected domain (at 0 or at ∞) with smooth boundary of both ends tend to 0 or ∞, and $f_{\mathbf{c}} : T \to N_R(\infty)$ (or $f_{\mathbf{c}} : T \to N_r(0)$) is a universal covering. We call such a T the exponential tract of $f_{\mathbf{c}}$.

Let T be a component of $f_{\mathbf{c}}^{-1}(N_R(\infty))$. If T is unbounded at ∞, then by (2), T is almost contained in a sector V_{2j} for some j when R is sufficiently large, and the boundary of T is asymptotic to the boundary of V_{2j} at ∞ because $f_{\mathbf{c}}(z)$ tends to zero along any ray in V_{2j-1} and in V_{2j+1}. If T is unbounded at 0, then by (3), T is almost contained in U_{2j-1} for some j for sufficiently small r, and the boundary of T is tangent to the boundary of U_{2j-1} at 0. Moreover, the growth of $f_{\mathbf{c}}$ in T is of exponential order.

The discussion of components of $f_{\mathbf{c}}^{-1}(N_r(0))$ is similar as of $f_{\mathbf{c}}^{-1}(N_R(\infty))$. We summarize the statement above as follows

Lemma 1: *$f_{\mathbf{c}} \in \mathcal{M}_{n,p,q}$ has exactly $2(p+q)$ exponential tracts (which are unbounded components of $f_{\mathbf{c}}^{-1}(N_R(\infty))$ and $f_{\mathbf{c}}^{-1}(N_r(0))$). p of them $T_1, T_2, \ldots, T_{2p}$ are unbounded at ∞ and other q of them $T_1', T_2', \ldots, T_{2q}'$ are unbounded at 0. They correspond alternatively to the components of $f_{\mathbf{c}}^{-1}(N_R(\infty))$ and $f_{\mathbf{c}}^{-1}(N_r(0))$. Each T_j $(j = 1, 2, \ldots, 2p)$ is almost contained in V_j which is defined in (4) and each T_j' $(j = 1, 2, \ldots, 2q)$ is nearly closed to U_j which is defined in (5). Furthermore, the exponential tracts are analytically dependent on the parameter* **c**.

We also denote T_j and T_j' by $T_j(\mathbf{c})$ and $T_j'(\mathbf{c})$ respectively.

3   Periodic Points And Critical Points

Let us consider $\mathcal{M} = \mathcal{M}_{n,p,q}$ as a submanifold of $\mathbf{C}^{p+q+1}$. Both notations $f_{\mathbf{c}} \in \mathcal{M}$ and $c \in \mathcal{M}$ are used.

We don't define again the dynamical terms here. All of them can be found, for example, in *[2, 4]*.

Consider periodic points of period d of the function $f_{\mathbf{c}} \in \mathcal{M}$. They are defined by the equation

$$f_{\mathbf{c}}^d(z) = z. \qquad (6)$$

The solution $z = z(\mathbf{c})$ of (6) is a multi-valued analytic function.

Denote N_d the set of algebraic singularities of $z(\mathbf{c})$, i.e.

$$N_d = \{\mathbf{c} \in \mathcal{M} | (f_{\mathbf{c}}^d)'(z(\mathbf{c})) = 1\}. \qquad (7)$$

It is well-known that N_d is an analytic subset of $\mathcal{M}$ and nowhere dense in $\mathcal{M}$.

Take a point $\mathbf{c}_0 \in \mathcal{M} \backslash N_d$. Then $(f_{\mathbf{c}_0}^d)'(z(\mathbf{c}_0)) \neq 1$. By the Implicit Function Theorem, every branch of $z(\mathbf{c})$ is single-valued in a small neighborhood of $\mathbf{c}_0$. We say that every branch of $z(\mathbf{c})$ can be analytically continued to any given simply connected neighborhood of $\mathbf{c}_0$ in $\mathcal{M} \backslash N_d$.

Lemma 2: *Let $c_0 \in \mathcal{M} \setminus N_d$. Given a neighborhood $D \subset \mathcal{M} \setminus N_d$ of c_0, then every branch $z_i(c)$ of $z(c)$ at c_0 can be analytically continued to the whole neighborhood D.*

Proof: Let $\gamma(t) = c_t$ $(0 \le t \le 1)$ be a curve in D, $\gamma(0) = c_0$. It is sufficient to prove that if $z_i(c)$ can be analytically continued along $\gamma(t)$ for $0 \le t < 1$, then $z_i(c)$ can be continued to $\gamma(1) = c_1$. Let $z_i(t) = z_i(c_t)$ and $f_t = f_{c_t}$. Two cases are possible:

1. There exists a sequence $t_n \to 1$ such that $z_i(t_n)$ tends to a point $z_i \in \mathbf{C}^*$ as $n \to \infty$. Then $f_{c_1}^d(z_i) = z_i$. Since $c_1 \in \mathcal{M} \setminus N_d$, $(f_{c_1}^d)'(z_i) \ne 1$. $z_i(c)$ can be continued to c_1 by the Implicit Function Theorem.

2. $z_i(t)$ tends to ∞ or 0 along γ as $t \to 1$. We say that this is impossible. We prove it only for the case $z_i(t) \to \infty$ as $t \to 1$.

Let $z_i^k(t) = f_t^k(z_i(t))$, $0 \le t < 1$, $0 \le k \le d$. Since $z_i^d(t) = z_i^0(t) = z_i(t) \to \infty$ as $t \to 1$ and $f_t(z_i^k(t)) = z_i^{k+1}(t)$, $k = 0, \ldots, d-1$, we deduce that $z_i^{d-1}(t)$, and then $z_i^k(t)$ $(0 \le k \le d-1)$, tends to ∞ or 0. Suppose that $z_i^{k+1}(t) \to \infty$ as $t \to 1$. For t closed to 1, $z_i^{k+1}(t) \in N_R(\infty) = \{z \in \mathbf{C}^* \,|\, |z| > R\}$, where R is sufficiently large. By Lemma 1,

$$z_i^k(t) \in T_{2j}(c_t) \quad \text{for some } 0 \le j < p, \text{ if } z_i^k(t) \to \infty \quad (t \to 1)$$

and

$$z_i^k(t) \in T'_{2j+1}(c_t) \quad \text{for some } 0 \le j < q, \text{ if } z_i^k(t) \to 0 \quad (t \to 1).$$

Suppose that $z_i^{k+1}(t) \to 0$ as $t \to 1$. For t closed to 1,

$$z_i^k(t) \in T_{2j+1}(c_t) \quad \text{for some } 0 \le j < p, \text{ if } z_i^k(t) \to \infty \ (t \to 1)$$

and

$$z_i^k(t) \in T'_{2j}(c_t) \qquad \text{for some } 0 \le j < q, \text{ if } z_i^k(t) \to 0 \ (t \to 1).$$

Since each exponential tract of f_c is almost contained in some sector of form V_l or U_l (see §2 for definition) and is continuously dependent on the parameter c by Lemma 1, the argument of $z_i^k(t)$ is bounded with bound independent of t closed to 1, where $0 \le k \le d-1$.

Suppose we denote $f_t(z) = z^n \exp(P_t(z) + Q_t(1/z))$. Consider

$$z_i^{k+1}(t) = (z_i^k(t))^n \exp\left(P_t(z_i^k(t)) + Q_t\left(\frac{1}{z_i^k(t)}\right)\right).$$

Take a branch of logarithm function

$$\log z_i^{k+1}(t) = n \log z_i^k(t) + P_t(z_i^k(t)) + Q_t\left(\frac{1}{z_i^k(t)}\right).$$

If $z_i^k(t) \to \infty$ as $t \to 1$,

$$|\log z_i^{k+1}(t)| \ge |P_t(z_i^k(t))| - n|\log z_i^k(t)| + O(1).$$

Since $\arg z_i^{k+1}(t)$ and $\arg z_i^k(t)$ are bounded, for t closed to 1, we have

$$|\log|z_i^{k+1}(t)|| \geq |P_t(z_i^k(t))| - n\log|z_i^k(t)| + O(1)$$
$$\geq \frac{a_p}{2}|z_i^k(t)| > 2|\log|z_i^k(t)||$$

where a_p is the coefficient of the highest term of $P_1(z)$. Similarly, if $z_i^k(t) \to 0$ as $t \to 1$, for t closed to 1, we have

$$|\log|z_i^{k+1}(t)|| \geq |Q_t(\frac{1}{z_i^k(t)})| + n\log|z_i^k(t)| + O(1)$$
$$\geq \frac{b_q}{2}\frac{1}{|z_i^k(t)|} > 2\log\frac{1}{|z_i^k(t)|} = 2|\log|z_i^k(t)||$$

where b_q is the coefficient of the highest term of $Q_1(z)$. It follows that

$$|\log|z_i(t)|| = |\log|z_i^d(z)|| \geq 2|\log|z_i^{d-1}(t)|| \geq \cdots \geq 2^d|\log|z_i(t)||.$$

This is a contradiction. $\qquad\qquad$ QED. $\qquad\qquad$ ◼

Corollary: $z(\mathbf{c})$ has only algebraic singularities.

Let $\lambda(\mathbf{c}) = (f_{\mathbf{c}}^d)'(z(\mathbf{c}))$ the multiplier of the periodic point $z(\mathbf{c})$.

Lemma 3: All branches of $\lambda(\mathbf{c})$ are non-constant.

Proof: Fixed $f = f_c \in \mathcal{M}$, $f(z) = z^n \exp(P(z) + Q(1/z))$. Consider the subfamily $f_w = wf \in \mathcal{M}$, where $w \in \mathbf{C}^*$. Let $z(w)$ denote the solution of $(f_w^d)(z) = z$ and $\lambda(w) = (f_w^d)'(z(w))$. It is sufficient to prove that $\lambda(w)$ is non-constant.

Suppose $\lambda(w) = \lambda$. Since $\lambda(w)$ is finite for $w \in \mathbf{C}^*$ from Corollary of Lemma 2, $\lambda \neq \infty$.

Let $z^k(w) = f_w^k(z(w))$, $0 \leq k \leq d$. Suppose that for $w_j \to 0$, there exists k such that $z^k(w_j)$ is bounded away from 0 and ∞. For simplicity, we assume $z^k(w)$ is bounded in a neighborhood of 0. Then $z^k(w)/w$ is in the neighborhood of ∞ and $|z^k(w)/w|$ tends to infinity at the rate of $1/|w|$ when $|w| \to 0$. Since $z^k(w) = wf(z^{k-1}(w))$, $z^{k-1}(w)$ is in one of exponential tracts of f and $|z^{k-1}(w)|$ tends to infinity at the rate of $|\log|w||^{1/p}$ or tends zero at the rate of $|\log|w||^{-1/q}$, where p and q are the degree of P and Q respectively. Hence, $|z^{k-1}(w)/w|$ is also tends to ∞ at the rate not faster then $1/|w|^2$. Note that $z^0(w) = z^d(w)$, it follows that $z^k(w) \to 0$ or ∞ for all $0 \leq k \leq d$ by induction. This is a contradiction.

Therefore, we can take a path w_t $(0 < t < 1)$ such that w_t tends to 0 as $t \to 0$ and $z^k(w_t)$ tends to 0 or ∞ for $0 \leq k \leq d$. It follows that $z^{k-1}(w_t)$ is in some exponential tract of f_{w_t}. Note that the exponential tract of f_w is closed to the exponential tract of f, the argument of $z^k(w_t)$ is bounded for all t near to 0.

Consider the covering $\pi : \mathbf{C} \to \mathbf{C}^*$, $\pi(z) = e^z$. Let $F(\zeta) = n\zeta + P(e^\zeta) + Q(e^{-\zeta})$ be one of lifts of f, where $\zeta = \log z$. Then the lift of f_w is $F_w(\zeta) = F(\zeta) + \log w$ and $F'_w(\zeta) = F'(\zeta)$. Fixed a branch of $\log$, let $\zeta(w) = \log z(w)$, $\zeta^k(w) = \log z^k(w)$ $(0 \le k \le d)$. Then $\zeta^k(w) = F_w^k(\zeta^{k-1}(w))$, $F_w^d(\zeta(w)) = \zeta(w)$. Hence

$$\lambda(w) = (f_w^d)'(z(w)) = (F_w^d)'(\zeta(w))$$

$$= \prod_{k=0}^{d-1} F'(\zeta^k(w))$$

$$= \prod_{k=0}^{d-1} \left(n\zeta^k(w) + P'(e^{\zeta^k(w)})e^{\zeta^k(w)} - Q'(e^{-\zeta^k(w)})e^{-\zeta^k(w)} \right).$$

Since $z^k(w_t) \to 0$ or ∞ and the argument of $z^k(w_t)$ is bounded, $|\mathrm{Re}\,\zeta^k(w_t)| \to \infty$ $(t \to 0)$ and $|\mathrm{Im}\,\zeta^k(w_t)|$ is bounded. Hence, every factor in the product above tends to infinity, so does $\lambda(w_t)$. This follows that $\lambda(w)$ is non-constant.QED. ∎

Let us consider critical points of $f_\mathbf{c} \in \mathcal{M}$. They are defined by the equation

$$(f_\mathbf{c})'(z) = 0. \qquad (8)$$

The solution $a(\mathbf{c})$ of (8) is a multi-valued analytic function on $\mathcal{M}$. Since the equation (8) is equivalent to the following algebraic equation

$$n + zP'_\mathbf{a}(z) - \frac{1}{z}Q'_\mathbf{b}(\frac{1}{z}) = 0 \qquad (9)$$

where $(\mathbf{a}, \mathbf{b}) = \mathbf{c}$.

Lemma 4: *The critical points $a(\mathbf{c})$ has only algebraic singularities.*

Proof: It follows immediately from (9) and that the coefficients a_p and b_q of $P_\mathbf{a}$ and $Q_\mathbf{b}$ are not zero. QED. ∎

Let S_0 denote the set of singularities of $a(\mathbf{c})$, then S_0 is an analytic subset of $\mathcal{M}$. It is closed and nowhere dense in $\mathcal{M}$. Therefore, if $\mathbf{c}_0 \in \mathcal{M} \setminus S_0$, $a(\mathbf{c})$ has at most $p + q$ single-valued analytical branches in a neighborhood of $\mathbf{c}_0$.

4 The Structural Stability

By using Mañé-Sad-Sullivan theory (*[10]* for rational functions and *[4]* for entire functions) and lemmas in §3, we prove here an analogue stability result for $f \in \mathcal{M}$.

The basis of the proof is the λ-lemma

Let M be a simply connected manifold, $\mathbf{c}_0 \in M$. A holomorphic motion of a set $X \subset \mathbf{C}$ over M is a map $\varphi : M \times X \to \mathbf{C}$ satisfying the following conditions:

a) The map $\mathbf{c} \to \varphi(\mathbf{c}, z)$ is analytic in M for every $z \in X$,

b) The map $\varphi_{\mathbf{c}} : z \to \varphi(\mathbf{c}, z)$ is injective for every $\mathbf{c} \in M$,

c) $\varphi_{\mathbf{c}_0} = id$.

Lemma: *(The λ-Lemma [10]) A holomorphic motion φ of a set X may be extended to a holomorphic motion of the closure $\bar{X}$. Furthermore, $\varphi_{\mathbf{c}} : \bar{X} \to \mathbf{C}$ is quasiconformal for any $\mathbf{c} \in M$.*

An extension of the λ-lemma shows that the holomorphic motion in fact may be extended to the whole complex plane *[12, 11]*.

Recall the dynamics of an individual holomorphic function f from $\mathbf{C}^*$ to itself. The Julia set of f is the set of points at which the iterate sequence $\{f^n\}$ is not normal. Denote by $J(f)$ the Julia set of f, $J(f)$ is non-empty and perfect and complete f-invariant. It coincides with the closure of repulsive periodic points of f. We refer to *[1, 5, 6]* for other dynamical properties of holomorphic functions from $\mathbf{C}^*$ to itself.

If $f_{\mathbf{c}} \in M = M_{n,p,q}$, then $f_{\mathbf{c}}$ has only finitely many singular points. Hence, all components of Fatou set of f are eventually periodic *[6]* and the periodic component can be classified into four types: (super)attractive basin, parabolic basin, Siegel disk and Herman ring ($f_{\mathbf{c}}$ has no essential parabolic basin *[8]*). Moreover, the number of attractive points $\leq p + q$.

A function $f_{\mathbf{c}_0} \in M$ is called J-stable if for all $\mathbf{c} \in M$ sufficiently closed to $\mathbf{c}_0$, there exists a homeomorphism $\varphi_{\mathbf{c}} : J(f_{\mathbf{c}_0}) \to J(f_{\mathbf{c}})$ depending continuously on $\mathbf{c}$ such that $f_{\mathbf{c}}|J(f_{\mathbf{c}}) = \varphi_{\mathbf{c}} \circ f_{\mathbf{c}_0}|J(f_{\mathbf{c}_0}) \circ \varphi_{\mathbf{c}}^{-1}$ (i.e. $f_{\mathbf{c}}$ is J-conjugate to $f_{\mathbf{c}_0}$). If $\varphi_{\mathbf{c}}$ above is a homeomorphism from $\mathbf{C}^* \to \mathbf{C}^*$ depending continuously on $\mathbf{c}$ such that $f_{\mathbf{c}} = \varphi_{\mathbf{c}} \circ f_{\mathbf{c}_0} \circ \varphi_{\mathbf{c}}^{-1}$, then we call $f_{\mathbf{c}_0}$ is structurally stable.

Consider the multi-valued analytic function $z_d = z_d(\mathbf{c}) : M \to \mathbf{C}^*$ defined by $f_{\mathbf{c}}^d(z_d(\mathbf{c})) = z_d(\mathbf{c})$ (see §3). Let N_d denote the set of algebraic singularities of $z_d(\mathbf{c})$ and put $N =$ the closure of $\bigcup_{d=1}^{\infty} N_d$ and $\mathcal{H} = M \setminus N$. The following theorem is an analog of the theorems of obtained in *[10]* for rational functions and in *[4]* for a class of entire functions.

Theorem 1: *$f \in M$ is J-stable if and only if $f \in \mathcal{H}$. The set $\mathcal{H}$ is open and dense in M.*

Proof: The proof is a modification of the proof of Theorem 9 in *[4]*. At first, we note that if $\mathbf{c} \in M$ then $\lambda_d(\mathbf{c}) = (f_{\mathbf{c}}^d)'(z_d(\mathbf{c})) = 1$.

Suppose $\mathbf{c}_0 \in \mathcal{H}$. Let $W \subset \mathcal{H}$ be a simply connected neighborhood of $\mathbf{c}_0$. By the Implicit Function Theorem, every branch $z_{d,i}(\mathbf{c})$ of $z_d(\mathbf{c})$ is well defined in a small neighborhood of $\mathbf{c}_0$ and by Lemma 2 it can be analytically continued to W. If $z_{d,i}(\mathbf{c}) = z_{d',j}(\mathbf{c})$ for some $\mathbf{c} \in W$, then $z_{d,i} \equiv z_{d',j}$. For otherwise, $\mathbf{c}$ is a singular point of $z_{dd'}$. Denote by $P_{\mathbf{c}}$ the set of periodic points of $f_{\mathbf{c}}$. Then the map $\varphi_{\mathbf{c}} : z_{d,i}(\mathbf{c}_0) \mapsto z_{d,i}(\mathbf{c})$ defines the holomorphic motion of $P_{\mathbf{c}_0}$ over W. By the λ-lemma this holomorphic motion may be extended to $\overline{P_{\mathbf{c}_0}}$ and conjugates $f_{\mathbf{c}_0}|\overline{P_{\mathbf{c}_0}}$ to $f_{\mathbf{c}}|\overline{P_{\mathbf{c}}}$. Since $J(f_{\mathbf{c}}) \subset \overline{P_{\mathbf{c}}}$, $J(f_{\mathbf{c}})$ is perfect and the attractive points are isolated, the map $\varphi_{\mathbf{c}}$ is them a conjugacy between $f_{\mathbf{c}_0}|J(f_{\mathbf{c}_0})$ and $f_{\mathbf{c}}|J(f_{\mathbf{c}})$ and the J-stability follows.

Now we show that $\mathcal{H}$ is dense in $\mathcal{M}$. Let $\mathbf{c}_0 \in N$. For any neighborhood W of $\mathbf{c}_0$, there exists $\mathbf{c}' \in W$ arbitrarily near to $\mathbf{c}_0$ such that $f_{\mathbf{c}'}$ has an periodic point $z_{d,i}(\mathbf{c}')$ with multiplier $\lambda_{d,i}(\mathbf{c}') = 1$. By Lemma 3, there exists $\mathbf{c}_1 \in N$ closed to $\mathbf{c}'$ such that the periodic point $z_{d,i}(\mathbf{c}_1)$ become attractive. Since attractive cycles are stable under perturbation, if $f_{\mathbf{c}_1}$ is also in N, the process can be repeated and the number of attractive cycles increases. Note that the number of attractive cycles is not more than $p + q$, the process breaks off after finite many steps. Therefore we find a $\mathbf{c} \in \mathcal{H}$ closed to $\mathbf{c}_0$.

On the other hand, J-stability implies that the Julia set is continuously depending on the parameter $\mathbf{c}$. Let $\mathbf{c}_0 \in N$, $\mathbf{c}'$ and $z_{d,i}(\mathbf{c}')$ as above. By Lemma 3, there exists $\mathbf{c}_2$ arbitrarily closed to $\mathbf{c}'$ such that $z_{d,i}(\mathbf{c}_2)$ is a centre of a Siegel disk. Therefore, $z_{d,i}(\mathbf{c}_2) \notin J(f_{\mathbf{c}_2})$. Again by Lemma 3, there exists $\mathbf{c}_3$ arbitrarily near to $\mathbf{c}_2$ such that the norm of multiplier of $z_{d,i}(\mathbf{c}_3)$ is greater than 1 and hence $z_{d,i}(\mathbf{c}_3) \in J(f_{\mathbf{c}_3})$. This contradicts the continuity of Julia sets. We complete the proof.   QED.   ∎

Let us consider the structural stability of $f_{\mathbf{c}} \in \mathcal{M}$. It is sufficient to consider $f_{\mathbf{c}} \in \mathcal{H} = \mathcal{M} \setminus N$. The structural stability of $f_{\mathbf{c}}$ is nearly relative to the orbit of critical points of $f_{\mathbf{c}}$. Note that every $f_{\mathbf{c}} \in \mathcal{H}$ has no rational neutral cycles by the definition of $\mathcal{H}$ and then $f_{\mathbf{c}} \in \mathcal{H}$ has no neutral cycles by Lemma 3. Therefore, according to the classification of periodic components of Fatou set, $f_{\mathbf{c}}$ has only (super)attractive periodic components and (or) a Herman ring ($f_{\mathbf{c}}$ has at most a Herman ring *[1]*).

Let S_0 denote the set of singularities of the multi-valued analytical function of critical points $a(\mathbf{c})$ (see §3). By Lemma 4 S_0 is closed and nowhere dense in $\mathcal{M}$. Therefore, we may consider the structural stability only for $f_{\mathbf{c}} \in \mathcal{H} \setminus S_0$. If $\mathbf{c}_0 \in \mathcal{H} \setminus S_0$, $a(\mathbf{c})$ has single-valued analytical branches $a_1(\mathbf{c}), \ldots, a_k(\mathbf{c})$ ($k \leq p + q$) in any simply connected neighborhood $W \subset \mathcal{H} \setminus S_0$ of $\mathbf{c}_0$. $a_j(\mathbf{c})$, $j = 1, 2, \ldots, k$, are all of critical points of $f_{\mathbf{c}}$.

Define equivalent relation by $x \sim y$ if and only if $x = y$ or x and y in the same leaf of the foliation of the Herman ring. We then define a subset $C \subset \mathcal{H} \setminus S_0$ by $\mathbf{c}_0 \in C$ if and only if there exists neighborhood $W_0 \subset \mathcal{H} \setminus S_0$ such that either $f_{\mathbf{c}}^m(a_j(\mathbf{c})) \sim f_{\mathbf{c}}^n(a_i(\mathbf{c}))$ for some $(m, j) \neq (n, i)$ and all $\mathbf{c} \in W_0$, or $f_{\mathbf{c}}^m(a_j(\mathbf{c})) \not\sim f_{\mathbf{c}}^n(a_i(\mathbf{c}))$ for all $\mathbf{c} \in W_0$, $(m, j) \neq (n, i)$, where $a_j(\mathbf{c})$ ($j = 1, 2, \ldots, k$) are the critical points of $f_{\mathbf{c}}$.

Theorem 2: *If $\mathbf{c} \in C$ then $f_{\mathbf{c}}$ is structurally stable. The set C is open and dense in $\mathcal{M}$.*

Proof: Since $\mathbf{c}_0 \in \mathcal{H}$, $f_{\mathbf{c}_0}$ is J-stable. Let $W \subset \mathcal{H} \setminus S_0$ be a neighborhood of $\mathbf{c}_0$. There exists homeomorphism $\varphi_{\mathbf{c}} : J(f_{\mathbf{c}_0}) \to J(f_{\mathbf{c}})$ conjugating $f_{\mathbf{c}_0}$ to $f_{\mathbf{c}} \in W$. The purpose is to extend $\varphi_{\mathbf{c}}$ to the Fatou set of $f_{\mathbf{c}_0}$. Suppose the critical points $a_1(\mathbf{c}_0), \ldots, a_l(\mathbf{c}_0)$ lie in the Fatou set of $f_{\mathbf{c}_0}$ and $a_{l+1}(\mathbf{c}_0), \ldots, a_k(\mathbf{c}_0)$ lie on the Julia set of $f_{\mathbf{c}_0}$. The J-stability of $f_{\mathbf{c}_0}$ ensures that $a_1(\mathbf{c}), \ldots, a_l(\mathbf{c}), a_{l+1}(\mathbf{c}), \ldots, a_k(\mathbf{c})$ have the same properties for all

$\mathbf{c} \in W$. Furthermore, if $f_{\mathbf{c}_0}^m(a_j(\mathbf{c}_0)) = f_{\mathbf{c}_0}^n(a_i(\mathbf{c}_0))$ for some $(m,j) \neq (n,i)$, $i, j \in [l+1, k]$, then $f_{\mathbf{c}}^m(a_j(\mathbf{c})) = f_{\mathbf{c}}^n(a_i(\mathbf{c}))$ for all $\mathbf{c} \in W$.

Let us show that C is dense in $\mathcal{M}$. It is sufficient to show that $\Lambda = W \setminus C$ is nowhere dense in W. If $f_{\mathbf{c}}$ has an attractive periodic point which is also a critical point, by Lemma 3, $\mathbf{c} \in \Lambda$ and such $\mathbf{c}$'s is nowhere dense in W. So we can assume that $f_{\mathbf{c}}$ has no superattractive periodic points.

Let $\Lambda(m,n,j,i) = \{\mathbf{c} \in W | f_{\mathbf{c}}^m(a_j(\mathbf{c})) \sim f_{\mathbf{c}}^n(a_i(\mathbf{c}))\}$. Recall $f_{\mathbf{c}}^m(a_j(\mathbf{c})) \sim f_{\mathbf{c}}^n(a_i(\mathbf{c}))$ means that either $f_{\mathbf{c}}^m(a_j(\mathbf{c})) = f_{\mathbf{c}}^n(a_i(\mathbf{c}))$ or $f_{\mathbf{c}}^m(a_j(\mathbf{c}))$ and $f_{\mathbf{c}}^n(a_i(\mathbf{c}))$ lies on the same leaf of the foliation of the Herman ring. For the later case, by Lemma III.6 in $[10]$, there exists homeomorphism $\varphi_{\mathbf{c}}$ dependent analytically on $\mathbf{c}$ such that $\varphi_{\mathbf{c}} \circ f_{\mathbf{c}}^m(a_j(\mathbf{c})) = e^{2\pi i \theta} \varphi_{\mathbf{c}} \circ f_{\mathbf{c}}^n(a_i(\mathbf{c}))$. Therefore, if $\Lambda(m,n,j,i) \neq \emptyset$ or $\neq W$, then $\Lambda(m,n,j,i)$ is a proper analytic subset of W and is closed and nowhere dense in W. $\Lambda =$ the closure of $\bigcup\{\Lambda(m,n,j,i) | \Lambda(m,n,j,i) \neq \emptyset$ or $\neq W, (m,j) \neq (n,i)\}$. By Lemma III.3 and Lemma III.6 in $[10]$, if $f_{\mathbf{c}}^m(a_j(\mathbf{c})) \sim f_{\mathbf{c}}^n(a_i(\mathbf{c}))$ for fixed j, i, then there exist minimal m_0 and n_0 such that $f_{\mathbf{c}}^{m_0+s}(a_j(\mathbf{c})) \sim f_{\mathbf{c}}^{n_0+s}(a_i(\mathbf{c}))$ for all $s \geq 0$ but $f_{\mathbf{c}}^{m_0+s}(a_j(\mathbf{c})) \not\sim f_{\mathbf{c}}^{n_0+t}(a_i(\mathbf{c}))$ if $s, t \geq 0$, $s \neq t$. This means that there exist M_0 and N_0 such that $\Lambda = \{\Lambda(m,n,j,i) | \Lambda(m,n,j,i) \neq \emptyset$ or $\neq W, (m,j) \neq (n,i), m \leq M_0, n \leq N_0\}$. Therefore, Λ is nowhere dense in W.

Let us show that if $\mathbf{c}_0 \in C$, then $f_{\mathbf{c}_0}$ is structurally stable. Let V_0 be the union of the $f_{\mathbf{c}_0}$-invariant subdomains of the attractive basins and the Herman ring of $f_{\mathbf{c}_0}$ defined in Lemma III.3 and Lemma III.6 in $[10]$, let W_0 be a simply connected neighborhood of $\mathbf{c}_0$ which satisfies the conditions of Lemma III.3 and Lemma III.6 in $[10]$. Using Lemma IV.1 in $[10]$, we can construct a holomorphic motion $\varphi_{\mathbf{c}} : V_0 \to V_{\mathbf{c}} = \varphi_{\mathbf{c}}(V_0)$ over W_0 such that

(a) $\varphi_{\mathbf{c}}$ maps $f_{\mathbf{c}_0}^m(a_j(\mathbf{c}_0))$ to $f_{\mathbf{c}}^m(a_j(\mathbf{c}))$, $m \geq 0, 1 \leq j \leq l$,

(b) $\varphi_{\mathbf{c}}$ conjugates $f_{\mathbf{c}_0}|V_0$ to $f_{\mathbf{c}}|V_{\mathbf{c}}$.

We now extend $\varphi_{\mathbf{c}}$ to $V = \bigcup_{n \geq 0} f_{\mathbf{c}_0}^{-n}(V_0)$. Let $\hat{V}_0$ be V_0 minus the intersection of forward orbits of critical points of $f_{\mathbf{c}_0}$ with V_0. Set $\hat{V} = \bigcup_{n \geq 0} f_{\mathbf{c}_0}^{-n}(\hat{V}_0)$, Then $\hat{V}$ is dense in V.

Let $z \in f_{\mathbf{c}_0}^{-n}(\hat{V}_0)$. Consider the functional equation

$$f_{\mathbf{c}}^n(\psi_z(\mathbf{c})) = \varphi_{\mathbf{c}}(f_{\mathbf{c}_0}^n(z)), \qquad \psi_z(\mathbf{c}_0) = z. \qquad (10)$$

By the Implicit Function Theorem and the fact that $(f_{\mathbf{c}_0}^n)'(z) \neq 0$, it has an analytic solution $\zeta = \psi_z(\mathbf{c})$ in a neighborhood of $\mathbf{c}_0$. We claim that ψ_z may be analytically extended to the whole domain W_0.

Let $\mathbf{c}_t$ $(0 \leq t \leq 1)$ be a path in W_0 such that ψ_z can be analytically continued along the path $\mathbf{c}_t$ $(0 \leq t < 1)$. Set $t \to 1$. Then the right hand of (10) tends to $\varphi_{\mathbf{c}_1}(f_{\mathbf{c}_0}^n(z)) \in \mathbf{C}^*$. Since $f_{\mathbf{c}}$ has only algebraic singularities in $\mathbf{C}^*$ and the possible asymptotic values of $f_{\mathbf{c}}$ are 0 and ∞, by (10), $\psi_z(\mathbf{c}_t)$ can not tend to 0 or ∞ as $t \to 1$. Therefore, we have $\psi_z(\mathbf{c}_1) \in \mathbf{C}^*$, If $(f_{\mathbf{c}_1}^n)'(\psi_z(\mathbf{c}_1)) \neq 0$ then by the Implicit Function Theorem, ψ_z can be analytically continued to $\mathbf{c}_1$. The claim is proved. If $(f_{\mathbf{c}_1}^n)'(\psi_z(\mathbf{c}_1)) = 0$, then $\psi_z(\mathbf{c}_1)$ is critical point of $f_{\mathbf{c}_1}^n$ and $\varphi_{\mathbf{c}_1}(f_{\mathbf{c}_0}^n(z))$ belongs to the critical orbit of $f_{\mathbf{c}_1}$. By (a), $f_{\mathbf{c}_0}^n(z)$ belongs to the critical orbit of $f_{\mathbf{c}_0}$ which contradicts that $z \in f_{\mathbf{c}_0}^{-n}(\hat{V}_0)$. We complete the proof of the claim.

Let $\varphi_{\mathbf{c}}(z) = \psi_z(\mathbf{c})$. we have extended $\varphi_{\mathbf{c}}$ to $\hat{V}$. If $f_{\mathbf{c}_0}$ has not Herman ring, then $\hat{\bar{V}} = \bar{V} = \mathbf{C}^*$. The application of λ-lemma completes the proof of the theorem. If $f_{\mathbf{c}_0}$ has a Herman ring A, since we can take the $f_{\mathbf{c}_0}$-invariant domain in A arbitrarily filled in A, we complete the proof of the theorem by applying the extension of λ-lemma *[12]*.

QED.

Acknowledgement: The author would like to thank professor Fuyao Ren for his encouragement.

References

[1] I.N.Baker, Wandering domains for maps of the punctured plane, *Ann. Acad. Sci. Fenn. Ser.A. I.Math.*, **12**(1987), 191-198.

[2] P.Blanchard, Complex analytic dynamics on the Riemann sphere, *Bull. Amer. Math. Soc.*, **11**(1984), 85-141.

[3] A.Douady and J.H.Hubbard, Étude dynamique des polynômes complexes, Publ. Math. d'Orsay, 84-2, 85-4.

[4] A.E.Eremenko and M.Yu.Lyubich, Dynamical properties of some classes of entire functions, Preprint of Institute for Math. Sciences, SUNY, 1990/4.

[5] Liping Fang, Complex analytic dynamics on $\mathbf{C}^*$, (in Chinese) *Acta Math. Sinica,* **34**(1991), 611-621.

[6] L.Keen, Topology and growth of a special class of holomorphic self-maps of $\mathbf{C}^*$, *Erg. Th. and Dyn. Sys.*, **9**(1989), 321-328, ·

[7] L.Keen Dynamics of holomorphic maps of $\mathbf{C}^*$, in Holomorphic Functions and Moduli I, ed. D.Drasin, Speringer-Verlag, New York, 1988.

[8] J.Kotus, Iterated holomorphic maps on the punctured plane, Preprint of Institute of Math. Polish Academy of Sciences, Warsaw, 1986.

[9] M.Yu.Lyubich, Some typical properties of the dynamics of rational maps, *Russian Math. Surveys,* **38**(1983), 154-155.

[10] R.Mañé, P.Sad and D.Sullivan, On the dynamics of rational maps, *Ann. Scient. Ecole Norm. Sup.*, **16**(1983), 193-217.

[11] Z.Slodkowski, Holomorphic motions and polynomial hulls, *Proc. Amer. Math. Soc.*, **111**(1991).

[12] D.Sullivan and W.Thurston, Extending holomorphic motions, *Acta Math.*

keywords: Holomorhpic dynamics, Topologically complete family, Structural stability, Julia set.

1991 Mathematical Subject Classifications: Primary 58F23, 58F08.

Sums of squares of real entire functions of one complex variable[1]

Lee A. Rubel

Department of Mathematics, University of Illinois
1409 West Green Street, Urbana, Illinois 61801, U.S.A.

Dedicated to the fond memory of Mahlon M. Day

Abstract: It is proved that every non–negative entire function of exponential type is the sum of two squares of real entire functions of exponential type. An extension to more general rates of growth is discussed.

1 Introduction

The functions we study are entire functions f of one complex variable. To say that f is "real" is to say that $f(x)$ is real if x is real. To say that f is "non–negative" is to say that $f(x) \geq 0$ if x is real. It is well–known that every non–negative polynomial is the sum of two squares of real polynomials. Moreover, every non-negative entire function is the sum of two squares of real entire functions. We will, as a prelude, sketch proofs of these simple facts— long ago, I saw these proofs, but I can't recall where. (A different proof for the polynomial case can be found in *[7]* p. 82, #44, solution on p. 275. For related papers, see the references at the end of this paper.) The proofs are easily adapted to handle the case, say, where f has order $\leq \rho$; we can then, under the additional hypothesis that f is non–negative, write $f = g^2 + h^2$ where g and h are real entire functions of order $\leq \rho$. However, the proof breaks down if we suppose that, say, f is of exponential type ($|f(z)| \leq ae^{b|z|}$) and demand the same of g and h. By modifying the proof, though, we show that the theorem is true in this context. The difference comes from Lindelöf's theorem (see *[8]*) that a sequence (z_n) of non–zero complex numbers is the exact zero–sequence of an entire function of exponential type if and only if both

(i) $$\sum_{|z_n| \leq r} 1 = O(r) \text{ and}$$

(ii) $$\sum_{|z_n| \leq r} \frac{1}{z_n} = O(1).$$

Condition (ii) is a *balance* condition and has no counterpart in the earlier cases considered.

---
[1] This research was partially supported by a grant from the National Security Agency

Proceedings of the International Conference on Complex
Analysis at the Nankai Institute of Mathematics, 1992 pp. 180-187
©INTERNATIONAL PRESS 1994

This leads us to entire functions of finite λ–type. Here, $\lambda(r)$ is an arbitrary "growth function"—that is, $\lambda(r)$ is defined and positive on $0 < r < \infty$, non-decreasing, and unbounded. We say that an entire function f is of finite λ–type to mean that the exist positive constants A and B such that

$$T(r, f) \leq A\lambda(Br) \text{ for all } r$$

or equivalently, that

$$|f(z)| \leq \exp A\lambda(B|z|) \text{ for all } z \in \mathbb{C}.$$

In this paper, we ask the question "If f is a non–negative entire function of finite λ–type, must we have $f = g^2 + h^2$ for two real entire functions g, h of finite λ–type?" In this generality, the question seems very hard.

At the end of the paper, we explore some of the difficulties that arise in trying to answer this question. First, from *[8]*, there is the result that a sequence $\mathcal{Z} = (z_n)$ of non–zero complex numbers is the exact sequence of zeros of an entire function of finite λ–type if and only if there exist finite constants A and B such that

(i') $N(r) \leq A\lambda(Br)$ and

(ii') $\left| \displaystyle\sum_{r < |z_n| \leq s} \frac{1}{z_n^k} \right| \leq \dfrac{A\lambda(Br)}{r^k} + \dfrac{A\lambda(Bs)}{s^k}.$

for $k = 1, 2, 3...$

Here, $N(r) = \int_0^r n(t)/t \, dt$ where $n(t) = \sum_{|z_n| \leq t} 1$. This can be considered to be an extension of Lindelöf's theorem. We describe three difficulties that arise if we attempt to generalize our proof from the case of finite exponential type, and we are able to resolve one difficulty and to partially resolve another one. The first difficulty is resolved by a lemma of independent interest that if $\sum A_n z^n$ is an entire function of finite λ type and if $|a_n| \leq |A_n|$ for $n = 0, 1, 2, \ldots$, then $\sum a_n z^n$ must also be an entire function of finite λ type.

It seems extremely difficult to solve the main problem on the sum of squares in its full generality. We wish to thank Joseph Miles for his helpful comments on our proofs.

2 Polynomials, unrestricted entire functions, entire functions of given order and functions of exponential type

Suppose $p = p(z)$ is a non–negative polynomial. To see that we can write $p = q^2 + r^2$ where q, r are real polynomials, observe first that the real zeros of p all have even order. We may write

$$p(z) = \ell^2(z) \prod_{\text{Im} z_n > 0} \left(1 - \frac{z}{z_n}\right)\left(1 - \frac{z}{\bar{z}_n}\right),$$

where ℓ is a real polynomial with the *real* zeros of p as its zeros, and the z_n are the non–real zeros of p. Note that since p is real on $\mathbb{R}$, $p(z) = \bar{p}(\bar{z})$ for all z, so that the zeros of p occur in conjugate pairs. Also, the real zeros of p all have *even* order. Hence

$$p(z) = P(z)\bar{P}(\bar{z})$$

where

$$P(z) = \ell(z) \prod_{\mathrm{Im}_{z_n}>0} \left(1 - \frac{z}{z_n}\right)$$

Finally

$$p(z) = \frac{[P(z) + \bar{P}(\bar{z})]^2}{2} + \frac{[iP(z) - i\bar{P}(\bar{z})]^2}{2}$$

is a sum of two squares of real polynomials. The same argument, mutatis mutandis, works for *entire* functions of unrestricted growth or for entire functions of given order, by writing the function as a Weierstrass product, and then grouping the complex zeros in the upper half–plane into a separate product.

As mentioned in the Introduction, this proof breaks down for non–negative entire functions f of exponential type. For let

$$f(z) = \prod_{n=1}^{\infty} \left(1 - \frac{z}{n + ni}\right) \left(1 - \frac{z}{-n + ni}\right) \left(1 - \frac{z}{n - ni}\right) \left(1 - \frac{z}{-n - ni}\right).$$

It is easy to verify, say by Lindelöf's theorem, that f is an entire function of exponential type. (Actually, there is a simple formula for f in terms of the sine function.) But, with (formally)

$$Q(z) = \prod_{n=1}^{\infty} \left(1 - \frac{z}{n + ni}\right) \left(1 - \frac{z}{-n + ni}\right).$$

we have a function that, by Lindelöf's theorem, is not of exponential type—its zero set is not balanced.

Here is how we overcome this difficulty. We suppose that f is a non–negative entire function of exponential type, and we wish to write $f = g^2 + h^2$ where g and h are real entire functions of exponential type. We may write

$$f(z) = Ae^{\alpha z}z^{2k} \prod \left(1 - \frac{z}{z_n}\right) e^{z/z_n}.$$

It is easy to see that $A \geq 0$ and that α is real. As before, we see that the real zeros of f have even order.

To keep the argument simple, let us suppose that

$$f(z) = \prod \left(1 - \frac{z}{z_n}\right) e^{z/z_n}$$

where all the z_n are non–real. By Lindelöf's theorem, we have

(i) $$\sum_{|z_n|\leq r} 1 = O(r)$$

(ii) $\displaystyle\sum_{|z_n|\le r}\frac{1}{z_n}=O(1).$

(Note that this is $\sum\frac{1}{z_n}$ and *not* $\sum\frac{1}{|z_n|}$.) We want to write

$$f=(g+ih)(g-ih)$$

where g and h are real entire functions of exponential type.

The idea is that the zeros z_n occur in conjugate pairs, and if z_n^+ is a zero in the upper half–plane, with $z_n^-=\overline{z_n^+}$ in the lower half–plane, then we put into $G=g+ih$ the factor $\left(1-\frac{z}{w_n}\right)e^{z/w_n}$, where w_n is either z_n^+ or z_n^-. (In case z_n occurs with multiplicity $2p$ or $2p+1$ for p a strictly positive integer, then we put into $g+ih$ also the factor $\left[\left(1-\frac{z}{z_n}\right)\left(1-\frac{z}{\overline{z_n}}\right)\right]^p$.) We let G be the product of these terms. Then we let $H=g-ih$ be the product of what is left. We call them $g+ih$ and $g-ih$ because we have $f=GH$ where $G=\bar{H}$ on the real axis—just look at the Taylor coefficients a_n and b_n of G and H and break them into real and imaginary parts. (Here, we could use our later lemma, or the direct fact, very easy to prove, that if $\sum A_n z^n$ is an entire function of exponential type and if $|a_n|\le|A_n|$, Then $\sum a_n z^n$ is again an entire function of exponential type. Just use Cauchy's estimate $|A_n|\le\frac{Ae^{Br}}{r^n}$ and choose $r=n$ and so on.)

It is clear that G and H are entire functions of order ≤ 1, and that their zeros satisfy (i), i.e. they have finite upper density. To prove that, say G has exponential type, we must verify (ii) (since (i) is automatic.)

Here, we are using the simple corollary of Lindelöf's theorem that if (z_n) satisfies (i) and (ii), and if $G(z)=\prod\left(1-\frac{z}{z_n}\right)e^{z/z_n}$ is the canonical product over (z_n), then $G(z)$ is an entire function of exponential type. For, because of (i) and (ii), (z_n) must be the exact zero sequence of some entire function $\mathcal{G}(z)$ of exponential type. But then $G(z)$ and $\mathcal{G}(z)$ are two entire functions of order ≤ 1 with the same zero sequences, so $G(z)=e^{\alpha z}\mathcal{G}(z)$ for some complex number α, and the result follows.

On considering the real and the imaginary parts of our sum, (ii) becomes

$$\sum_{|w_n|\le r}\frac{x_n}{x_n^2+y_n^2}=O(1)\tag{1}$$

$$\overset{*}{\sum_{|w_n|\le r}}\frac{\pm|y_n|}{x_n^2+y_n^2}=O(1),\tag{2}$$

for suitable choices of $\pm$, depending on n. The asterisk on the sum in (2) indicates that multiple zeros contribute simply. Now (1) follows from

$$\sum_{|w_n|\le r}\frac{x_n}{x_n^2+y_n^2}=\frac{1}{2}\sum_{|z_n|\le r}\frac{x_n}{x_n^2+y_n^2}\tag{3}$$

and the last sum is $O(1)$ because f is of exponential type.

But in (2),

$$\frac{|y_n|}{x_n^2 + y_n^2} \to 0 \quad \text{as} \quad n \to \infty \tag{4}$$

and it is a fact from elementary calculus that we may choose the $\pm$ so that (2) holds; indeed, we may even make the series

$$\sum \frac{\pm |y_n|}{x_n^2 + y_n^2}$$

converge. Since it is clear from the construction that $\bar{G}(x) = H(x)$ for x real, the proof is complete.

3   A lemma on majorization of coefficients

We prove that if $F(z) = \sum A_n z^n$ is an entire function of finite λ–type and if $|a_n| \le |A_n|$ for $n = 0, 1, 2, \ldots$, then $f(z) = \sum a_n z^n$ is also an entire function of finite λ–type. This clearly follows from the following lemma.

Lemma 3.1: *If $F(z) = \sum A_n z^n$ is entire and if $f(z) = \sum a_n z^n$ with $|a_n| \le |A_n|$ for all n, then*

$$M_f(r) \le 2M_F(2r),$$

where, if g is an entire function,

$$M_g(r) = \max\{|g(z)| : |z| \le r\}.$$

Proof: We first prove it for $r = \frac{1}{2}$. The general result then follows by taking $f_1(z) = f(\rho z)$ and $F_1(z) = F(\rho z)$ and applying the case $r = \frac{1}{2}$ to f_1 and F_1. Now $a_n = A_n B_n$ where $|B_n| \le 1$. In other words, $f(z) = (F * B)(z)$ is the Hadamard convolution, with $B(z) = \sum B_n z^n$; on $\{|z| = \frac{1}{4}\}$,

$$f(z) = \frac{1}{2\pi i} \int_{|w|=\frac{1}{2}} F(w) B\left(\frac{z}{w}\right) \frac{dw}{w},$$

as is easily checked by looking at the power series. Hence

$$|f(z)| \le M_F\left(\frac{1}{2}\right) \frac{1}{2\pi} \int_{|w|=\frac{1}{2}} \left|B\left(\frac{z}{w}\right)\right| \left|\frac{dw}{w}\right|.$$

But $|z| = \frac{1}{4}$ and $|w| = \frac{1}{2}$ implies that $|z/w| = \frac{1}{2}$ so that

$$\left|B\left(\frac{z}{w}\right)\right| \le \sum |B_n| \left|\frac{z}{w}\right|^n \le \sum \left|\frac{z}{w}\right|^n = \frac{1}{1 - \left|\frac{z}{w}\right|} = 2.$$

Hence, on $|z| = \frac{1}{4}$,

$$|f(z)| \le 2M_F\left(\frac{1}{2}\right),$$

and the result is proved

4   Two more difficulties

In the general case, we must choose one zero from each conjugate pair of zeros, and still satisfy the condition (ii$'$) mentioned in the introduction. We now show how to do this for a *single* value of k. To do it for all k simultaneously seems beyond our reach. For simplicity of notation, we take $k = 1$, but for any one k, a simple modification does the job. We suppose that $\lambda(0+) > 0$. In the notation of our second section, we want

$$|S| = \left| \sum_{r < |z_n| \le s} \frac{\epsilon_n |y_n|}{|z_n^2|} \right| \le \frac{A\lambda(B_r)}{r} + \frac{A\lambda(B_s)}{s}.$$

We now choose the $\epsilon_n = \pm 1$ once and for all. Order the $|y_n|/|z_n^2|$ in decreasing order and then choose the ϵ_n to alternate ± 1 in that order. Now because $|y_n|/|z_n|^2 \le 1/|z_n|$, we have

$$S = \sum_{\substack{\frac{1}{s} \le \frac{|y_n|}{|z_n|^2} < \frac{1}{r}}} \frac{\epsilon_n |y_n|}{|z_n|^2} - \sum_{\substack{\frac{|y_n|}{|z_n|^2} < \frac{1}{r} \\ |z_n| \le r}} \frac{\epsilon_n |y_n|}{|z_n|^2} + \sum_{\substack{\frac{|y_n|}{|z_n|^2} < \frac{1}{s} \\ |z_n| \le s}} \frac{\epsilon_n |y_n|}{|z_n|^2}$$

$$= S_1 - S_2 + S_3.$$

Since S_1 is an alternating series of terms that decrease in modulus, we have, for a finite constant c,

$$|S_1| < \frac{c}{s} \le \frac{A\lambda(Bs)}{s}$$

since λ is increasing and $\lambda(0+) > 0$.

Also,

$$|S_2| \le \frac{1}{r} \sum_{|z_n| \le r} 1 \le \frac{A\lambda(Br)}{r}$$

by (i$'$) of the introduction, and likewise

$$|S_3| \le \frac{1}{s} \frac{A\lambda(Bs)}{s}$$

and the result is proved. We have no real idea how to choose the $\epsilon_n = \pm 1$ to work for more than one k at a time in (ii$'$).

Finally, we mistakenly thought that if $\lambda(r) = r^{k(r)}$, and $k(r)$ stays within a number $\delta < 1$ of a particular integer k_0, and if (z_n) satisfies (i$'$), then it must automatically satisfy (ii$'$) for every positive integer $k \ne k_0$. This was suggested to us by Lindelöf's theorem, but is just plain wrong, as the following example shows.

We construct a growth function $\lambda(t)$ with

$$t^{\frac{1}{2}}(1 + \log t) \le \lambda(t) \le t^{\frac{3}{2}}(1 + \log t)$$

for all large t and a real positive sequence (z_n) satisfying (i$'$)—indeed

$$\sum_{|z_n| \le r} 1 = [\lambda(r)]$$

—that fails the balance condition (ii$'$) for $k = 2$; i.e.

$$\sum_{r < z_n \leq s} \frac{1}{z_n^2} \leq \frac{A\lambda(Br)}{r^2} + \frac{A\lambda(Bs)}{s^2}$$

for all r, s is FALSE for all finite A and B. We construct $\lambda(t)$ in blocks. First, we construct a growth function $\bar{\lambda}(t)$. Let r_n be a rapidly increasing sequence of positive numbers, let $t_n = r_n \log r_n$, $u_n = r_n \log^2 r_n$, $s_n = u_n^3$. For $t_n \leq t \leq u_n$, let $\bar{\lambda}(t) = t^{\frac{3}{2}}$. For $u_n \leq t \leq s_n$, let $\bar{\lambda}(t) = s_n^{\frac{1}{2}}$. for all other values of t, let $\bar{\lambda}(t) = t^{\frac{1}{2}}$, except that $\bar{\lambda}(t) = 1$ for $0 < t \leq 1$.

Now

$$\sum_{r_n < z_n \leq s_n} \frac{1}{z_n^2} \geq \int_{t_n}^{u_n} \frac{1}{t^2} d[\lambda(t)]$$

$$\sim \frac{3}{2} \frac{1}{t^{\frac{3}{2}}} \Big|_{t_n}^{u_n} \sim \frac{3}{2} \frac{1}{(r_n \log r_n)^{\frac{1}{2}}}.$$

Now we let $\lambda(t) = \bar{\lambda}(t)(1 + \log t)$ for $t \geq 1$ with $\lambda(t) = 1$ for $0 < t \leq 1$.

It is easy to check though, that aside from factors that grow at worst like $\log r_n$, $A\lambda(Br_n)/r_2^n$ behaves like $r_n^{-\frac{3}{2}}$ and $A\lambda(Bs_n)/s_2^n$ behaves like $r_n^{-\frac{9}{2}}$ so that

$$\left(\frac{1}{r_n \log r_n}\right)^{\frac{1}{2}} \leq \frac{A\lambda(Br_n)}{r_n^2} + \frac{A\lambda(Bs_n)}{s_n^2}$$

is FALSE, for any finite A and B when n is large. This ends the paper.

References

[1]   L. I. Il'evskii "Solutions of functional equations in various classes of entire functions" (Russian) *Teor. Funkcii Funkcional. Anal. i Priložen*, No. 34 (1980) pp. 52–64, ii

[2]   M. G. Krein, B. Ya. Levin, and A. A. Nudel'man "On a special representation of polynomials that are positive on a system of closed intervals" (Ten Papers Translated from the Russian) *Amer. Math. Soc. Translations, series 2* 142 (1989) pp. 33–60

[3]   M. G. Krein "On inverse problems of the theory of filters and λ–zones of stability" *Dokl. Akad. Nauk SSSR* 93 (1953) pp. 767–770 (Russian) MR 15, 874

[4]   M. G. Krein and A. A. Nudel'man "On representation of entire functions positive on the real axis, or on a semiaxis, or outside a finite interval" (Eleven Papers in Analysis) *Amer. Math. Soc. Translations, series 2* 127 (1986) pp. 17–32

[5]   B. Ja. Levin "Distribution of Zeros of Entire Functions" *Translations of Mathematical Monographs, vol. 5, Amer. Math. Soc.* 5 (1964) (Appendix V, pp. 437–443)

[6]   V. A. Marčenko and I. V. Ostrovskii "A characterization of the spectrum of Hill's operator" *Math. USSR Sbornik* 26 (1975) (No. 4, 493–554)

[7]   G. Pólya and G. Szegö *Aufgaben und Lehrsätze aus der Analysis*, Springer–Verlag, 2 (1954)

[8]   Lee A. Rubel and B. A. Taylor "A Fourier series method for meromorphic and entire functions" *Bull. Soc. Math. France* 96 (1968) pp. 53–96

Univalent and Starlike Integral Operators and Certain Associated Families of Linear Operators

H.M.Srivastava

Department of Mathematics and Statistics University of Victoria Victoria, British Columbia V8W 3P4,Canada

1 Introduction

Operational techniques based upon various classes of linear operators are becoming increasingly useful in *Geometric Function Theory* which is the study of the relationship between the analytic properties of its image domain.On the other hand, an immensely useful class of special functions(namely,the generalized hypergeometric function) played a rather crucial role in Louis de Branges' recent proof of the celebrated Bieberbach, Robertson, and Milin conjectures in the theory of analytic and univalent functions.These latter developments in an area other than the so-called traditional areas of applications of generalized hypergeometric functions have naturally provided a new impetus for the study of such an important class of special functions.With this points in view, we illustrate the usefulness (in the study of univalent, starlike, and convex generalized hypergeometric functions) of certain families of linear operators which are defined in terms of (for example) fractional derivatives and fractional integrals, Hadamard product of convolution, and so on. We also consider severral inclusion theorems associated with the Hardy space of analytic functions whose derivative has a positive real part.

2 Subclasses of Analytic Functions

Let **A** denote the class of functions $f(z)$ *normalized* by

$$f(z) = z + \sum_{n=2}^{\infty} a_n z^n, \qquad (2.1)$$

which are *analytic* in the *open* unit disk

$$\mathbf{U} := \{z : z \in \mathbf{C} \text{ and } |z| < 1\}.$$

Also let **S** denote the class of all functions in **A** which are univalent in **U**. We denote by $\mathbf{S}^*(\alpha)$ and $\mathbf{K}(\alpha)$ the subclasses of **S** consisting of all functions which are, respectively, *starlike* and *convex of order* α in **U** $(0 \leq \alpha < 1)$, that is,

$$\mathbf{S}^*(\alpha) := \left\{ f : f \in \mathbf{S} \quad \text{and} \quad \Re\left(\frac{zf'(z)}{f(z)}\right) > \alpha \quad (0 \leq \alpha < 1; z \in \mathbf{U})\right\} \quad (2.2)$$

Proceedings of the International Conference on Complex
Analysis at the Nankai Institute of Mathematics, 1992 pp. 188-208
©INTERNATIONAL PRESS 1994

and

$$\mathbf{K}(\alpha) := \{f : f \in \mathbf{S} \text{ and } \Re e \left(1 + \frac{zf''(z)}{f'(z)} \right) > \alpha \quad (0 \leq \alpha < 1; z \in \mathbf{U})\} \quad (2.3)$$

It follows readily from the definitions (2.2) and (2.2) that

$$f(z) \in \mathbf{K}(\alpha) \Leftrightarrow zf'(z) \in \mathbf{S}^*(\alpha) \qquad (0 \leq \alpha < 1), \tag{2.4}$$

whose special case, when $\alpha = 0$, is the familiar Alexander theorem (*cf.*, *e.g.*, Duren*[9]*, p. 43, Theorem 2.12. We note also that

$$\mathbf{K}(\alpha) \subset \mathbf{S}^*(\alpha) \subset \mathbf{S} \qquad (0 \leq \alpha < 1), \tag{2.5}$$

$$\mathbf{S}^*(\alpha) \subseteq \mathbf{S}^*(0) \equiv \mathbf{S}^* \qquad (0 \leq \alpha < 1), \tag{2.6}$$

and

$$\mathbf{K}(\alpha) \subseteq \mathbf{K}(0) \equiv \mathbf{K} \qquad (0 \leq \alpha < 1), \tag{2.7}$$

where $\mathbf{S}^*$ denotes the class of all functions in $\mathbf{A}$ which are starlike (with respect to the origin) in $\mathbf{U}$.

In statements like those involved in the definitions (2.2) and (2.3), and in analogous situations throughout this paper, it should be understood that functions such as

$$\frac{zf'(z)}{f(z)} \quad \text{and} \quad \frac{zf''(z)}{f'(z)},$$

which have *removable singularities* at $z = 0$, have had these singularities removed.

For the functions $f_j(z)$ defined by

$$f_j(z) = \sum_{n=0}^{\infty} a_{j,n+1} z^{n+1} \quad (j = 1, 2), \tag{2.8}$$

we denote by $f_1 \star f_2(z)$ the *Hadamard product* or *convolution* of the functions $f_1(z)$ and $f_2(z)$, that is,

$$f_1 \star f_2(z) = \sum_{n=0}^{\infty} a_{1,n+1} a_{2,n+1} z^{n+1}. \tag{2.9}$$

Thus, following the work of Ruscheweyh *[41]*, a function $f(z) \in \mathbf{A}$ is said to be *prestarlike of order* α $(\alpha \leq 1)$ if and only if

$$\begin{cases} \dfrac{z}{(1-z)^{2(1-\alpha)}} \star f(z) \in \mathbf{S}^*(\alpha) \ (\alpha < 1) \\ \Re e(\dfrac{f(z)}{z}) > \dfrac{1}{2} \qquad (\alpha = 1; z \in \mathbf{U}), \end{cases} \tag{2.10}$$

and we denote by $\Im(\alpha)$ the subclass of $\mathbf{A}$ consisting of all prestarlike functions of order α in $\mathbf{U}$.

Next, with a view to introducing another interesting family of analytic functions, we recall the concept of subordination between analytic functions. Given two functions $f(z)$ and $g(z)$, which are analytic in $\mathbf{U}$, the function $f(z)$ is said to be *subordinate* to $g(z)$ if there exists a function $h(z)$, analytic in $\mathbf{U}$ with

$$h(0) = 0 \quad \text{and} \quad |h(z)| < 1 \tag{2.11}$$

such that

$$f(z) = g(h(z)) \quad (z \in \mathbf{U}). \tag{2.12}$$

We denote this subordination by

$$f(z) \prec g(z). \tag{2.13}$$

In particular, if $g(z)$ is univalent in $\mathbf{U}$, the subordination 2.13 is *equivalent* to (*cf.* Goodman *[12]* (p.85)

$$f(0) = g(0) \quad \text{and} \quad f(\mathbf{U}) \subset g(\mathbf{U}). \tag{2.14}$$

The concept of subordination between analytic functions can be traced back to Lindelöv *[21]*, although Littlewood *[22, 23]* and Rogosinski *[39, 40]* introduced the term and established the basic results involving subordination. Making use of this concept, we have

Definition 1:   (cf. Janowski *[14]*.) For $-1 \leq B < A \leq 1$, a function $p(z)$, analytic in $\mathbf{U}$ with $p(0) = 1$, is said to belong to the class $\mathcal{P}(A, B)$ if

$$p(z) \prec \frac{1 + Az}{1 + Bz} \quad (-1 \leq B < A \leq 1). \tag{2.15}$$

Definition 2:   (cf. Janowski *[14]*.) A function $f(z) \in \mathbf{A}$ is said to be in the class $\mathbf{S}^*(A, B)$ if and only if

$$\frac{z f'(z)}{f(z)} \in \mathcal{P}(A, B) \quad (-1 \leq B < A \leq 1). \tag{2.16}$$

We note from Definition 2 that

$$\mathbf{S}^*(1, -1) \equiv \mathbf{S}^*. \tag{2.17}$$

More generally, we recall

Definition 3:   (cf. Noor *[29]*.) A function $f(z) \in \mathbf{A}$ is said to be in the class $\mathbf{B}(A, B; \alpha)$ if and only if

$$\left(\frac{f(z)}{z} \right)^{\alpha - 1} f'(z) \in \mathcal{P}(A, B) \quad (\alpha \geq 0; -1 \leq B < A \leq 1). \tag{2.18}$$

Clearly, we have the following relationships:

$$\mathbf{B}(A, B; 0) = \mathbf{S}^*(A, B) \quad (-1 \leq B < A \leq 1). \tag{2.19}$$

and

$$\mathbf{B}(1, -1; \alpha) = \mathbf{B}_1(\alpha) \quad (\alpha \geq 0), \tag{2.20}$$

where $\mathbf{S}^*(A, B)$ is given by Definition 2, and $\mathbf{B}_1(\alpha)$ is a subclass of Bazilevič functions, which was introduced and studied by Singh *[45]*. Yet another subclass of analytic functions is given by

Definition 4: A function $f(z) \in \mathbf{A}$ is said to be in the class $\Re(\gamma)$ if it satisfies the inequality:

$$\Re\{f'(z)\} > \gamma \quad (0 \leq \gamma < 1; z \in \mathbf{U}). \tag{2.21}$$

Evidently, we have

$$\Re(\gamma) \subseteq \Re(0) \equiv \Re(0 \leq \gamma < 1). \tag{2.22}$$

The class $\Re$ was studied rather systematically by MacGregor *[25]* who did indeed refer to numerous earlier works (by, for example, Alexander *[2]*, Wolff *[60]*, Noshiro *[30]*, Warschawski *[58]*,Tims *[57]*, and Herzog and Piranian *[13]*) investigating various properties of functions whose derivative has a positive real part. In fact, a more general class of functions than those satisfying the inequality:

$$\Re e f(z) > 0 \quad (z \in \mathbf{U}) \tag{2.23}$$

is the class of close-to-convex functions considered by Kaplan *[16]*. (See also Duren *[9]*) More recently, several interesting subclasses of $\mathbf{A}$ associated with the class $\Re(\gamma)$ were considered elsewhere by (among others) Sarangi and Uralegaddi *[44]*, Owa and Uralegaddi *[37]*, and Srivastava and Owa *[52]*.

Finally, let $\mathcal{H}^p$ $(0 < p \leq \infty)$ denote the *Hardy space* of analytic functions $f(z)$ in $\mathbf{U}$, and define the *integral means* $M_p(r, f)$ by

$$M_p(r, f) = \begin{cases} (\frac{1}{2\pi} \int_0^{2\pi} |f(re^{i\theta})|^p d\theta)^{1/p} & (0 < p < \infty) \\ \sup_{0 \leq \theta < 2\pi} |f(re^{i\theta})| & (p = \infty). \end{cases} \tag{2.24}$$

Definition 5: A function $f(z)$, analytic in $\mathbf{U}$, is said to belong to the Hardy space $\mathcal{H}^p$ $(0 < p \leq \infty)$ if

$$\lim_{r \to 1-} \{M_p(r, f)\} < \infty \quad (0 < p \leq \infty). \tag{2.25}$$

For $0 \leq p \leq \infty$, $\mathcal{H}^p$ is a *Banach space* with the norm $\| f \|_p$ defined by (*cf., e.g.,* Duren *[9]*, p.23; see also Koosis *[19]*)

$$\| f \|_p = \lim_{r \to 1-} \{M_p(r, f)\} \quad (0 \leq p \leq \infty). \tag{2.26}$$

Furthermore, $\mathcal{H}^\infty$ is the familiar class of *bounded* analytic functions in $\mathbf{U}$, whereas $\mathcal{H}^2$ is the class of power series $\sum a_n z^n$ with

$$\sum |a_n|^2 < \infty. \tag{2.27}$$

3   The Hypergeometric Function and Its Generalization

In Geometric Function Theory, which indeed is (as we remarked in Section 1) the study of the relationship between the analytic properties of a given function $f(z)$ and the geometric properties of its image domain

$$\mathbf{D} = f(\mathbf{U}), \tag{3.1}$$

it is an extremely difficult open problem to find a (useful) set of conditions on the coefficients $a_n(n \in \mathbb{N}_0 = \{0, 1, 2, \ldots\})$ that are both *necessary* and *sufficient* for the function $f(z)$ to be in the class $\mathbf{S}$. One of the several partial results in connection with this problem is provided by de Branges' theorem (*cf.*, *e.g.*,[7]) which asserts the truth of the **Milin conjecture** of 1971:

$$f(z) \in \mathbf{S} \quad \text{and} \quad \log\left(\frac{f(z)}{z}\right) = 2\sum_{n=1}^{\infty} \gamma_n z^n$$

$$\Rightarrow \sum_{k=1}^{n}(n - k + 1)\left(k|\gamma_k|^2 - \frac{1}{k}\right) \leq 0 \quad (n \in \mathbb{N} = \mathbb{N}_0 \setminus \{0\}). \tag{3.2}$$

In fact, in view of the second Lebedev-Milin inequality (*cf.* [9](p.143), it is not difficult to show that (3.2) implies the **Robertson conjecture** of 1936:

$$f(z) \text{ is odd and in } \mathbf{S}$$

$$\Rightarrow \sum_{k=1}^{n} |a_{2k-1}|^2 \leq n \quad (n = 2, 3, 4, \ldots; \ a_1 \equiv 1), \tag{3.3}$$

which, in turn, implies the celebrated **Bieberbach conjecture** of 1916:

$$f(z) \in \mathbf{S} \Rightarrow |a_n| \leq n \quad (n = 2, 3, 4, \ldots), \tag{3.4}$$

where the equality holds true for all integers $n \geq 2$ only if

$$f(z) = K_\varphi(z) = \frac{z}{(1 - ze^{i\varphi})^2}$$

$$= \sum_{n=1}^{\infty} ne^{i(n-1)\varphi} z^n \quad (\varphi \text{ real}), \tag{3.5}$$

$K_\varphi(z)$ being a rotation of the Koebe function:

$$K(z) \equiv K_0(z) = \sum_{n=1}^{\infty} nz^n = \frac{z}{(1 - z)^2} \tag{3.6}$$

The key ingredients in the Branges' proof of the Milin conjecture (3.2), and hence also of the Robertson conjecture (3.3) and the Bieberbach conjecture 3.4, include Löwner's differential equation(*cf.* [9],p.83) and the following nonnegativity result due to Askey and Gasper [3], p.713, Theorem 3:

$$\sum_{k=0}^{n} P_k^{(\alpha,0)}(x) \geq 0 \quad (\alpha \geq -2; -1 < x \leq 1), \tag{3.7}$$

where $P_n^{(\alpha,\beta)}(x)$ denotes the classical Jacobi polynomial of index or order (α,β) and degree n in x (*cf.* Szegö *[56]*). In fact, we have

$$P_n^{(\alpha,\beta)}(x) = \sum_{k=0}^{n} \binom{n+\alpha}{n-k}\binom{n+\beta}{k}\left(\frac{x-1}{2}\right)^k \left(\frac{x+1}{2}\right)^{n-k}, \qquad (3.8)$$

where, in terms of *Gamma functions*,

$$\binom{\lambda}{\mu} = \frac{\Gamma(\lambda+1)}{\Gamma(\lambda-\mu+1)\Gamma(\mu+1)} = \binom{\lambda}{\lambda-\mu} \quad (\lambda,\mu \in \mathbb{C}),$$

so that

$$\binom{\lambda}{0} = 1 \quad \text{and} \quad \binom{\lambda}{n} = \frac{\lambda(\lambda-1)\ldots(\lambda-n+1)}{n!} \quad (\lambda \in \mathbb{C}; n \in \mathbb{N}). \qquad (3.9)$$

Equivalently, (3.8) may be written in the form:

$$P_n^{(\alpha,\beta)}(x) = \binom{n+\alpha}{n} {}_2F_1(-n,\alpha+\beta+n+1;\alpha+1;\frac{1-x}{2}), \qquad (3.10)$$

in terms of the *Gaussian case*

$$l - 1 = m = 1$$

of the *generalized hypergeometric function* ${}_lF_m$ defined below.

Definition 6: Let λ_j $(j = 1,\ldots,l)$ and μ_j $(j = 1,\ldots,m)$ be complex numbers such that

$$\mu_j \neq 0, -1, -2, \ldots \quad (j = 1,\ldots,l).$$

Then the generalized hypergeometric function ${}_lF_m(z)$ is defined by

$$\begin{aligned}
{}_lF_m(z) &\equiv {}_l!F_m(\lambda_1,\ldots,\lambda_l;\mu_1,\ldots,\mu_m;z) \\[1mm]
&= {}_lF_m\left[\begin{array}{c} \lambda_1,\ldots,\lambda_l; \\ \\ \mu_1,\ldots,\mu_m; \end{array} z \right] \\[1mm]
&:= \sum_{n=0}^{\infty} \frac{(\lambda_1)_n \ldots, (\lambda_l)_n}{(\mu_1)_n, \ldots, (\mu_m)_n} \frac{z^n}{n!} \quad (l \leq m+1),
\end{aligned} \qquad (3.11)$$

where $(\lambda)_n$ denotes the *Pochhammer symbol* defined, again in terms of Gamma functions, by

$$(\lambda)_n := \frac{\Gamma(\lambda+n)}{\Gamma(\lambda)} = \begin{cases} 1 & (n=0) \\ \lambda(\lambda+1)\ldots(\lambda+n-1) & (n \in N). \end{cases} \qquad (3.12)$$

We note in passing that

$$z \, {}_lF_m(\lambda_1,\ldots,\lambda_l;\mu_1,\ldots,\mu_m;z) \in \mathbf{A} \qquad (3.13)$$

since the $_lF_m$ series in (3.11) converges absolutely for (*cf.,e.g.*, Erdélyi *et al.* *[10]*, Chapter 4)

(i) $|z| < \infty$ if $l < m + 1$;
(ii) $z \in U$ if $l = m + 1$;
(iii) $z \in \partial U = \{z : z \in \mathbb{C}$ and $|z| = 1\}$ if $l = m + 1$,
provided further that

$$\Re e\left(\sum_{j=1}^{m} \mu_j - \sum_{j=1}^{m} \lambda_j\right) > 0,$$

unless (of course) the series terminates.

Making use of the hypergeometric representation (3.10), it is not difficult to rewrite the inequality (3.7) in the generalized hypergeometric form *[3]*,p.717, Equation (3.1):

$$\frac{(\alpha + 2)_n}{n!} \, {}_3F_2 \left[\begin{array}{c} -n, \alpha + n + 2, \frac{1}{2}(\alpha + 1); \\ \\ \alpha + 1, \frac{1}{2}(\alpha + 3); \end{array} \; x \right] \geq 0 \tag{3.14}$$

$$(0 \leq x < 1; \; \alpha \geq -2; \; n \in \mathbb{N}_0).$$

The theory of special functions has so far remained unavoidable in proving the aforementioned conjectures in Geometric Function Theory (see also Aleksandrov *[1]*). Even the relatively more recent attempt by Weinstein *[59]* to prove the Bieberbach conjecture (3.4) *directly* is based rather heavily upon the addition theorem for the Legendre (or spherical) polynomials $P_n(x)$, where [*cf.* Equation (3.10)]

$$P_n(x) = P_n^{(}0, 0)(x) = {}_2F_1(-n, \; n + 1; \; 1; \; \frac{1 - x}{2}). \tag{3.15}$$

All these developments using special functions in an area other than the so-called traditional areas of applications of generalized hypergeometric functions (*cf.*, *e.g.*, *[50, 49]*, and *[47]*) have provided a new impetus for the study of the generalized hypergeometric functions, especially in connection with the various subclass of analytic functions which we enumerated in the preceding section.

4   Associated Families of Linear Operators

The various linear operators (whose usefulness, in the study of such subclasses of analytic functions as those defined in Section 1, will be considered in this paper) are given below:

I. Carlson-Shaffer Operator. The *Carlson-Shaffer operator* $\mathcal{L}(a, c)$ is defined by the convolution (*cf.* Carlson and Shaffer *[6]*; see also Owa *et al.* *[36]*):

$$\mathcal{L}(a, c)f(z) := \varphi(a, c; z) \star f(z) \quad (f(z) \in \mathbf{A}), \tag{4.1}$$

where $\varphi(a, c; z)$ is an *incomplete Beta function* defined by

$$\varphi(a, c; z) := \sum_{n=0}^{\infty} \frac{(a)_n}{(c)_n} z^{n+1} = z \, {}_2F_1(1, a; c; z) \tag{4.2}$$

$$(c \neq 0, -1, -2, \ldots; z \in \mathbf{U}).$$

The operator $\mathcal{L}(a, c)$ maps $\mathbf{A}$ onto itself. Furthermore, if we let

$$a \neq 0, -1, -2, \ldots,$$

then $\mathcal{L}(c, a)$ is an inverse of $\mathcal{L}(a, c)$. Observe also that (*cf.* *[33]*, p.1067)

$$\mathbf{K}(\alpha) = \mathcal{L}(1, 2)\mathbf{S}^*(\alpha); \quad \mathbf{S}^*(\alpha) = \mathcal{L}(2, 1)\mathbf{K}(\alpha); \tag{4.3}$$

$$(0 \leq \alpha < 1).$$

II. Generalized Bernardi-Libera-Livingston Integral Operator. An interesting generalization of the *Bernardi- Libera- Livingston integral operator*, denoted here by $\mathbf{J}_\gamma$, is defined by

$$\mathbf{J}_\gamma f(z) := \frac{\gamma + 1}{z^\gamma} \int_0^z t^{\gamma - 1} f(t) dt \quad (\gamma > -1; f(z) \in \mathbf{A}), \tag{4.4}$$

which, for various further constraints on the parameter γ, was used recently by several authors (see, *e.g.*, *[53]*, pp.66,154,181, and 338; see also *[4, 20, 24]*).

III. Miller-Mocanu-Reade Integral Operator. The *Miller-Mocanu-Reade integral operator* $\mathbf{I}$ is defined (for suitable analytic functions $\varphi(z)$ and $\phi(z)$, and for suitable constants $\alpha, \beta \neq 0, \gamma$, and δ) by

$$\mathbf{I}f(z) := \left(\frac{\beta + \gamma}{z^\gamma \phi(z)} \int_0^z \{f(t)\}^\alpha \varphi(t) t^{\delta - 1} dt \right) \quad (f(z) \in \mathbf{A}), \tag{4.5}$$

which, in the special case when

$$\alpha = \beta = 1, \quad \delta = \gamma, \quad \text{and} \quad \varphi(z) = \phi(z) = 1,$$

reduced at once to the generalized Bernardi-Libera-Livingston operator $\mathbf{J}_\gamma (\gamma > -1)$ (see Miller *et al.* *[26, 27]*).

IV. Operators of Fractional Calculus. Numerous operators of *fractional calculus* (that is, *fractional integral* and *fractional derivative*) have indeed been studied in the literature (*cf.,e.g.,[11]*, Chapter 13, *[28, 31, 42, 43]*, *[48]*, p.21 *et seq.,[51]*, Chapter 5, and *[32]*). We choose to recall here the following operators of fractional calculus.

Definition 7: (Fractional Integral Operator). The *fractional integral of order* λ is defined, for a function $f(z)$, by

$$D_z^{-\lambda} f(z) := \frac{1}{\Gamma(\lambda)} \int_0^z \frac{f(\zeta)}{(z - \zeta)^{1 - \lambda}} d\zeta \quad (\lambda > 0), \tag{4.6}$$

where $f(z)$ is an analytic function in a simply-connected region of the z-plane containing the origin, and the multiplicity of $(z - \zeta)^{\lambda - 1}$ is removed by requiring $\log(z - \zeta)$ to be *real* when $z - \zeta > 0$.

Definition 8: (Fractional Derivative Operator). The *fractional derivative of order* λ is defined, for a function $f(z)$, by

$$D_z^\lambda f(z) := \begin{cases} \dfrac{1}{\Gamma(1-\lambda)} \dfrac{d}{dz} \displaystyle\int_0^z \dfrac{f(\zeta)}{(z-\zeta)^\lambda} d\zeta & (0 \le \lambda < 1) \\[2ex] \dfrac{d^n}{dz^n} D_z^{\lambda-n} f(z) & (n \le \lambda < n+1; n \in \mathbb{N}), \end{cases} \tag{4.7}$$

where $f(z)$ is constrained, and the multiplicity of $(z-\zeta)^{-\lambda}$ is removed, as in Definition 7.

Definition 9: (Generalized Fractional Integral Operator). Under the hypotheses of Definition 7, the *generalized fractional integral of order* λ is defined, for a function $f(z)$, by

$$I_{0,z}^{\lambda,\mu,\nu} f(z) := \frac{z^{-\lambda-\lambda}}{\Gamma(\lambda)} \int_0^z (z-\zeta)^{\lambda-1} f(\zeta) \, {}_2F_1(\lambda+\mu, -\nu; \lambda; 1 - \frac{\zeta}{z}) d\zeta \tag{4.8}$$

$$(\lambda > 0; k > \max\{0, \mu - \nu\} - 1),$$

provided further that

$$f(z) = O(|z|^k) \quad (z \to 0). \tag{4.9}$$

It follows readily from Definition 7 and Definition 9 that

$$D_z^{-\lambda} f(z) := I_{0,z}^{\lambda,-\lambda,\nu} f(z) \quad (\lambda > 0). \tag{4.10}$$

Furthermore, since

$${}_2F_1(a,b;b;z) = {}_1F_0(a;-;z) = (1-z)^{-a} \quad (z \in \mathbf{U}), \tag{4.11}$$

we have relationship:

$$I_{0,z}^{\lambda,\mu,-\lambda} f(z) := D_z^{-\lambda} z^{-\lambda-\mu} f(z) \quad (\lambda > 0). \tag{4.12}$$

The fractional calculus operator D_z^λ, given by Definition 7 and Definition 8, is related rather closely to the Carlson-Shaffer operator $\mathcal{L}(a,c)$ defined by (4.1); in fact, we have

$$\mathcal{L}(2,c)f(z) = \Gamma(c)z^{2-c} D_z^{2-z} f(z) \quad (c \ne 0, -1, -2, \ldots) \tag{4.13}$$

or, equivalently,

$$D_z^\lambda f(z) = \frac{z^{-\lambda}}{\Gamma(2-\lambda)} \mathcal{L}(2, 2-\lambda) f(z) \quad (\lambda \ne 2, 3, 4, \ldots) \tag{4.14}$$

On the other hand, operator $I_{0,z}^{\lambda,\mu,\nu}$ is a generalization of the fractional integral operator which was studied by Saigo *[42]* and applied subsequently by Srivastava and Saigo *[54]* in solving various boundary value problems involving the Euler-Darboux equation:

$$\frac{\partial^2 u}{\partial x \partial y} - \frac{1}{x - y}\left(\beta\frac{\partial u}{\partial x} - \alpha\frac{\partial u}{\partial x}\right) = 0 \quad (\alpha; \beta > 0; \alpha + \beta < 1). \tag{4.15}$$

Definition 10:   (Generalized Fractional Derivative Operator). Under the hypotheses of Definition 8, the *generalized fractional derivative of order* λ is defined, for a function $f(z)$, by

$$J_{0,z}^{\lambda,\mu,\nu} f(z) := \frac{1}{\Gamma(1 - \lambda)}\frac{d}{dz}\left\{z^{\lambda-\mu}\int_0^z (z - \zeta)^{-\lambda}\right.$$

$$\left. \cdot {}_2F_1\left(\mu - \lambda, -\nu; 1 - \lambda; 1 - \frac{\zeta}{z}\right)f(\zeta)d\zeta\right\} \tag{4.16}$$

$$(0 \leq \lambda < 1; k > \max\{0, \mu - \nu\} - 1),$$

where k is given, as before, by the order estimate (4.9).

It follows readily from Definition 10 that

$$J_{0,z}^{\lambda,\lambda,\nu} f(z) = D_z^{\lambda} f(z) \quad (0 \leq \lambda < 1), \tag{4.17}$$

where the fractional calculus operator D_z^{λ} is, in fact, given by Definition 7 and 8 for all values of λ (see, *e.g.*, *[53]*, p.343). Furthermore, in terms of Gamma functions, we have

$$J_{0,z}^{\lambda,\mu,\nu} z^{\rho} = \frac{\Gamma(\rho + 1)\Gamma(\rho - \mu + \nu + 2)}{\Gamma(\rho - \mu + 1)\Gamma(\rho - \lambda + \nu + 2)} z^{\rho-\mu} \tag{4.18}$$

$$(\rho + 2 \geq \mu - \nu).$$

(See also *[46, 55, 35]*.)

5   Operational Techniques Involving Subclasses of the Class A

By applying the Carlson-Shaffer operator $\mathcal{L}(a, c)$ defined by (4.1), Owa and Srivastava *[33]* proved

Theorem 1: *For the generalized hypergeometric function* ${}_lF_m(z)$ *defined by (3.11), let*

$$\left|\frac{z_l F''{}_m(z)}{{}_lF'_m(z)}\right| < (1 - \alpha)^{-1}\left(1 - \frac{3}{2}\alpha + \alpha^2\right) \tag{5.1}$$

$$(0 \leq \alpha \leq \frac{1}{2}; \ \lambda_1\ldots\lambda_l \neq 0; \ z \in \mathbf{U}).$$

Then

$$z \ {}_{l+1}F_{m+1}(\lambda_1 + 1,\ldots,\lambda_l + 1, 1; \mu_1 + 1,\ldots,\mu_m + 1, 2; z) \in \mathbf{S} * (\alpha) \tag{5.2}$$

$$(0 \leq \alpha \leq \frac{1}{2}).$$

For the generalized Bernardi-Libera-Livingston operator $\mathbf{J}_\gamma$ defined by (4.4), it is known that

$$f(z) \in \mathbf{S}^* \Rightarrow \mathbf{J}_\gamma f(z) \in \mathbf{S}^* \, (0 \le \gamma \le 1). \tag{5.3}$$

Making use of this last inclusion property (5.3) and the definition (4.4), it is not difficult to apply Theorem 1 (with $\alpha = 0$) *iteratively* in order to deduce

Theorem 2: *For the generalized hypergeometric function* ${}_lF_m(z)$ *defined by (3.11), let*

$$\left| \frac{z \, {}_lF''m(z)}{{}_lF'_m(z)} \right| < 1 \tag{5.4}$$

$$(\lambda_1 \ldots \lambda_l \ne 0; z \in \mathbf{U}).$$

Then

$$z \, {}_{l+s+1}F_{m+s+1} \left[\begin{array}{c} \lambda_1 + 1, \ldots, \lambda_l + 1, 1, \sigma_1 + 1, \ldots, \sigma_s + 1; \\ \\ \mu_1 + 1, \ldots, \mu_l + 1, 1, \sigma_1 + 2, \ldots, \sigma_s + 2; \end{array} z \right] \in \mathbf{S}^* \tag{5.5}$$

$$(0 \le \sigma_j \le 1; \; j = 1, \ldots, s).$$

From amongst the various special cases of Theorem 2, which are worthy of note, we consider the case when

$$\sigma_j = 1 \quad (j = 1, \ldots, s).$$

The assertion (5.5) reduces, in this case, to the inclusion relation:

$$z \, {}_{l+s}F_{m+s} \left[\begin{array}{c} \lambda_1 + 1, \ldots, \lambda_l + 1, 1, 2, \ldots, 2; \\ \\ \mu_1 + 1, \ldots, \mu_l + 1, 3, 3, \ldots, 3; \end{array} z \right] \in \mathbf{S}^*, \tag{5.6}$$

provided that the relevant hypotheses of Theorem2 hold true. The following result is analogous to the inclusion relation (5.3); it holds true for the substantially more general Miller-Mocanu-Reade operator $\mathbf{I}$ defined by (4.5).

Theorem 3: *Let the functions* $f(z)$ *and* $\varphi(z)$ *be in the class* $\mathbf{S}^*(\rho)$ *$(0 \le \rho < 1)$. Then the function $F(z)$ defined by [cf. Equation (4.5) with* $\Phi(z) = 1, \beta = \alpha+1$, *and* $\delta = \gamma$]

$$F(z) := \left(\frac{\gamma + \alpha + 1}{z^\gamma} \int_0^z \{f(t)\}^\alpha \varphi(t) t^{\gamma-1} dt \right)^{1/(\alpha+1)}$$

$$= z + \sum_{n=2}^\infty b_n z^n \tag{5.7}$$

$$(\gamma > 0; \alpha > 0)$$

is also in the class $\mathbf{S}^*(\rho)(0 \le \rho < 1)$.

The proof of Theorem 3, detailed elsewhere by Kim *et al.* *[17]*, would make use of a number of results associated with the function spaces $\mathcal{P}(A, B)$ and $\mathbf{B}(A, B; \alpha)$ (see Definition 1 and Definition 3, respectively), and with the Miller-Mocanu-Reade integral operator $\mathbf{I}$ defined by (4.5). As a matter of fact, by applying the integral operator $\mathbf{I}$, Kim *et al.* *[17]* also proved Theorem 4 below, and also Theorem 5 in the *special* case when $n = 1$ and $\alpha \in \mathbb{N}$.

Theorem 4: *Let the functions $f(z)$ and $\varphi(z)$ be in the class $\mathbf{S}^*(A, B)$ given by Definition 2. Then the function $F(z)$ defined by (5.7) is also in the class $\mathbf{S}^*(A, B)$.*

Theorem 5: *Let the function $f(z)$ be in the class $\mathbf{B}(A, B; \alpha)$ given by Definition 3. Then the function $G_n(z)$ defined by*

$$G(z) := \left(\frac{\gamma + \alpha + n}{z^\gamma} \int_0^z \{f(t)\}^\alpha t^{\gamma + n - 1} dt \right)^{1/(\alpha + n)}$$

$$= z + \sum_{n=2}^\infty c_n z^n \tag{5.8}$$

$$(\gamma > -\alpha - n; \ \ n \in \mathbb{N})$$

(that is, by (5.7) with $\varphi(z) = z^n$, $\gamma > -\alpha - n$, and $n \in \mathbb{N}$) is in the class $\mathbf{B}(A, B; \alpha + n)$.

Next we turn to an application of the foregoing operational techniques involving the classes $\mathcal{F}(\alpha)$ and $\mathbf{K}(\alpha)$, that is, the classes of prestarlike and convex functions of order α. Indeed it is easily verified from the definitions 2.10 and (4.1) that

$$\mathcal{F}(\alpha) = \mathcal{L}(1, 2 - 2\alpha)\mathbf{S}^*(\alpha) \quad (\alpha < 1). \tag{5.9}$$

Moreover, we have

$$\mathcal{F}(1) = \left\{ f : f \in \mathbf{A} \ \text{and} \ \Re e \left(\frac{f(z)}{z} \right) > \frac{1}{2} \ \ (z \in \mathbf{U}) \right\} \tag{5.10}$$

In order to present a connection theorem involving the classes $\mathcal{F}(\alpha)$ and $\mathbf{K}(\alpha)$, we introduced the operator $\Omega_z^{\lambda, \mu, \nu}$ defined by

$$\Omega_z^{\lambda, \mu, \nu} f(z) = \frac{\Gamma(2 - \mu)\Gamma(3 - \lambda + \nu)}{\Gamma(3 - \mu + \nu)} z^\mu \mathbf{J}_{0,z}^{\lambda, \mu, \nu} f(z) \tag{5.11}$$

$$(f(z) \in \mathbf{A}),$$

where the generalized fractional derivative operator $\mathbf{J}_{0,z}^{\lambda, \mu, \nu}$ is given by Definition 10.

In view of the formula (4.18), it is fairly straightforward to relate the operator $\Omega_z^{\lambda, \mu, \nu}$ with the Carlson-Shaffer operator $\mathcal{L}(a, c)$ as follows:

$$\Omega_z^{\lambda, \mu, \nu} f(z) = \mathcal{L}(2, 2 - \mu)\mathcal{L}(3 - \mu + \nu, 3 - \lambda + \nu)f(z) \tag{5.12}$$

$$(0 \leq < 1; \mu - \nu < 3; f(z) \in \mathbf{A}),$$

Making use of the relationships (5.12) and (4.3) , and also the following inclusion relation (due essentially to Carlson and Shaffer *[6]*):

$$\mathcal{L}(2 - 2\beta, 2 - 2\alpha)\mathbf{S}^*(\alpha) \subset \mathbf{S}^*(\beta) \subset \mathbf{S}^*(\alpha), \tag{5.13}$$

$$(0 \le \alpha \le \beta < 1),$$

it can be shown that

$$\mathcal{L}(3 - \lambda + \nu, 3 - \mu + \nu)\Omega_z^{\lambda,\mu,\nu}\mathbf{K}(\tfrac{1}{2}) \subset \mathbf{S}^*(\tfrac{1}{2}), \tag{5.14}$$

$$(0 \le \lambda < 1; \ \mu - \nu < 3; \ 0 \le \mu < 1),$$

Finally, if we rewrite a special case $(\beta = \tfrac{1}{2})$ of (5.13) in the form:

$$\mathcal{L}(1, 2 - \alpha)\mathbf{S}^*(\tfrac{\alpha}{2}) \subset \mathbf{S}^*(\tfrac{1}{2}) \subset \mathbf{S}^*(\tfrac{\alpha}{2}) \tag{5.15}$$

$$(0 \le \alpha < 2),$$

we shall obtain the following connection theorem involving the classes $\mathcal{F}(\alpha)$ and $\mathbf{K}(\alpha)$ (*cf.* Owa and Srivastava *[34]*):

Theorem 6: *For the classes* $\mathbf{K}(\alpha)$ *and* $\mathcal{F}(\alpha)$ *defined by (2.3) and (2.10), respectively,*

$$\mathcal{L}(3 - \lambda + \nu, 3 - \mu + \nu)\Omega_z^{\lambda,\mu,\nu}\mathbf{K}(\tfrac{\mu}{2}) = \mathcal{F}(\tfrac{\mu}{2}) \tag{5.16}$$

$$((0 \le \lambda < 1; \ \mu - \nu < 3; \ 0 \le \mu < 2).$$

In its special case when $\mu = 0$ and $\mu = 1$, the assertion (5.16) of Theorem 6 would simplify considerably, and we have

$$\mathcal{L}(3 - \lambda + \nu, 3 + \nu)\Omega_z^{\lambda,0,\nu}\mathbf{K} = \mathcal{F}(0) \tag{5.17}$$

$$((0 \le \lambda < 1; \ \nu > -3)$$

and

$$\mathcal{L}(3 - \lambda + \nu, 2 + \nu)\Omega_z^{\lambda,1,\nu}\mathbf{K}(\tfrac{1}{2}) = \mathcal{F}(\tfrac{1}{2}) \tag{5.18}$$

$$((0 \le \lambda < 1; \nu > -2)$$

respectively.

6   Generalized Hypergeometric Functions Associated with the Hardy Space

Several inclusion theorems associated with the Hardy space of analytic functions (see Definition 5) were proven recently for various classes of generalizedhypergeometric functions whose derivative has a positive real part (see Definition 4). In this section we aim at developing a relatively simpler proof of a unification (and generalization) of these inclusion theorems.

We begin by recalling the following inclusion theorem which was proven by Jung *et al.* *[15]* by applying the *one-parameter* family of integral operators defined by (4.4):

Theorem 7:   *Let the function*

$$z \, {}_lF_m(\lambda_1, \ldots, \lambda_l; \mu_1, \ldots, \mu_m; z) \quad (l \le m + 1)$$

be in the class $\mathcal{R}$ *defined by (2.22).*
   Then the function

$$z \, {}_{l+s}F_{m+s} \left[\begin{array}{c} \lambda_1, \ldots, \lambda_l, \alpha_1 + 1, \ldots, \alpha_s + 1; \\[1ex] \mu_1, \ldots, \mu_m, \alpha_1 + 2, \ldots, \alpha_s + 2; \end{array} \; z \right]$$

is in $\mathcal{K}^\infty$ *at least for* $\alpha_j > 0$ $(j = 1, \ldots, s)$.

Another inclusion theorem for generalized hypergeometric functions, involving the class $\mathcal{R}(\gamma)$ given by Definition 4, is contained in

Theorem 8:   *Let the function*

$$z \, {}_lF_m(\lambda_1, \ldots, \lambda_l; \mu_1, \ldots, \mu_m; z) \quad (l \le m + 1)$$

be in the class $\mathcal{R}(\gamma)$ $(0 \le \gamma < 1)$.
   Then the function

$$z \, {}_{l+s}F_{m+s} \left[\begin{array}{c} \lambda_1, \ldots, \lambda_l, 2, \ldots, 2; \\[1ex] \mu_1, \ldots, \mu_m, \alpha_1 + 2, \ldots, \alpha_s + 2; \end{array} \; z \right]$$

is in $\mathcal{K}^\infty$ *for* $\alpha_j \in N$ $(j = 1, \ldots, s)$.

The proof of Theorem 8 by Kim *et al.* *[18]* makes use of the generalized fractional integral operator $I_{0,z}^{\lambda,\mu,\nu}$ given by Definition 9.
   We now give a simple and direct proof of the following unification (and generalization) of Theorem 7 and Theorem 8, *without* using the integral operators $\mathbf{J}_\gamma$ and $I_{0,z}^{\lambda,\mu,\nu}$ defined by (4.4) and (4.8), respectively.

Theorem 9:   *Let the function*

$$z \, {}_lF_m(\lambda_1, \ldots, \lambda_l; \mu_1, \ldots, \mu_m; z) \quad (l \le m + 1)$$

be in the class $\mathcal{R}(\gamma)$ $(0 \le \gamma < 1)$. *Suppose also that the function* $\Phi(z)$ *is defined, in terms a generalized hypergeometric function, by*

$$\Phi(z) = z \, {}_{l+s}F_{m+s} \left[\begin{array}{c} \lambda_1, \ldots, \lambda_l, \alpha_1, \ldots, \alpha_s; \\[1ex] \mu_1, \ldots, \mu_m, \beta_1, \ldots, \beta_s; \end{array} \; z \right] \quad (l \le m + 1; s \in N) \qquad (6.1)$$

for (real or complex) parameters $\alpha_1, \ldots, \alpha_s$ and $\beta_1, \ldots, \beta_s$ such that

$$\beta_j \neq 0, -1, -2, \ldots \quad (j = 1, \ldots, s).$$

Then

$$\Phi(z) \in \mathcal{K}^\infty \tag{6.2}$$

and, more precisely,

$$|\Phi(z)| < \infty \quad (z \in \bar{\mathbf{U}} = \mathbf{U} \cup \partial\mathbf{U} = \{z : z \in \mathbb{C} \ \text{ and } \ |z| \leq 1\}), \tag{6.3}$$

provided that

$$\Re\left(\sum_{j=1}^{s} \beta_j - \sum_{j=1}^{s} \alpha_j\right) > 0. \tag{6.4}$$

In place of the integral operators $\mathbf{J}_\gamma$ and $I_{0,z}^{\lambda,\mu,\nu}$ (which were applied by Jung *et al.* *[15]* and Kim *et al.* *[18]* to prove Theorem 7 and Theorem 8, respectively), our proof of Theorem 9 is based upon the coefficient inequality asserted by the following (*cf.* MacGregor *[25]*, p.533, Theorem 1)

Lemma   *Let the function $f(z)$ be in the class $\mathcal{R}(\gamma)$ $(0 \leq \gamma < 1)$.*
   Then

$$|a_n| \leq \frac{2}{n} \quad (n \in \mathbb{N}^* = \mathbb{N}\backslash\{1\}). \tag{6.5}$$

Proof. Since $0 \leq \gamma < 1$, the hypothesis $f(z) \in \mathcal{R}(\gamma)$ implies that

$$\Re\{f'(z)\} > \gamma \geq 0 \quad (z \in \mathbf{U}).$$

Thus the assertion 6.5 of the Lemma can be deduced fairly readily by setting $g(z) = f'(z)$ in the following well-known (*rather classical*) result due to Constantin Carathéodory (1873-1950): *If*

$$g(z) = 1 + \sum_{n=1}^{\infty} b_n z^n$$

is analytic in the open unit disk $\mathbf{U}$,*and if*

$$\Re\{g(z)\} > 0 \quad (z \in \mathbf{U}),$$

then

$$|b_n| \leq 2 \quad (n \in \mathbb{N}).$$

(*cf.* Carathéodory *[5]*; see also Pólya and Szegö *[38]*, pp.150 and 355, Problem 235.)

Proof of Theorem 9. For the sake of convenience, we put

$$\Omega_n = \frac{(\lambda_1)_n \ldots (\lambda_l)_n}{(\mu_1)_n \ldots (\mu_m)_n} \quad (n \in \mathbb{N}_0 = \mathbb{N} \cup \{0\}). \tag{6.6}$$

Then, by the assertion (6.5) of the Lemma, the hypothesis

$$z \, {}_lF_m(\lambda_1,\ldots,\lambda_l;\mu_1,\ldots,\mu_m;z) = z + \sum_{n=2}^{\infty} \frac{\Omega_{n-1}}{(n-1)!} z^n \in \mathcal{R}(\gamma)$$

$$(0 \le \gamma < 1)$$

implies that

$$\left| \frac{\Omega_{n-1}}{(n-1)!} \right| \le \frac{2}{n} \quad (n \in \mathbb{N}^*). \tag{6.7}$$

From the definition (3.12) and Stirling's asymptotic expansion for the Gamma function (*cf.,e.g.*, Erdélyi *et al.* *[10]*, p.47, Section 1.18), it is not difficult to show for

$$\Delta_n = \frac{(\alpha_1)_n \ldots (\alpha_s)_n}{(\beta_1)_n \ldots (\beta_s)_n} \quad (n \in \mathbb{N}_0)$$

with *fixed* parameters α_j and β_j $(j=1,\ldots,s)$ that

$$\Delta_n = K^{-1} n^{-\omega}[1 + O(n^{-1})] \quad (n \to \infty), \tag{6.8}$$

where, for convenience,

$$K = \frac{\Gamma(\alpha_1)\ldots\Gamma(\alpha_s)}{\Gamma(\beta_1)\ldots\Gamma(\beta_s)}$$

and

$$\omega = \sum_{j=1}^{s} \beta_j - \sum_{j=1}^{s} \alpha_j. \tag{6.9}$$

Now, for the function $\Phi(z)$ defined by 6.1, we readily have

$$|\Phi(z)| \le |z| + \sum_{n=2}^{\infty} \left| \frac{\Omega_{n-1}}{(n-1)!} \right| |\Delta_n||z|^n, \tag{6.10}$$

which, for $z \in \bar{\mathbf{U}}$, yields

$$|\Phi(z)| \le 1 + \sum_{n=2}^{\infty} |c_n|, \tag{6.11}$$

where

$$|c_n| = \left| \frac{\Omega_{n-1}}{(n-1)!} \right| |\Delta_n|, \quad (n \in \mathbb{N}^*). \tag{6.12}$$

Applying the results (6.7) and (6.8), we find from (6.12) that

$$|c_n| < \frac{2M}{|K|} \frac{1}{n^{1+\Re e(\omega)}} \quad (n \ge N \in \mathbb{N}; M > 0), \tag{6.13}$$

which proves that the power series for the function $\Phi(z)$ converges absolutely for *each* $z \in \bar{\mathbf{U}}$, provided that the real part of ω defined by (6.9) is positive, that is, that the condition (6.4) of Theorem 9 is satisfied.

This evidently completes our direct proof of *both* the assertions (6.2) and (6.3) of Theorem 9.

In its special case when $\gamma = 0$ *and*

$$\beta_j = \alpha_j + 1 \quad (j = 1, \ldots, s), \tag{6.14}$$

the assertion (6.2) of Theorem 9 would correspond to Theorem 7 *without* the inequalities required there to be satisfied by the parameters $\alpha_1, \ldots, \alpha_s$. Furthermore, a special case of the assertion (6.2) of Theorem 9 when

$$\alpha_j = 2 \quad \text{and} \quad \beta_j = \alpha_j + 2 \quad (j = 1, \ldots, s), \tag{6.15}$$

would yield Theorem 8 with the relatively less stringent condition:

$$\Re(\alpha_1 + \ldots + \alpha_s) > 0.$$

We should like to conclude by remarking that, under the aforementioned special cases (6.14) and (6.15), our proof of the assertion (6.3) of Theorem 9 would show that not only are the functions (involved in Theorem 7 and Theorem 8) bounded, but their power series are also absolutely convergent, for *each* $z \in \partial \mathbf{U}$.

7   Acknowledgements

This paper incorporates the text of a lecture presented on **June 21, 1992** at the **International Conference on Complex Analysis** held (in the memory of **Professor King-Lai Hiong** on the occasion of his birth centenary) at Tianjin (People's Republic of China) on June 19-23, 1992. The author wishes to thank the Nankai Institute of Mathematics (Nankai University), Beijing Institute of Technology, and Fudan University for having invited and supported his visits to these institutions, as well as his participation in the Nankai Institute's **Special Year in Complex Analysis.** The present investigation was supported, in part, by the Natural Sciences and Engineering Research Council of Canada under Grant OGP0007353.

References

[1] I.A.Aleksandrov, L.de Branges' proof of the I.M.Milin conjecture and the L.Bieberbac conjecture (Russian), *Sibirsk.Mat.Zh.* 28(1987), no.2,7-20,223; English trns.,*Siberian Math.J.* 28(1987),178-191.

[2] J.W.Alexander, Functions which map the interior of the unit circle upon simple regions, *Ann.of Math.*(2)17(1915-1916),12-22.

[3] R.Askey and G.Gasper, Positive Jacobi polynomial sums. II, *Amer.J.Math.* 98(1976), 709-737.

[4] S.D.Bernardi, Convex and starlike univalent functions, *Trans. Amer. Math. Soc.* 135(1969), 429-446.

[5] C.Carathéodory, Über den Variabilitätsbereich der Fourier'schen Konstanten von positiven harmonischen Funktionen, *Rend. Circ. Mat. Palermo* 32(1911),193-217.

[6]   B.C.Carlson and D.B.Shaffer, Starlike and prestarlike hypergeometric functions, *SIAM J.Math.Anal.* 15(1984), 737-745.

[7]   L.de Branges, A proof of the Bieberbach conjecture, *Acta Math.* 154(1985), 137-152.

[8]   P.L.Duren, *Theory of H p Spaces*, A Series of Monographs and Textbooks in Pure and Applied Mathematics, Vol.38, Academic Press, New York and London, 1970.

[9]   P.L.Duren, *Univalent Functions*, Grundlehren der Mathematischen Wissenschaften 259, Springer-Verlag, New York, Berlin,Heidelberg, and Tokyo, 1983.

[10]   A.Erdélyi, W.Magnus, F.Oberhettinger, and F.G.Tricomi, *Higher Transcendental Functions*, Vol.I, McGraw-Hill Book Company, New York, London, and Toronto, 1953.

[11]   A.Erdélyi, W.Magnus, F.Oberhettinger, and F.G.Tricomi, *Tables of Integral Transforms*, Vol.II, McGraw-Hill, New York, London, and Toronto, 1954.

[12]   A.W.Goodman, *Univalent Functions*, Vol.I,Polygonal Publishing House, Washington,New Jersey,1983.

[13]   F.Herzog and Piranian, On the univalence of functions whose derivative has a positive real part, *Proc. Amer.Math. Soc.* 2(1951), 625-633.

[14]   W.Janowski, Some extremal problems for certain families of analytic functions. I, *Ann.Polon.Math.* 28(1973), 297-326.

[15]   I.B.Jung, Y.C.Kim,and H.M.Srivastava, The Hardy space of analytic functions associated with certain one-parameter families of integral operators, *J.Math.Anal. Appl.*(to appear).

[16]   W.Kaplan, Close-to-convex schlicht functions, *Michigan Math.J.* 1(1952), 169-185.

[17]   Y.C.Kim, K.S.Lee, and H.M.Srivastava, Some families of univalent and starlike integral operators, in *Current Topics in Analytic Function Theory* (H.M.Srivastava and S.Owa, Editors), pp.118-129, World Scientific Publishing Company, Singapore, New Jersey, London, and Hong Kong,1992.

[18]   Y.C.Kim,Y.S.Park,and H.M.Srivastava, A class of inclusion theorems associated with some fractional integral operators, *Proc.Japan Acad.Ser. A Math Sci.* 67(1991),313-318.

[19]   P.Koosis, *Introduction to H_p Spaces*, London Mathematical Society Lecture Note Series, Vol.40, Cambridge University Press, Cambridge, London,and New York,1980.

[20]   R.J.Libera, Some classes of regular univalent functions, *Proc. Amer. Math. Soc.* 16(1965), 755-758.

[21]   E.Lindelöf, Mémoire sur certaines inégaliés dans la théorie des fonctions monogénes et sur quelques propriétés nouvelles de ces fonctions dans le voisinage d'un point singulier essentiel, *Ann.Soc.Sci.Fenn.* 35(1909), no.7,1-35.

[22]   J.E.Littlewood, On Equalities in the theory of functions, *Proc.London Math. Soc.* (2)23(1925), 481-519.

[23]   J.E.Littlewood, *Lectures on the Theory of Functions*, Oxford University Press, Oxford and London, 1944.

[24]   A.E.Livingston, On the radius of univalence of certain analytic functions, *Proc. Amer.Math.Soc.* 17(1966), 352-357.

[25]   T.H.MacGregor, Functions whose derivative has a positive realpart, *Trans. Amer.Math.Soc.* 104(1962), 532-537.

[26]   S.S.Miller and P.T.Mocanu, Classes of univalent integral operators, *J.Math. Anal.Appl.* 157(1991), 147-165.

[27]   S.S.Miller, P.T.Mocanu, and M.O.Reade, Starlike integral operators, *Pacific J.Math.* 79(1978), 157-168.

[28]   K.Nishimoto, *Fractional Calculus*, Vols.I,II,and III, Descartes Press, Koriyama, 1984, 1987, and 1989.

[29]   K.I.Noor, On some univalent integral operators, *J.Math. Anal. Appl.* 128(1987), 586-592.

[30]   K.Noshiro, On the theory of schlicht functions, *J.Fac. Sci. Hokkaido Univ.* (1)2(1934-1935), 129-155.

[31]   K.B.Oldham and Spanier, *The Fractional Calculus: Theory and Applications of Differentiation and Integration to Arbitrary Order*, Academic Press, New York and London, 1974.

[32]   S.Owa, On the distortion theorems. I, *Kyungpook Math.J.* 18(1978), 53-59.

[33]   S.Owa and H.M.Srivastava, Univalent and starlike generalized hypergeometric functions, *Canad. J.Math.* 39(1987). 1057-1077.

[34]   S.Owa and H.M.Srivastava, An application of a certain fractional derivative operator, *Proc. Japan Acad.Ser.A Math.Sci.* 66(1990),307-311.

[35]   S.Owa, M.Saigo, and H.M.Srivastava, Some characterization theorems for starlike and convex functions involving a certain fractional integral operator, *J.Math. Anal. Appl.* 140(1989), 419-426.

[36]   S.Owa, H.M.Srivastava, and C.Y.Shen, Applications of a certain linear operator defined by a Hadamard product or convolution, *Utilitas Math.* 33(1988), 173-181.

[37]   S.Owa and B.A.Uralegaddi, An application of the fractional calculus, *J.Karnatak Univ. Sci.* 32(1987), 194-201.

[38]   G.Pólya and G.Szegö, *Problems and Theorems in Analysis*, Vol.I *(Series, Integral Calculus, Theory of Functions)*, Translated from the German by D.Aeppli, Springer-Verlag, New York, Heidelberg, and Berlin, 1972.

[39]   W.Rogosinski, On subordinate functions, *Proc. Cambridge Philos. Soc.* 35(1939), 1-26.

[40]   W.Rogosinski, On the coefficients of subordinate functions, *Proc. London Math.Soc.* (2)48(1945),48-82.

[41]   S.Ruscheweyh, Linear operators between classes of prestarlike functions, *Comment. Math. Helv.* 52(1977), 497-509.

[42]   M.Saigo, A remark on integral operators involving the Gauss hypergeometric functions, *Math. Rep. College General Ed.Kyushu Univ.* 11(1978), 135-145.

[43]   S.G.Samko, A.A.Kilbas, and O.I.Marichev, *Integrals and Derivatives of Functional Order and Some of Their Applications* (Russian), Nauka i Tekhnika, Minsk,1987.

[44]   S.M.Sarangi and B.A.Uralegaddi, The radius of convexity and starlikeness for certain classes of analytic functions with negative coefficients. I, *Atti Acad.Naz. Lincei Rend. Cl. Sci. Fis. Mat. Natur.* (8) 65(1978), 38-42.

[45] R.Singh, On Bazilevič functions, *Proc. Amer. Math.Soc.* 38(1973), 261-271.

[46] N.S.Sohi, Distortion theorems involving certain operators of fractional calculus on a class of p-valent functions, In *Fractional Calculus and Its Applications* (k.Nishimoto, Editor), pp. 245-252, College of Engineering (Nihon University), Koriyama, 1990.

[47] J.B.Seaborn, *Hypergeometric Functions and Their Applications*, Springer Texts in Applied Mathematics, Vol.8, Springer-Verlag, New York, Berlin, Heidelberg, and London, 1991.

[48] H.M.Srivastava and R.G.Buschman, *Theory and Applications of Convolution Integral Equations*, Kluwer Academic Publishers, Dordrecht, Boston, and London, 1992.

[49] H.M.Srivastava and P.W.Karlsson, *Multiple Gaussian Hypergeometric Series*,Halsted Press (Ellis Horwood Limited, Chichester), John Wiley and Sons, New York, Chichester, Brisbane, and Toronto, 1985.

[50] H.M.Srivastava and B.K.K.Kashyap, *Special Functions in Queuing Theory and Related Stochastic Processes*, Academic Press, New York and London, 1982.

[51] H.M.Srivastava and H.L.Manocha, *A Treatise on Generating Functions*, Halsted Press (Ellis Horwood Limited, Chichester), John Wiley and Sons, New York, Chichester, Brisbane, and Toronto,1984.

[52] H.M.Srivastava and S.Owa, Certain classes of analytic functions with varying arguments, *J. Math. Anal. Appl.* 136(1988), 217-228.

[53] H.M.Srivastava and S. Owa (Editors), *Univalent Functions, Fractional Calculus, and Their Applications*, Halsted Press (Ellis Horwood Limited, Chichester), John Wiley and Sons, New York, Chichester, Brisbaine, and Toronto,1989.

[54] H.M.Srivastava and M.Saigo, Multiplication of fractional calculus operators and boundary value problems involving the Euler-Darboux equation, *J.Math.Anal.Appl.* 121(1987),325-369.

[55] H.M.Srivastava, M.Saigo, and S.Owa, A class of distortion theorems involving certain operators of fractional calculus, *J.Math.Anal.Appl.* 131(1988), 412-420.

[56] G.Szegö, *Orthogonal Polynomials*, Fourth Edition, American Mathematical Society Colloquium Publications, Vol.23, American Mathematical Society, Providence, Rhode Island, 1975.

[57] S.R.Tims, A theorem on functions schlicht in convex domains, *Proc. London Math. Soc.* (3) 1(1951), 200-205.

[58] S.Warschawski, On the higher derivatives at the boundary in conformal mapping, *Trans. Amer. Math, Soc.* 38(1935), 310-340.

[59] L.Weinstein, The Bieberbach conjecture, *Internat. Math. Res. Notices* 5(1991), 61-64.

[60] J.Wolff, L'integrale d'une fonction holomorphe et à partie réelle positive dans un demiplan est univalente, *C.R.Acad. Sci. Paris* 198(1934), 1209-1210.

Continuous and discrete Newton's algorithms

Lei Tan

UMPA, UMR 128 CNRS,École Normale Supérieure de Lyon
46 allée d'Italie,69364 Lyon cedex 07,FRANCE

Abstract: In the case that the continuous Newton's root-finding algorithm for a cubic complex polynomial has a saddle connection, we show that the discrete version of the algorithm, considered as a complex dynamical system, does not "converge" (its sense will be precised below), as the step h of discreteness tends to zero. The same result is given for all real polynomials of any degree with real saddle connections. We also show that while the step h tends to zero, the set of bad cubic polynomials for the discrete algorithm tends to a well defined and thin limit set, namely the set of cubic polynomials with saddle connections. With a convenient parametrization this limit set is just a connected graph made of two parts of circles and a segment.

1 Introduction

This work is motivated by works of Flexor and Sentenac (*[6]*) and Sauper (*[12]*, especially the computer generated picture figure 7).

Denote by $\overline{\mathbb{C}}$ the Riemann sphere, viewed as a one-point (the point ∞) compactification of the complex plane $\mathbb{C}$. Let $P : \mathbb{C} \to \mathbb{C}$ be a complex polynomial. We define a vector field $V = V_P$ by $V_P(z) = -\frac{P(z)}{P'(z)}\frac{\partial}{\partial z}$. It is called the *Newton's field* of P. For $h > 0$, we denote by $N_h = N_{h,P}$ the rational map $z \mapsto z - h\frac{P(z)}{P'(z)}$. It is called the *relaxed Newton's method* of P with *step h*. Note that the mapping N_1 with step 1 is the original Newton's method of P. These are continuous and discrete versions of a root-finding algorithm. In this paper we describe several relationships between V and N_h, especially when $h \to 0$. We make the assumption that all roots of P are simple. Denote by $X = \{x|P(x) = 0\}$ the set of roots of P, and by $C = \{c|P'(c) = 0\}$ the set of critical points of P.

Let us state in this section some well known results:

Lemma 1.1: *Properties of the Newton's field:*

1. *The field V is well defined in the complement of C, with singularities on $\{\infty\} \cup X$. The point ∞ is a source, each point of X is a sink. An appropriate rescaling field of V is well defined in a neighborhood of C with each point of C as a* hyperbolic saddle *singularity.*

Proceedings of the International Conference on Complex
Analysis at the Nankai Institute of Mathematics, 1992 pp. 208-220
©INTERNATIONAL PRESS 1994

2. *The image field of V by the differential of P is well defined in $\mathbb{C} - P(C)$, and coinsides with the radial field $-w\frac{\partial}{\partial w}$.*

3. *For each $z \in \mathbb{C} - (C \cup X)$, we denote by $\phi(t, z)$ the local flow of the field V with initial value z. There is a maximal interval $]t_z^-, t_z^+[$ such that $\phi(t, z)$ is well defined on it. Moreover, $\lim_{t \to t_z^-} \phi(t, z)$ is equal to either ∞ or a point of C, and $\lim_{t \to t_z^+} \phi(t, z)$ is equal to either a point of X or a point of C. If $\lim_{t \to t_z^-} \phi(t, z) = \infty$, then $t_z^- = -\infty$. If $\lim_{t \to t_z^+} \phi(t, z) \in X$, then $t_z^+ = +\infty$.*

4. *The set $T_P = \{z, \lim_{t \to t_z^-} \phi(t, z) \in C\} \cup C \cup X$ is a connected topological tree.*

5. *For each $x \in X$, there is a finite but non-zero number of trajectories coming from points of C, and landing on x.*

6. *Denote by S_P the set $\{z, \lim_{t \to t_z^+} \phi(t, z) \in C\} \cup C \cup \{\infty\}$. Then each connected component L in the complement of S_P is unbounded, simply connected, containing a unique root $x_L \in X$. And the restriction of P from L to $P(L)$ is a homeomorphism. Moreover L coincides with the set $\{z \in \mathbb{C} - C, \lim_{t \to t_z^+} \phi(t, z) = x_L\}$, or the attracting basin of x_L. The bijection $L \leftrightarrow x_L$ is sometimes written by $L_x \leftrightarrow x$.*

For the proof of the lemma, refer for example to *[10]* and *[18]*.

The set T_P is called Newton's graph or Shub-Smale tree of P. One can get many interesting informations about it in *[3]*, *[9]*, *[11]*, *[12]*, *[16]* for example.

Lemma 1.2: *Properties of the relaxed Newton's method:*

1. *For $0 < h < 2$, the set of fixed points of N_h is $\{\infty\} \cup X$, with ∞ as a repulsive fixed point and each $x \in X$ as an attracting fixed point. The eigenvalue of N_h at x is $1 - h$ (i.e $N_h'(x) = 1 - h$). It is independent of the choice of x in X.*

2. *For $0 < h < 2$, the Julia set J_h of N_h is connected (where the Julia set is defined to be the set of $z \in \overline{\mathbb{C}}$ such that the sequence of iterations of N_h is not normal in any neighborhood of z).*

Proof:

We can get 1 from a simple calculation. The point 2 is a consequence of a result of Shishikura. He proved in *[13]* that the Julia set J_1 for $h = 1$ is connected. But J_h is homeomorphic of J_1 (by a quasi-conformal surgery, cf. *[14]*). So J_h is connected as well. ∎

Lemma 1.3: *A relationship between V and N_h: Set $q = \#C$. For $c \in C$, let $k(c)$ be the multiplicity of P' at c (i.e. $P(z) = P(c) + K(z - c)^{k(c)+1} + \cdots$, with K a constant). Then N_h has $q + \sum_{c \in C} k(c)$ critical points $c_{h,i}$ (counting with multiplicity). To each $c \in C$ and each trajectory γ of the field V coming out from c correspond a unique critical point $c_{h,\gamma}$ of N_h. The set $\{c_{h,\gamma}\}_{0 < h \leq h_0}$ forms a continuous curve, asymptotic to γ as $h \to 0$, with $\lim_{h \to 0} c_{h,\gamma} = c$.*

Proof: It is a consequence of the Weierstrass preparation theorem. cf. *[6]*. ∎

For $0 < h < 2$ and $x \in X$ denote by $B_h(x) = \{z, N_h^n(z) \to x\}$ the attracting basin of x for N_h, and $A_h(x)$ the *immediate basin*, i.e. the connected component of $B_h(x)$ containing x. Set

$$BV_h = \{z \in \overline{\mathbb{C}} \mid N_h^n(z) \text{ does not converge to a root of } P\},$$

the set of bad initial values of N_h, considered as a root-finding algorithm. A theorem of Fatou applied to our case gives (cf. *[5],[4]*) :

Lemma 1.4: *If the critical points of N_h are either contained in $\cup_{x \in X} A_h(x)$, or strictly preperiodic, then BV_h has empty interior and zero Lebesgue measure.*

2   Consequences of results in $[6]$

A recent work of Flexor and Sentenac illustrates an important relationship between the continuous and discrete Newton's algorithm. In this section we state their result and some straightforward consequences.

We say that a polynomial P *has a connection* if the vector V_P has a saddle connection, i.e. a trajectory connecting one saddle to another saddle. In case that P has no connection, each trajectory coming from a point of C will land on a root of P (according to Lemma 1.1). Here is the result of Flexor and Sentenac :

Theorem 2.1: *([6]) Let P be a polynomial having no connection. There exists $h_P > 0$, such that for each $0 < h < h_P$, all critical points of $N_{h,P}$ are contained in $\cup_{x \in X} A_h(x)$. As a consequence, BV_h has empty interior and zero Lebesgue measure.*

One of their arguments (Lemma 5, *[6]*) uses the uniform convergence, over a compact set, of the Euler piece-wise linear approximations of the local flow of V_P. The same argument gives us (cf. as well *[7],[11]*)

Lemma 2.2: *Let P be any polynomial (having connections or not). For each compact subset K of a connected component L of $\overline{\mathbb{C}} - S_P$, there exists $l > 0$, such that for any $0 < h < l$, $K \subset A_h(x_L)$.*

An immediate consequence of this lemma is :

Corollary 2.3: *Let P be any polynomial. Give $\overline{\mathbb{C}} \approx S^2$ the spherical metric and the normalized Lebesgue measure m (i.e. $m(S^2) = 1$). We have*

$$J_h \subset BV_h \subset \overline{\mathbb{C}} - \cup_{x \in X} A_h(x).$$

And for any $\varepsilon > 0$, there exists $l > 0$, such that for $0 < h < l$, the set $\overline{\mathbb{C}} - \cup_{x \in X} A_h(x)$ is contained in the ε-neighborhood of S_P. Thus,

$$\lim_{h \to 0} m(J_h) = \lim_{h \to 0} m(BV_h) = \lim_{h \to 0} m(\overline{\mathbb{C}} - \cup_{x \in X} A_h(x)) = 0.$$

One can prove easily as well

$$\lim_{h\to 0} d(J_h, S_P) = \lim_{h\to 0} d(BV_h, S_P) = \lim_{h\to 0} d(\overline{\mathbb{C}} - \cup_{x\in X} A_h(x), S_P) = 0 \ ,$$

where d denotes the Hausdorff distance in the space of compact subsets of $\overline{\mathbb{C}}$. Theorem 2.1 can be generalized to the case of a family of polynomials.

Lemma 2.4: *Let Λ be a topological space with $\lambda_0 \in \Lambda$. Let $\{P_\lambda\}_{\lambda\in\Lambda}$ be a family of same degree polynomials, depending continuously on λ. Assume that none of the P_λ has connections. Assume in addition that one can parametrize continuously the critical points $c_i(\lambda)$ of P_λ, such that each $c_i(\lambda)$ has a multiplicity k_i which is independent of λ. Then the constant $h_{P_\lambda} = h_\lambda$ described in theorem 2.1 can be chosen as a constant in a neighborhood of λ_0.*

Proof: Let us check the proof of the theorem again, but this time taking into account the parameter λ. Let $c_i(\lambda_0) = c$ be a non-root critical point of P_{λ_0}, with multiplicity $k_i = k$. Let $c(\lambda)$ be the critical point of P_λ such that the mapping $\lambda \mapsto c(\lambda)$ is continuous, with $c(\lambda_0) = c$ and with multiplicity k. The choice of $c(\lambda)$ is unique by assumption. First let us normalize the polynomials. Set

$$b_\lambda = \left(-\frac{kk!P_\lambda(c_\lambda)}{P_\lambda^{(k+1)}(c_\lambda)} \right)^{\frac{1}{k+1}} , \quad \psi_\lambda : z \mapsto \frac{z - c(\lambda)}{b_\lambda},$$

with a continuous choice on λ of the $(k+1)$-th root. And set

$$Q_\lambda(z) = \frac{P_\lambda \circ \psi_\lambda^{-1}(z)}{P_\lambda(c_\lambda)} = 1 - \frac{k}{k+1} z^{k+1} + \cdots .$$

For each h, the mapping $N_{P_\lambda,h}$ is analytically conjugate to $N_{Q_\lambda,h}$. The six lemmas in *[6]* remain valid for $N_{Q_\lambda,h}$, with the constants independent of λ. This proves that for $h < l$ all critical points of $N_{Q_\lambda,h}$ which stay in a neighborhood of 1 are attracted by roots of Q_λ. Similarly all critical points of $N_{P_\lambda,h}$ in a neighborhood of $c(\lambda)$ are attracted by the roots of P_λ. We can apply the same process to each critical point $c_i(\lambda_0)$ of P_{λ_0}. And finally we take h_{λ_0} as the minimum of the l's. Then for all $0 < h < h_{\lambda_0}$, and for all λ in a neighborhood of λ_0, all critical points of $N_{P_\lambda,h}$ are attracted by roots of P_λ.   ∎

3   Polynomials with connections in the cubic case

We restrict ourself now to cubic polynomials in order to study the converse of theorem 2.1. We will see that once a polynomial has connections, its relaxed Newton's method $N_{h,P}$ behaves quite chaotically as $h \to 0$.

For $a \in \mathbb{C}$, set

$$P = P_a : z \mapsto (z + \frac{1}{2} - a)(z + \frac{1}{2} + a)(z - 1).$$

We will see in the next section that any cubic polynomial with connections is conjugate to some P_a, with $a \in]0, \frac{\sqrt{3}}{2}i[$ (here $]0, \frac{\sqrt{3}}{2}i[$ denotes the set $\{iy | 0 < y < \frac{\sqrt{3}}{2}\}$).

Theorem 3.1: *Assume $a \in]0, \frac{\sqrt{3}}{2}i[$. Then the polynomial P_a has a connection. For $0 < h < 2$, one critical point x_h of $N_h = N_{h,P_a}$ is contained in the complement of $\cup_{x \in X} A_h(x)$, the three other critical points are contained in $\cup_{x \in X} A_h(x))$. There exists a sequence $h_{1,n} \to 0$ such that $x_{h_{1,n}}$ is periodic for $N_{h_{1,n}}$. As a consequence, the set of bad initial values $BV_{h_{1,n}}$ has non-empty interior. On the other hand, there exists an other sequence $h_{2,n} \to 0$ such that $BV_{h_{2,n}}$ has empty interior and zero measure.*

Before proving the theorem, remark at first that the two critical points $\pm\sqrt{\frac{1}{4} + \frac{a^2}{3}}$ of P_a are real when $a \in]0, \frac{\sqrt{3}}{2}i[$. Set $c_- = -\sqrt{\frac{1}{4} + \frac{a^2}{3}}$ and $c_+ = \sqrt{\frac{1}{4} + \frac{a^2}{3}}$. Assume $a \in]0, \frac{\sqrt{3}}{2}i[$. Here are three preliminary lemmas whose proof will be postponed to the section 5.

Lemma 3.2: *The behavior of N_h on $]c_-, c_+[$ is :*
1. *For all $h \in]0, 2[$, we have*
$$\lim_{x \to c_-^+} N_h(x) = \lim_{x \to c_+^-} N_h(x) = -\infty,$$
and there is a unique critical point x_h of N_h in the interval $]c_-, c_+[$. Moreover, for all $h, l \in]0, 2[$, $h < l$, for all $x \in]c_-, c_+[$, we have $N_l(x) < N_h(x) < x$.
2. *For $h \in]0, 1]$, we have $x_h \geq 0$. The mapping N_h is strictly increasing in $]c_-, x_h[$, and strictly decreasing in $]x_h, c_+[$. Moreover, x_h is decreasing with respect to h, and $x_h \to c_+$ when $h \to 0$.*
3. *If there are a positive integer n and a value $l > 0$ such that $N_l^k(x_l) > c_-$ for any $k = 1, 2, \cdots, n$, then for all $0 < h < l$, we have $N_h^k(x_h) > N_l^k(x_l) > c_-$, $k = 1, \cdots, n$.*

Lemma 3.3: *The mapping $N_h(x)$ on $] - \infty, c_-[$ is :*
1. *increasing with respect to x, and $N_h :] - \infty, c_-[\to] - \infty, +\infty[$ is bijective,*
2. *asymptotic to $y = \frac{3-h}{3}x$ while $x \to -\infty$.*
3. *asymptotic to $x = c_-$ while $x \to c_-$.*
4. *increasing with respect to h (when the value of x remains fixed).*

Lemma 3.4: *For n sufficiently large, there exists $h > 0$ such that $N_h^k(x_h) > c_-$, for $k = 1, 2, \cdots, n$.*

Proof of Theorem 3.1 : Let $a \in]0, \frac{\sqrt{3}}{2}i[$. The polynomial P_a has a unique real root, which is the point 1. The other two roots are $-\frac{1}{2}+a, -\frac{1}{2}-a = \overline{-\frac{1}{2} + a}$. For $x \in]c_-, c_+[$, the scale $-\frac{P(x)}{P'(x)}$ of the vector $V(x)$ is negative. There exists then a trajectory γ of the Newton's flow from c_+ to c_-, which is exactly the real interval $]c_-, c_+[$. Since the coefficients of N_h are functions of a^2, they are all real. The mapping $N_h(x)$ preserves then the real axis $\mathbb{R}$. By Lemma 3.2, one critical point x_h of N_h is in $]c_-, c_+[$. We can conclude then $x_h \notin A_h(-\frac{1}{2} + a) \cup A_h(-\frac{1}{2} - a)$, because by symmetry $A_h(-\frac{1}{2} - a)$ and $A_h(-\frac{1}{2} + a)$ are complex conjugate to each other, thus both disjoint from $\mathbb{R}$. We claim as well $x_h \notin A_h(1)$. Because $A_h(1)$ is symmetric with respect to $\mathbb{R}$, simply connected, thus $A_h(1) \cap \mathbb{R}$ is connected. On the other hand, we have $x_h < c_+ < 1$, with $1 \in A_h(1)$ and $c_+ \notin A_h(1)$ (note that $N_h(c_+) = \infty$ and $N_h(\infty) = \infty$). (In fact $A_h(1) \cap \mathbb{R} =]c_+, +\infty[$. cf. [19]).

For n sufficiently large, set

$$h_n = \sup\{h > 0, N_h^k(x_h) > c_-, k = 1, 2, \cdots, n\}.$$

In the case $h = 1$, we have $x_1 = 0$. There exists $n_0 > 0$ such that $N_1^k(0) > c_-$ for $k = 1, \cdots, n_0$, and $N_1^{n_0+1}(0) \leq c_-$. Thus $h_{n_0} \geq 1$. It is clear that $h_{n_0+1} \leq h_{n_0}$. Moreover $N_h^{n_0+1}(x_h) < c_-$ for $h_{n_0+1} < h \leq 1$; $N_h^{n_0+1}(x_h) = c_-$ for $h = h_{n_0+1}$ and $N_h^{n_0+1}(x_h) > c_-$ for $0 < h < h_{n_0+1}$.

Assume $n \geq n_0 + 1, h \in]h_{n+1}, h_n[$. Then $N_h^n(x_h) > c_-$ and $N_h^{n+1}(x_h) < c_-$. Hence $N_h^{n+2}(x_h)$ is well defined and is continuous with respect to h. We show now that it is also decreasing with respect to h. In fact : for $h, l \in]h_{n+1}, h_n[$, and $h < l$:

$$N_h^{n+2}(x_h) - N_l^{n+2}(x_l) =$$

$$(N_h(N_h^{n+1}(x_h)) - N_l(N_h^{n+1}(x_h))) + (N_l(N_h^{n+1}(x_h)) - N_l(N_l^{n+1}(x_l))) .$$

The above value is positive by Lemma 3.3.4 and Lemma 3.2.3. When $h \to h_n^-$, we have $N_h^n(x_h) \to c_+^+$, $N_h^{n+1}(x_h) \to -\infty$, and $N_h^{n+2}(x_h) \to -\infty$. When $h \to h_{n+1}^+$, we have $N_h^{n+1}(x_h) \to c_-^-$ and $N_h^{n+2}(x_h) \to +\infty$. There exists then $h_{1,n} \in]h_{n+1}, h_n[$ such that $N_{h_{1,n}}^{n+2}(x_{h_{1,n}})$ is a preimage of $x_{h_{1,n}}$. In other words, the point $x_{h_{1,n}}$ is periodic of period $n + 3$.

Concerning the second sequence $h_{2,n}$ required by the theorem, one can simply set $h_{2,n} = h_n$. In this case $x_{h_{2,n}}$ is strictly preperiodic, and one can apply Lemma 1.4 to $BV_{h_{2,n}}$. We have to show now $h_n \to 0$. If $h_n \to h_0 \neq 0$, then $N_{h_0}^n(x_{h_0}) > c_-$ for all n. We have either $N_{h_0}^n(x_{h_0}) \to x > c_-$ or $N_{h_0}^n(x_{h_0}) \to c_-$. In the former case the limit point x must be a fixed point of N_{h_0}, but $N_{h_0}(x) < x$ for all $x \in]c_-, c_+[$, according to Lemma 3.2. In the last case we must have $N_{h_0}^{n+1}(x_{h_0}) \to -\infty$, which gives a contradiction.

4　A study in the parameter space

We study a stability problem in the space of cubic polynomials.

Let P, Q be two polynomials of degree d. It is easy to see that the following conditions are equivalent : Fix $h \in]0, 2[$,

1.　$N_{h,Q} = \psi^{-1} \circ N_{h,P} \circ \psi$, where ψ is a Möbius transformation.
2.　$N_{h,Q} = \psi^{-1} \circ N_{h,P} \circ \psi, \psi : z \mapsto az + b, a, b \in \mathbb{C}$
3.　The two trees T_P and T_Q described in Lemma 1.1 are isotopic (cf. for example [10])

4. $Q(z) = K \cdot P(az + b),\ a, b, K \in \mathbb{C}$.

Hence for any cubic polynomial P, its Newton's method $N_{h,P}$ is conjugate to $N_{h,a} = N_{h,P_a}$ for some P_a in the form $z \mapsto (z + \frac{1}{2} + a)(z + \frac{1}{2} - a)(z - 1)$. Moreover $N_{h,a}$ is conjugate to $N_{h,a'}$ iff $G(a) = G(a')$, where G is the rational map

$$a \in \mathbb{C} \mapsto -\frac{(\frac{3}{4} + a^2)^3}{(\frac{1}{4} - a^2)^2} \ .$$

Thus for any $h \in\ ]0, 2[$, the parameter a-plane can be considered as a covering space, of the space of the conformal conjugacy classes of relaxed Newton's method with step h, of all cubic polynomials.

One can determine all cubic polynomials with connections in a-plane, as showed in *[12]*, proposition 7 (with a slightly different parametrization):

Proposition 4.1: (Sauper) *The polynomial P_a has a connection iff $G(a) \in\ ] - \frac{27}{4}, 0[$.*

Remark that the segment $]0, \frac{\sqrt{3}}{2} i[$ is a fundamental domain of $G^{-1}(] - \frac{27}{4}, 0[)$ (compare with Theorem 3.1).

According to the result of Flexor and Sentenac, if P_a has no connection, the set of bad initial values $BV_{h,a}$ has zero measure as soon as the step h becomes sufficiently small. But for a fixed value of h, the set $BV_{h,a}$ can very well have non-zero measure. In this case the polynomial P_a is considered to be "bad" for the step h. We then define the set of bad polynomials in the parameter space: given h, set

$$BP_h = \{a \in \mathbb{C},\ BV_{h,a} \text{ has non-zero measure}\}$$

and the set of quasi-bad polynomials

$$QP_h = \{a \in \mathbb{C},\ \text{a critical point of } N_{h,a} \text{ is not contained in} \bigcup_{x, P_a(x)=0} A_{h,a}(x)\}$$

(recall that $A_{h,a}$ denotes the immediate attracting basin of x for the rational map N_{h,P_a}).

Denote by $\Phi = \text{closure}(G^{-1}(] - \frac{27}{4}, 0[)$, the set of P_a's with connections. A simple calculation shows

$$\Phi = \{iy | -\frac{\sqrt{3}}{2} \leq y \leq \frac{\sqrt{3}}{2}\} \bigcup \{-\frac{1}{2} + e^{i\theta} \mid \frac{\pi}{3} \leq \theta \leq \frac{5\pi}{3}\} \bigcup \{\frac{1}{2} + e^{i\theta} \mid -\frac{2\pi}{3} \leq \theta \leq \frac{2\pi}{3}\} \ .$$

We have

Lemma 4.2:
$$\Phi \subset QP_h,\ BP_h \subset QP_h \text{ for all } 0 < h < 2$$
$$\bigcap_{0<l<2} \bigcup_{0<h<l} BP_h = \bigcap_{0<l<2} \bigcup_{0<h<l} QP_h = \bigcap_{0<h<2} QP_h = \Phi.$$

Proof: The fact $\Phi \subset QP_h$ is a consequence of Proposition 4.1 and theorem 3.1 ; the fact $BP_h \subset QP_h$ follows from Fatou lemma (Lemma 1.4).

The equality $\bigcap_{0<h<2} QP_h = \Phi$ follows from Theorem 2.1. The inclusion $\Phi \subset \bigcap_{0<l<2} \bigcup_{0<h<l} BP_h$ is garanteed by the existence of the sequence $h_{1,n}$ in Theorem 3.1. Finally the inclusion $\bigcap_{0<l<2} \bigcup_{0<h<l} BP_h \subset \bigcap_{0<l<2} \bigcup_{0<h<l} QP_h \subset \Phi$ is again a consequence of Theorem 2.1. ∎

We can even get more:

Theorem 4.3: *Let m be the normalized Lebesgue measure on $\overline{\mathbb{C}}$.*
1. *For each $0 < h < 2$, $m(QP_h) \neq 0$.*
2. *$d(QP_h, \Phi) \to 0$ when $h \to 0$ (where d denotes the Hausdorff distance).*
3. *$m(QP_h) \to 0$ when $h \to 0$.*

Proof:

For 1, one can either make a quasi-conformal surgery (cf. *[7]*), or construct, as in Theorem 3.1 a sequence $a_n \in \,]0, \frac{\sqrt{3}}{2}i[$, with $a_n \to 0$ (the value of h is fixed this time) such that x_h is periodic for N_{h,a_n}. The point a_n is then an interior point of QP_h. The limit function of N_{h,a_n} as $a_n \to 0$ is no longer the identity, but a rational map of degree two:

$$z \mapsto \frac{(6-2h)z^2 - (3-h)z + h}{3(2z-1)}.$$

For 2 we note at first that the critical points of P_a for $a \in \mathbb{C} - \{\pm 1/2\}$ are simple. We can then apply Lemma 2.4 and choose the constant h_{P_a} depending continuously on a. Thus for each compact set $K \subset \overline{\mathbb{C}} - \Phi$, there exists $l > 0$, such that for each $0 < h < l$, we have $K \cap QP_h = \emptyset$. We can then arrange h and ε so that QP_h is contained in the ε-neighborhood of Φ.

The point 3 is an immediate consequence of 2. ∎

The computer generated picture fig. 7 in the paper of Sauper (*[12]*) is quite relevant to the above two results.

5 Proof of lemmas

Proof of Lemma 3.2.
1. Since $\frac{\partial^2 N(h,x)}{\partial x^2} = h\frac{d^2 N(1,x)}{dx^2} < 0$ for $x \in \,]c_-, c_+[$ and $h \in \,]0, 2[$, the mapping $\frac{\partial}{\partial x} N(h,x)$ is decreasing with respect to x. Since $\lim_{x \to c_-^+} N_h(x) = \lim_{x \to c_+^-} N_h(x) = -\infty$, there exists a unique critical point x_h of N_h in $]c_-, c_+[$. For $h \in \,]0, 2[$ and $x \in \,]c_-, c_+[$, we have $N_h(x) - x = -h\frac{P(x)}{P'(x)} < 0$, and $\frac{\partial N_h(x)}{\partial h} = -\frac{P(x)}{P'(x)} < 0$. We get 1.

2. We have $\frac{\partial}{\partial h}\frac{\partial}{\partial x} N(h,x) = \frac{d}{dx} N(1,x) - 1$. It is negative for $x \in [0, c_+[$. But when $h = 1, x = 0$, we have $\frac{\partial}{\partial x} N(1,x)|_{x=0} = 0$. Thus $\frac{\partial}{\partial x} N(h,x)|_{x=0} > 0$ for all $h \in \,]0, 1[$. On the other hand

$$\lim_{x \to c_+} \frac{\partial}{\partial x} N(h,x) = \lim_{x \to c_+} (1-h) + h\frac{PP''}{P'^2} = -\infty.$$

We have then $x_h \in \,]0, c_+[$. If $h < l < 1$, then $\frac{\partial}{\partial x} N(h, x_h) > \frac{\partial}{\partial x} N(l, x_l) = 0$ and then $x_h > x_l$. To prove the convergence $x_h \to c_+$ as $h \to 0$ it is sufficient to note that $N_h(x)$ is strictly increasing on $]c_-, x_h[$ and strictly decreasing on $]x_h, c_+[$, moreover on each compact subset of $]c_-, c_+[$, the mappings $\{N_h(x)\}|_{0<h<2}$ converge uniformly to the identity mapping x, while $h \to 0$.

3. Let us do it by induction. At first $x_h > x_l$. Set $\varepsilon_k = N_l^k(x_l) - N_h^k(x_h)$, $k = 1, 2, \cdots, n$. Then

$$\varepsilon_k = N_l(N_l^{k-1}(x_l)) - N_h(N_l^{k-1}(x_l)) + N_h(N_l^{k-1}(x_l)) - N_h(N_h^{k-1}(x_h)) = I_1 + I_2 \,,$$

with $I_1 < 0$ since $N_l(x) < N_h(x)$ for $x \in \,]c_-, c_+[$, and $N_l^{k-1}(x_l) \in \,]c_-, c_+[$. We have $I_2 < 0$ as well. Since $\varepsilon_{k-1} < 0$, so $c_- < N_l^{k-1}(x_l) < N_h^{k-1}(x_h) \le x_h$, and N_h is decreasing on $]c_-, x_h[$.

Proof of Lemma 3.3. Only 1 and 4 need to be verified. To prove 1, we have

$$\frac{\partial N_h(x)}{\partial x} = 1 - h + h \frac{P P''}{P'^2} > 0$$

for $x \in \,] - \infty, c_-[$. To prove 4, we have

$$\frac{\partial N_h(x)}{\partial h} = -\frac{P}{P'} > 0$$

for $x \in \,] - \infty, c_-[$.

Proof of Lemma 3.4. Given a, $0 < a < c_+$. For each η such that $0 < \eta < a$, there exists $h_\eta > 0$, such that for each $0 < h < h_\eta$ and each $x \in [-a, a]$, we have $|N_h(x) - x| < \eta$ (uniform convergence). Denote by Q the mapping: $x \mapsto x - \eta$. Fix η such that $a - (n+1)\eta > -a$. For each $x \in [a - \eta, a]$ and $k = 1, 2, \cdots, n$, we have $Q^k(x) = x - k\eta$, and $-a < Q^k(x) < Q^{k-1}(x)$. Assume $0 < h < h_\eta$. We have $Q(x) < N_h(x) < x$ for all $x \in [-a, a]$. Assume $x \in [a - \eta, a]$. We will show by induction that $N_h^k(x) > -a$ and $N_h^k(x) > Q^k(x)$, for $k = 1, 2, \cdots, n$. In fact, $N_h^k(x) - Q^k(x)$ can be considered as the sum of two parts $I_1 + I_2$, with $I_1 = N_h(N_h^{k-1}(x)) - Q(N_h^{k-1}(x))$ and $I_2 = Q(N_h^{k-1}(x)) - Q(Q^{k-1}(x))$. We have $I_1 > 0$ since $N_h^{k-1}(x) > -a$, and $I_2 > 0$ since $N_h^{k-1}(x) > Q^{k-1}(x)$. As a consequence, $N_h^k(x) > Q^k(x) > -a$.

According to Lemma 3.2, there exists l with $0 < l < h_\eta$ such that for each $0 < h < l$, we have $x_h \in \,]a, c_+[$. Assume now $0 < h < l$. For each $x \in [a, x_h]$, we have $N_h(x) \ge N_h(a) > a - \eta$ (because N_h is increasing). Since $N_h^k(x_h)$ is decreasing with respect to k, there exists a minimal integer $k_0 \ge 1$ such that $N_h^{k_0-1}(x_h) > a$ and $a - \eta < N_h^{k_0}(x_h) \le a$. We conclude from the above paragraph that $N_h^k(N_h^{k_0}(x_h)) > -a$, $k = 1, 2, \cdots, n$. In particular, $N_h^k(x_h) > -a > c_-$, $k = 1, 2, \cdots, n$.

6   About higher degree real polynomials

We ask ourself now the same question as in sections 2 and 3 for real polynomials of any degree, assuming that their Newton's flow on $\mathbb{R}$ is known. We can prove Theorem 3.1 similarly to those polynomials, but the proof of Theorem 2.1 is no more available (we don't have an analogue of Fatou Lemma on $\mathbb{R}$). Fortunately there is a theorem of Barna ($[1]$), which shows that for $h = 1$, and P a real polynomial with only real roots, the set $BV_{h,P} \cap \mathbb{R}$ has zero measure. The idea of the proof can also be applied to the case that not all roots of P are real, but we use $N_{h,P}$ instead of $N_{1,P}$, and we let h tend to 0. More precisely :

Theorem 6.1: *Let P be a real polynomial, of degree $d \geq 3$. Denote by X the set of roots of P and by C the set of non-root critical points of P. The following conditions are equivalent :*

1. *Either $\#C \cap \mathbb{R} \geq \#X \cap \mathbb{R} \geq 1$, or $\#X \cap \mathbb{R} = 0$ and $\#C \cap \mathbb{R} \geq 2$.*

2. *P has connections on $\mathbb{R}$.*

3. *There exists $h_n \to 0$ such that $BV_{P,h_n} \cap \mathbb{R}$ is different from $\mathbb{R}$ but with non-empty interior.*

4. *There exists $h_n \to 0$ such that $BV_{P,h_n} \cap \mathbb{R}$ is different from $\mathbb{R}$ but with non-zero measure.*

For the case $\#X \cap \mathbb{R} = 0, \#C \cap \mathbb{R} \leq 1$ we have $\#C \cap \mathbb{R} = 1$ and $BV_{h,P} \cap \mathbb{R} = \mathbb{R} \subset J_{N_{h,P}}$ for each $0 < h < 2$ (where J_N denotes the Julia set of N).

Proof: The Newton's flow and the relaxed Newton's method $N_{h,P}$ preserve $\mathbb{R}$ when P is real.

$1 \Leftrightarrow 2$ and $3 \Rightarrow 4$ are evident.

$2 \Rightarrow 3$ follows from the same proof as theorem 3.1.

$4 \Rightarrow 2$. We will prove that if P has no connection on $\mathbb{R}$, then either $BV_{h,P} = \mathbb{R}$ for each $0 < h < 2$, or $BV_{h,P} \cap \mathbb{R}$ has zero measure for sufficiently small h. Assume at first $\#X \cap \mathbb{R} = 0$, $\#C \cap \mathbb{R} \leq 1$. In this case the degree of P is even and the degree of P' is odd. Hence $\#C \cap \mathbb{R} = 1$. Since $N_{h,P}$ preserves $\mathbb{R}$, no point on the real axis can be captured by a non-real root. This shows $\mathbb{R} \subset BV_{h,P}$.

Assume now $\#X \cap \mathbb{R} > 0$. Because P has no connection on $\mathbb{R}$, for h sufficiently small, the graph of $N_{h,P}$ has the shape as shown in Figure 6.1. In this case one can easily prove that the union of the immediate basins of $x \in X \cap \mathbb{R}$ contains the set $\{y \mid |N'_{h,P}(y)| \leq 1\}$. Hence on $BV_{h,P} \cap \mathbb{R}$, the Newton's method $N_{h,P}$ is expanding. Since $BV_{h,P} \cap \mathbb{R}$ is invariant by $N_{h,P}$, it must have zero measure (Barna's argument).   ∎

To end this paper we say one more word about higher degree non-real polynomials. We have seen that $N_{h,P}$ is analytically conjugate to $N_{h,Q}$ iff there exists $a, b, c \in \mathbb{C}$ such that $P(z) = cQ(az + b)$. If Q maps a segment $I \subset \mathbb{C}$ into a ray, one can find a, b, c so that $P(z) = cQ(az + b)$ has real coefficients. In particular if Q has a connection which is a segment, we can make P real with real connections, and then apply the above results.

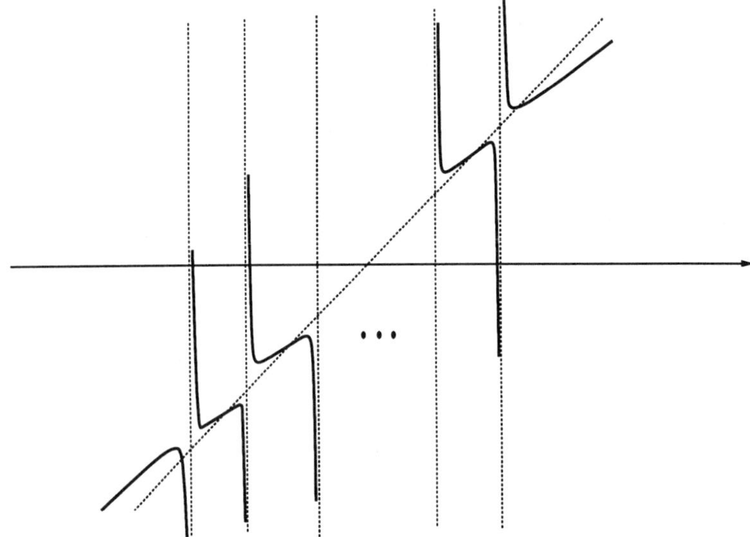

Figure 6.1: The graph of $N_{h,P}$

Acknowledgement. The author would like to thank A. Marin for helpful discussions, and E. Ghys for having read carefully the manuscript.

References

[1]   B. Barna, *Über die Divergenzpunkte des Newtonschen Verfahrens zur Bestimmung von Wurzeln algebraischer Gleichungen*, III, Publ. Math. Debrecen 8, 1961, 193-207

[2]   P. Blanchard, *Complex analytic dynamics on the Riemann sphere*, Bull. AMS, Vol 11, 1, July 1984, 85-141

[3]   D. Braess, *Über die Einzugsbereiche der Nullstellen von Polynomen beim Newton-Verfahren*, Numer. Math. 29, 1977, 123-132

[4]   A. Douady and J.H. Hubbard, *Etude dynamique des polynômes complexes*, II, Publ. Math. d'Orsay, 85-02, 1985

[5]   P. Fatou, *Sur les équations fonctionnelles*, Bull. Soc. Math. France 47, 1919, 161-271: 48, 1920, 33-94; 48, 1920, 208-314

[6]   M. Flexor and P. Sentenac, *Algorithmes de Newton généralisés*, CRAS 308, no.14, 1989, 445-448

[7]   F. von Haeseler and H. Kriete, *The relaxed Newton's method for polynomials*, preprint Universität Bremen, no.234, 1990

[8]   H. T. Jongen, P. Jonker and F. Twilt, *The continuous, desingularized Newton method for meromorphic functions*, Acta Applic. Math. 13, 1988, 81-121

[9]   J. Kahn, *Newtonian graphs for families of complex polynomials*, Jour. of Complexity, 7, 1991, 425-442

[10]   A. Marin, *Les arbres de Shub-Smale*, Séminaire de Géométrie algébrique réelle (J.J. Risler), Publ. Univ. Paris VII, 24, 1986

[11] H.-G. Meier, *The relaxed Newton-iteration for rational functions: The limiting case*, preprint Rhein.-Westf. Techn. Hocheschule Aachen, no.19, January 1990

[12] D. Saupe, *Discrete versus continuous Newton's method: A case study*, Acta Appl. Math. 13, 1988, 59-80

[13] M. Shishikura, *The connectivity of the Julia set and fixed points*, preprint IHES/M/90/37, 1990

[14] M. Shishikura, *On the quasi-conformal surgery of the rational functions*, Ann. scient. Éc. Norm. Sup., 4 série, t. 20, 1987, 1-29

[15] S. Smale, *The fundamental theorem of algebra and complexity theory*, Bull. A.M.S., volume 4, number 1, 1981, 1-36

[16] S. Smale, *On the efficiency of algorithms of analysis*, Bull. A.M.S., volume 13, number 2, 1985, 87-121

[17] M. Shub and S. Smale, *Computational complexity, On the geometry of polynomials and a theory of cost: Part I*, Ann. scient. Éc. Norm. Sup., 4 série, t. 18, 1985, 107-142

[18] M. Shub, D. Tischler and R. F. Williams, *The Newtonian graph of a complex polynomial*, Siam J. Math. Anal. Vol. 19, No. 1, January 1988, 246-256

[19] Tan L. *Cubic Newton's method of Thurston's type*, preprint ENS Lyon, no.26, 1990

On Generalization of Beurling-Ahlfors'Extension

Jian Wang and Zhimin Gong

Department of Mathematics, Xiangtan University
Xiangtan,Hunan,411105, P.R.China

Abstract: In this paper,a class of $Q.C.$extension will be constructed and an estimate of the maximal dilatation of them be given.The $Q.C.$ extensions of this kind include Beurling-Ahlfors'extension as a special one.

1 Introduction and Statement of Results

Let h be a ρ-quasi-symmetric function,that is ,h is a increasing homeomorphism from R onto itself and

$$(1.1) \qquad \frac{1}{\rho} \le \frac{h(x + t) - h(x)}{h(x) - h(x - t)} \le \rho$$

for all $x \in R$ and $t \ne 0$.

Denote by H_ρ the family of all such h.

The Beurling-Ahlfors'construction

$$(1.2) \qquad \begin{cases} \mu(x, y) = \frac{1}{2} \int_0^1 [h(x + ty) + h(x - ty)] dt \\ \nu(x, y) = \frac{r}{2} \int_0^1 [h(x + ty) - h(x - ty)] dt \\ f(x, y) = \mu(x, y) + i\nu(x, y) \end{cases}$$

gives a $Q.C.$ extension of h,and for $r{=}1$,Beurling and Ahlfors[1] obtained the first estimate of the maximal dilatation of f such as $K \le 2\rho(\rho + 1)$,before some other mathematicians [2],[3] improved the estimate of this kind.But the purpose of this paper is to construct a family of $Q.C.$ extension which include Beurling-Ahlfors'extension,and discuss the maximal dilatation.

let us replace dt in (1.2) with $d\mu(t)$ and $d\nu(t)$ respectively,both of which are nonnegative measures.More precisly,we construct two functions for fixed $h \in H_\rho$ as follows

$$(1.3) \qquad \begin{cases} \mu(h)(x, y) = \int_0^\infty [h(x + ty) + h(x - ty)] d\mu(t) \\ \nu(h)(x, y) = \int_0^\infty [h(x + ty) - h(x - ty)] d\nu(t) \end{cases}$$

We set $f_{\mu,\nu}(h) = \mu(h) + i\nu(h)$ and call it the generating function of $d\mu$ and $d\nu$. If $d\mu = \varphi(t)dt, d\nu = \psi(t)dt$,we denote $f_{\mu,\nu}(h)$ by $f_{\varphi,\psi}(h)$. Specially if $\varphi = \frac{1}{2}\chi_{[0,1]}, \psi = r\varphi$,then $f_{\varphi,\psi}$ is none other than Beurling-Ahlfors'extension.

Proceedings of the International Conference on Complex
Analysis at the Nankai Institute of Mathematics, 1992 pp. 220-225

Naturally, we ask what kind of measures $d\mu$ and $d\nu$ can make $f_{\mu,\nu}$ to be $Q.C.$ extensions of h for every $h \in H_\rho$?

Definition   Suppose $d\mu \geq 0$ and $\int_0^\infty d\mu = \frac{1}{2}$,a measure $d\nu \geq 0$ is said to be a quasi-conjugate-measure of $d\mu$ if $f_{\mu,\nu}$ are $Q.C.$ extensions of h for every $h \in H_\rho$.

Denote by $[d\mu]$ the family of all such $d\nu$. If $d\mu = \varphi dt$, we will denote by $[\varphi]$ the set of functions ψ such that $d\nu = \psi dt \in [d\mu]$.

Now the problem is to determine the conditions on $d\mu$,under which the set $[d\mu]$ is nonempty.We always assume $d\mu \geq 0$ and $\int_0^\infty d\mu = \frac{1}{2}$.

Denote by $K_{\mu,\nu}(h)$ the maximal dilatation of $f_{\mu,\nu}(h)$,Let

$$K_{\mu,\nu} = \sup_{h \in H_\rho} K_{\mu,\nu}(h)$$

If $d\mu = \varphi(t)dt, d\nu = \psi(t)dt$,we will prefer denote $K_{\mu,\nu}$ by $K_{\varphi,\psi}$.

The main results of ours follows

Theorem 1: *Suppose $\varphi \geq 0$ is a piecewise continous function of compact support and $\int_0^\infty \varphi(t)dt = \frac{1}{2}$.Let $n \geq 0$ be a integer and r be a positive number, then $\psi(t) = rt^n\varphi(t) \in [\varphi]$,and $f_{\varphi,\psi}$ are $K_{\varphi,\psi}$-Q.C.extensions of h for every $h \in H_\rho$,where $K_{\varphi,\psi}$ is a finite constant depending on φ and ψ.*

Remark.   We have only to suppose $\varphi \in L^\infty(0,\infty)$,other conditions unchanged,then the conclusions of theorem 1 remain true.

Theorem 2: *Suppose $\varphi \geq 0$ as in the remark above, Let $P(t) = \sum_{m=0}^{n} a_m t^m$, and $a_m \geq 0$, $a_m \not\equiv 0$, then $\psi = \varphi P \in [\varphi]$, and $f_{\varphi,\psi}$ are $K_{\varphi,\psi}$-Q.C.extensions of h for every $h \in H_\rho$, where $K_{\varphi,\psi}$ is a finite constant depending on φ and ψ.*

2   Proof of Theorems

We prove theorem 1 first. By hypotheses of the theorem,$f_{\varphi,\psi}(h)(x,0) = h(x)$. It is easy to check by the technique of mollification

$$K_{\varphi,\psi} = \sup_{h \in H_\rho \cap C^\infty} K_{\varphi,\psi}(h)$$

Suppose now $h \in H_\rho \cap C^\infty$,then we obtain from (1.3) that

$$(2.1) \quad \begin{cases} \mu_x(x,y) = \int_0^\infty [h'(x+ty) + h'(x-ty)]d\mu \\ \mu_y(x,y) = \int_0^\infty [th'(x+ty) - th'(x-ty)]d\mu \\ \nu_x(x,y) = \int_0^\infty [h'(x+ty) - h'(x-ty)]d\nu \\ \nu_y(x,y) = \int_0^\infty [th'(x+ty) + th'(x-ty)]d\nu \end{cases}$$

Since $h(at + b)$ is still in H_ρ for all $a > 0$ and $b \in R$, we have only to calculate the maximal dilatation of $f_{\mu,\nu}(h)$ at $x = 0, y = 1$. Set

$$J_{\varphi,\psi}(h) = \mid \partial_z f_{\varphi,\psi}(h)(0,1) \mid^2 - \mid \bar{\partial}_z f_{\varphi,\psi}(0,1) \mid^2 = (\mu_x\nu_y - \mu_y\nu_x)(0,1)$$

$$I_{\varphi,\psi}(h) = (\mu_x^2 + \mu_y^2 + \nu_x^2 + \nu_y^2)(0,1)$$

It is easy to show

(2.2)
$$K_{\varphi,\psi} = \sup_{h \in H_\rho \cap C^\infty} F\left(\frac{I_{\varphi,\psi}(h)}{2J_{\varphi,\psi}(h)}\right)$$

where $F(t) = t + \sqrt{t^2 - 1}$.

We know by the technique of mollification that if $f_{\varphi,\psi}(h)$ are K-$Q.C.$ extensions of h for every $h \in H_\rho \cap C^\infty$, then $f_{\varphi,\psi}$ are also K-$Q.C.$ extensions of h for every $h \in H_\rho$. So, in order to prove the theorem, it is enough to show $K_{\varphi,\psi} < +\infty$ and

(2.3)
$$J_{\varphi,\psi}(h) > 0$$

for all $h \in H_\rho \cap C^\infty$. Since $\psi = rt^n\varphi$, a direct calculation from (2.1) shows

(2.4)
$$J_{\varphi,\psi}(h) = J_1(h) + J_2(h) + J_1(h^*) + J_2(h^*)$$

where

$$\begin{cases} J_1(h) = r \int_0^\infty h'(-t)\varphi dt \int_0^\infty h'(t)t^{n+1}\varphi dt + r \int_0^\infty h'(-t)t\varphi dt \int_0^\infty h'(t)t^n\varphi dt \\ J_2(h) = r \int_0^\infty h'(t)\varphi dt \int_0^\infty h'(t)t^{n+1}\varphi dt - r \int_0^\infty h'(t)t\varphi dt \int_0^\infty h'(t)t^n\varphi dt \\ h^*(t) = -h(-t) \end{cases}$$

By the following inequality for any integer $n \geq 0$ and any nonnegative measure $d\mu$,

$$\int d\mu \int t^{n+1} d\mu \geq \int t d\mu \int t^n d\mu$$

we can easily prove $J_2(h) \geq 0$, and so (2.3) is true for all $h \in H_\rho \cap C^\infty$. By (2.1) and (2.4), $J_{\varphi,\psi}(h)$ is larger than

$$r \int_0^\infty h'(t)\varphi dt \int_0^\infty h'(-t)t^{n+1}\varphi dt + r \int_0^\infty h'(-t)\varphi dt \int_0^\infty h'(t)t^{n+1}\varphi dt$$

$$+r \int_0^\infty h'(-t)t\varphi dt \int_0^\infty h'(t)t^n\varphi dt + r \int_0^\infty h'(t)t\varphi dt \int_0^\infty h'(-t)t^n\varphi dt$$

Obviously,

$$I_{\varphi,\psi}(h) \leq 2\left(\int_0^\infty h'(t)\varphi dt\right)^2 + 2r^2\left(\int_0^\infty h'(-t)t^{n+1}\varphi dt\right)^2$$

$$+2\left(\int_0^\infty h'(-t)\varphi dt\right)^2 + 2r^2\left(\int_0^\infty h'(t)t^{n+1}\varphi dt\right)^2$$

$$+2r^2\left(\int_0^\infty h'(t)t^n\varphi dt\right)^2 + 2\left(\int_0^\infty h'(-t)t\varphi dt\right)^2$$

$$+2r^2\left(\int_0^\infty h'(-t)t^n\varphi dt\right)^2 + 2\left(\int_0^\infty h'(t)t\varphi dt\right)^2$$

Noticing that if $h \in H_\rho$ then $h^*(t) = -h(-t) \in H_\rho$, let $G(t) = t + \frac{1}{t}$ and we can get

$$(2.5) \qquad \sup_{h \in H_\rho \cap C^\infty} \frac{I_{\varphi, rt^n \varphi}}{2J_{\varphi, rt^n \varphi}} \leq \frac{\max(1, r^2)}{r} \sup_{h \in H_\rho \cap C^\infty} \max(G(A), G(B))$$

where $A = \dfrac{\int_0^\infty h'(t)\varphi dt}{\int_0^\infty h'(-t)t^{n+1}\varphi dt}$ and $B = \dfrac{\int_0^\infty h'(t)t\varphi dt}{\int_0^\infty h'(-t)t^n \varphi dt}$

By hypotheses of the theorem, there exists an interval $(\alpha, \beta) \subset supp\varphi$ such that $\varphi(t) \geq C > 0$ for some constant C,and we can assume $supp\varphi \subset [0, T]$, and $\beta - \alpha \geq \frac{T}{2^k}$,then

$$(2.6) \qquad \begin{cases} \int_0^\infty h'(t)t^{n+1}\varphi dt \leq BT^{n+1}[h(T) - h(0)] \\ \int_0^\infty h'(-t)t^{n+1}\varphi dt \leq BT^{n+1}[h(0) - h(-T)] \\ \int_0^\infty h'(t)t^{n+1}\varphi dt \geq \frac{C2^{n+1}}{(\rho+1)^{k+1}}[h(T) - h(0)] \\ \int_0^\infty h'(-t)t^{n+1}\varphi dt \geq \frac{C2^{n+1}}{(\rho+1)^{k+1}}[h(0) - h(-T)] \end{cases}$$

By (2.5) and (2.6) ,$K_{\varphi, rt^n \varphi}$ is less than

$$\frac{2B}{Cr}\rho(\rho + 1)^{k+1} \max(1, r^2) \max(\frac{1}{\alpha^{n+1}} + T^{n+1}, \frac{T}{\alpha^n} + \frac{T^n}{\alpha})$$

where $B = \max \varphi(t)$, and k is a positive integer satisfying $\beta - \alpha \geq \frac{T}{2^k}$ and may be taken as small as possible . This completes the proof.

Proof of theorem 2. A quick glance at the proof above shows $\psi = rt^n \varphi(t) \in [\varphi]$ .So,to prove theorem 2 it is sufficient to show that if $\psi_i \in [\varphi]$ for $i = 1, 2$,then $\psi = \psi_1 + \psi_2 \in [\varphi]$. In fact, for each $h \in H_\rho \cap C^\infty$ we have

$$J_{\varphi, \psi_1 + \psi_2}(h) = J_{\varphi, \psi_1}(h) + J_{\varphi, \psi_2}(h) > 0$$

$$I_{\varphi, \psi_1 + \psi_2}(h) \leq 2I_{\varphi, \psi_1}(h) + 2I_{\varphi, \psi_2}(h)$$

thus

$$\frac{I_{\varphi, \psi_1 + \psi_2}(h)}{2J_{\varphi, \psi_1 + \psi_2}(h)} \leq \frac{I_{\varphi, \psi_1}(h) + I_{\varphi, \psi_2}(h)}{J_{\varphi, \psi_1}(h) + J_{\varphi, \psi_2}(h)} \leq \max(\frac{I_{\varphi, \psi_1}(h)}{J_{\varphi, \psi_1}(h)}, \frac{I_{\varphi, \psi_2}(h)}{J_{\varphi, \psi_2}(h)})$$

and

$$K_{\varphi, \psi_1 + \psi_2} \leq 4 \max(K_{\varphi, \psi_1}, K_{\varphi, \psi_2}) < +\infty$$

the proof is completed.

3 Some Open Questions

It would be interesting to consider the following extremal problem

$$(3.1) \qquad K = \inf_{(\varphi, \psi)} K_{\varphi, \psi}$$

1. Does (3.1) has any solutions?

2. If (φ, ψ) is a solution of (3.1) ,what conditions should φ and ψ satisfy?

3. Is Beurling-Ahlfors'extension just a solution to (3.1)?

    Another problem is to determine whether or not there are many enough $Q.C.$ extensions such as $f_{\varphi,\psi}$.

4. Does there exist a $Q.C.$ extension of h such that it can not be approximated by the $Q.C.$ extensions of h such as $f_{\varphi,\psi}$?

References

[1] Ahlfors,L., Lecture on the Quasiconformal mapping,71-74.

[2] Li Zhong, On Beurling-Ahlfors'extension, Acta Math.Sinica,26(3)1988.

[3] Tan De-lin , On estimates of dilatation of Beurling-Ahlfors'extension, Chinese Ann.of Math.,7A(1),1986.

On the Problems of Meromorphic Derivatives

Yuefei Wang

Institute of Mathematics and Computing Center, Academia Sinica
Beijing 100080, P. R. China

Abstract: In this paper, we shall obtain the following results. (i) A precise estimate of the total deficiencies of meromorphic derivatives. (ii) A general criterion for normal families. (iii) A new proof of a conjecture of Hayman.

1 Introduction and Main Results

In the theory of value distribution, the research concerning the derivatives is an important and interesting topic. In this paper we shall deal with the following three problems. The notations and definitions of the theory will be freely used throughout the paper. (See Hayman *[7]* or Lo Yang *[13]*).

(A) Defects of Meromorphic Derivatives

Suppose that $f(z)$ is a transcendental meromorphic function in the plane. The most important and classical result is that $f(z)$ has vanishing deficiency for all but at most a countable set of values $a \in \overline{\mathbb{C}}$ and the total deficiency does not exceed two (*[13]*). The upper bound two is sharp in general. The deficiency problem is to find precise estimates of the total deficiencies for special classes of functions. This problem is still open for meromorphic functions of order bigger than 1.

For the derivatives of the function $f(z)$, a series of precise estimates of the total deficiency have been obtained by Hayman *[7]*, Mues *[15]*, and Lo Yang *[14]*. In *[14]* Lo Yang got a much better estimate for meromorphic derivatives. For any positive integer k, we have

$$\sum_{a \in \mathbb{C}} \delta(a, f^{(k)}) \leq 1 + \frac{1}{2k+1}.$$

However, the bound is not sharp. In fact Mues *[16]* conjectured that for a transcendental meromorphic function f, the inequality

$$(1.1) \qquad \sum_{a \in \mathbb{C}} \delta(a, f^{(k)}) \leq 1$$

Proceedings of the International Conference on Complex
Analysis at the Nankai Institute of Mathematics, 1992 pp. 225-235
©INTERNATIONAL PRESS 1994

holds for any positive integer k.

Recently Lo Yang and the author *[15]* first proved that (1.1) holds for all positive integers with at most finitely many exceptions of k, and then the author *[23]* affirmed further that the conjecture holds with at most four exceptions of k. A simple example shows that the upper bound one is attainable. Let $f(z) = e^z$. Then we have for $k = 1, 2, \cdots,$

$$\sum_{a \in \mathbb{C}} \delta(a, f^{(k)}) = 1.$$

It seems four exceptions do not exist, but we can not prove this assertion up to now.

Now we shall prove

Theorem 1: *Let $f(z)$ be a transcendental meromorphic function in the plane. Then for every positive integer k, there exist a number $d(k)$, $0 < d(k) \leq 1$, such that the inequality*

$$(1.2) \qquad \sum_{a \in \mathbb{C} \backslash \{0\}} \delta(a, f^{(k)}) \leq 1 - d(k)$$

holds with at most one exception of k.

We remark that one exception may exist. For some positive integer k_0, let $f(z) = e^z + z^{k_0}$. Then we have

$$\sum_{a \in \mathbb{C} \backslash \{0\}} \delta(a, f^{(k_0)}) = 1.$$

The better estimate for $\sum_{a \in \mathbb{C} \backslash \{0\}} \delta(a, f^{(k)})$ shows whether $\delta(a, f^{(k)})$ equals zero or not, the total deficiency of $f^{(k)}(z)$ at all non-zero finite values is always less than one for every positive integer k (posssibly one exception).

It is an interesting problem to determine the precise value of $d(k)$.

(B) A General Criterion for Normality

We now consider meromorphic functions in a domain $D \subset \mathbb{C}$, which satisfy the condition for all $z \in D$,

$$(1.3) \qquad\qquad (f^n)^{(k)}(z) \neq 1,$$

where n and k are positive integers.

The only meromorphic functions in the plane which fulfill condition (1.3) with $n \geq k + 1$ (except $n = 2, k = 1$) are constants. This was proved by Hayman *[6]* for $n \geq 4, k = 1$; by Mues *[17]* for $n = 3, k = 1$; by Hennekemper *[10]* for $n \geq k + 3$; and by the author *[23]* in the general case. In *[23]* we even eliminated the restriction $n \geq k+1$ and proved that the meromorphic functions in the plane which satisfy (1.3) with $n \geq 3, k \geq 1$ must be rational.

Corresponding to the Picard-type theorem, we shall prove the following general criterion for normal families.

Theorem 2: *If F is a family of meromorphic functions in a domain D and every $f \in F$ satisfies (1.3) with $n \geq k + 1$ (except $n = 2, k = 1$), then F is normal in D.*

Remarks: (i) Several criteria contained in Theorem 2 were previously proved by Gu *[22]* for $n \geq 4, k = 1$, by Pang *[21]* for $n = 3, k = 1$ and by Schwick *[17]* for $n \geq k + 3$ respectively.

(ii) The condition $n \geq k + 1$ in Theorem 2 is necessary though the Picard-type theorem holds for $n \geq 3, k \geq 1$. Example. Let $F = \{2mz\}_{m \in \mathbb{Z}}$ and $D : |z| < 1$. It is clear that the family F is not normal at $z = 0$, but if $n \leq k$, we always have for every $f \in F, z \in D$,

$$(f^n)^{(k)}(z) \neq 1.$$

(iii) Only one case "$n = 2, k = 1$" remains open for this kind of problems concerning the functions satisfying (1.3).

(C) A Conjecture of Hayman

In 1967, Hayman *[8]* posed the following conjecture. Suppose that $f(z)$ is meromorphic in the plane and that $f(z)f^{(k)}(z) \neq 0$, for some $k \geq 2$. Then $f(z) = e^{az+b}$ or $(Az + B)^{-n}$.

This conjecture was firstly proved by Frank in *[1]* for $k \geq 3$. Afterwards quite a few further results have been obtained (*[2, 4, 11, 19]*,etc.). Langley proved the conjecture for $k = 2$ very recently *[12]*. The main methods in their proofs are to use the techniques and arguments in the differential equations to deal with the solution systems of some complex differential equations.

By a different method based on an inequality which was established in *[23]*, we shall provide a simple proof to the conjecture. Further we shall prove the following theorem, from which the conjecture holds immediately.

Theorem 3: *Let $f(z)$ be meromorphic in the plane. Then for any integer $k \geq 3$, we have*

$$(1.4) \qquad T(r, f'/f) = O(\overline{N}(r, 1/f) + N(r, 1/f^{(k)}) + \log r), \quad r \overline{\in} E,$$

where E is a set with finite linear measure.

2   Lemmas

In this section we suppose that $f(z)$ is a transcendental meromorphic function in the plane and, k, n and m are positive integers. For the proofs of the theorems, we need the following lemmas.

Lemma 1: *(Frank and Kennekemper [3]).*

$$(2.1) \quad For \quad k \geq 1, \qquad m(r, f^{(k)}/f) = O(m(r, f'/f)) + S(r, f'/f),$$

$$(2.2) \quad For \quad n > k \geq 1, \quad m(r, f^{(n)}/f^{(k)}) = O(m(r, f'/f)) + S(r, f'/f),$$

$$(2.3) \quad For \quad k \geq 2, \quad m(r, f'/f) = O(\overline{N}(r, 1/f)) + \overline{N}(r, 1/f^{(k)}) + S(r, f'/f),$$

The next lemma will play an important role in the proofs. For our purpose, we shall state it in two forms.

Lemma 2: *(Wang [23]). For $k \geq 0, \varepsilon > 0$, we have*

$$(2.4) \quad (k-2)\overline{N}(r, f) + N(r, \frac{1}{f}) \leq 2\overline{N}(r, \frac{1}{f}) + N(r, \frac{1}{f^{(k)}}) + \varepsilon T(r, f) + S(r, f),$$

$$(2.5) \quad (k-2-\varepsilon)\overline{N}(r, f) \leq (1+\varepsilon)\overline{N}(r, \frac{1}{f}) + N(r, \frac{1}{f^{(k)}}) + O(m(r, \frac{f'}{f})) + S(r, \frac{f'}{f}).$$

Proof: (2.4) was already proved in *[23]*. Now we prove (2.5). From *[23]*, we have

$$(2.6) \quad \begin{aligned} (k-2-\varepsilon)\overline{N}(r, f) + (1-\varepsilon)N(r, \frac{1}{f}) &\leq 2\overline{N}(r, \frac{1}{f}) + N(r, \frac{1}{f^{(k)}}) \\ &+ O\left[\sum_{m \geq j \geq 1} m(r, \frac{f^{(j)}}{f}) + \sum_{2m \geq i > k} m(r, \frac{f^{(i)}}{f^{(k)}})\right], \end{aligned}$$

where the integer $2m$ is the degree of the Wronskian defined in *[22]*.

Lemma 1 gives for $m \geq j \geq 1, 2m \geq i > k$,

$$(2.7) \quad m(r, \frac{f^{(j)}}{f}) = O(m(r, \frac{f'}{f})) + S(r, \frac{f'}{f})$$

and

$$(2.8) \quad m(r, \frac{f^{(i)}}{f^{(k)}}) = O(m(r, \frac{f'}{f})) + S(r, \frac{f'}{f})$$

Therefore (2.5) follows from (2.6), (2.7) and (2.8).

We also need the following known results.

Lemma 3: *(Frank and Weissenborn [5]). For $k \geq 0, \varepsilon > 0$ we have*

$$k\overline{N}(r, f) \leq N(r, \frac{1}{f^{(k)}}) + (1+\varepsilon)N(r, f) + S(r, f^{(k)}).$$

Lemma 4: *(Hayman and Miles [9]). For a real number $K > 1$, there exists a set $M(K)$ of upper logarithmic density at most*

$$d(K) = \min\{(2e^{K-1} - 1)^{-1}, (1 + e(K-1))\exp(e(1-K))\},$$

such that for every positive integer k,

$$\varlimsup_{\substack{r \to \infty \\ r \bar{\in} M(K)}} \frac{T(r,f)}{T(r, f^{(k)})} \leq 3eK.$$

To prove the normal criterion, we use the following lemma, which is an extension of the principle originated with Zalcman *[24]*.

Lemma 5: *(Pang [21]) If a family F of meromorphic functions is not normal in the unit disk U, then for any real number β with $0 \leq \beta < 1$, there exist*
    (i) a real number $r_0, 0 < r_0 < 1$; (ii) complex numbers $z_j, |z_j| < r_0$; (iii) functions $f_j \in F$; and (iv) positive numbers ρ_j, $\lim_{j\to\infty} \rho_j = 0$, such that
$\rho_j^{-\beta} f_j(z_j + \rho_j \zeta)$ *converges uniformly on any compact subset of $\mathbb{C}$ to a nonconstant meromorphic function $g(\zeta)$.*

3   Proof of Theorem 1

Let $\triangle_k = \sum\limits_{a \in \mathbb{C}\backslash\{0\}} \delta(a, f^{(k)}), k = 1, 2, \cdots$. We need only to prove that for any positive integers $n, m(n > m)$, either $\triangle_n \leq 1 - d_n$ for some $d_n, 0 < d_n \leq 1$ or $\triangle_m \leq 1 - d_m$ for some $d_m, 0 < d_m \leq 1$.

It is obvious when $\triangle_m = 0$. Now we suppose that $\triangle_m > 0$. Hence there exists a finite set $\mathbb{C}_0$ in $\mathbb{C}\backslash\{0\}$, such that

$$\begin{aligned}
\frac{\triangle_m}{2} &\leq \sum_{a \in \mathbb{C}_0} \delta(a, f^{(m)}) \\
&\leq \varlimsup_{r \to \infty} \sum_{a \in \mathbb{C}_0} \frac{m(r, \frac{1}{f^{(m)} - a})}{T(r, f^{(m)})} \\
&\leq \varlimsup_{r \to \infty} \frac{[m(r, \frac{1}{f^{(n)}}) + S(r, f^{(m)})]}{T(r, f^{(m)})}.
\end{aligned}$$

Since $T(r, f^{(n)}) \leq (n - m + 1)T(r, f^{(m)}) + S(r, f^{(m)})$, we have

$$\delta_0 = \varlimsup_{\substack{r \to \infty \\ r \bar{\in} E_0}} \frac{m(r, \frac{1}{f^{(n)}})}{T(r, f^{(n)})} \geq \frac{\triangle_m}{2(n - m + 1)} > 0,$$

where E_0 is a set in $\mathbb{R}^+$ with finite measure.

We distinguish two cases.

(I). There exist a number $\sigma_0, 0 < \sigma_0 \leq 1$ and a set S_1 in $\mathbb{R}^+$ with infinite measure such that for $r \in S_1$,

$$\sigma_0 T(r, f^{(m)}) + (n - m)\overline{N}(r, f) \leq N(r, \frac{1}{f^{(n)}}).$$

Then we have for arbitrary finite set $\mathbb{C}_1$ in $\mathbb{C}$,

$$\sum_{a \in \mathbb{C}_1} m(r, \frac{1}{f^{(m)} - a}) \leq m(r, \frac{1}{f^{(n)}}) + S(r, f^{(m)})$$

$$\leq T(r, f^{(n)}) - N(r, \frac{1}{f^{(n)}}) + S(r, f^{(m)})$$

$$\leq T(r, f^{(m)}) + (n - m)\overline{N}(r, f) - N(r, \frac{1}{f^{(n)}}) + S(r, f^{(m)})$$

$$\leq (1 - \sigma_0)T(r, f^{(m)}) + S(r, f^{(m)}), \qquad (r \in S_1).$$

This means $\sum_{a \in \mathbb{C}} \delta(a, f^{(m)}) \leq 1 - \sigma_0$. Therefore $\triangle_m \leq 1 - d_m$ if we set $d_m = \sigma_0$.

(II) For arbitrary small positive number σ,

$$(3.1) \qquad N(r, \frac{1}{f^{(n)}}) \leq (n - m)\overline{N}(r, f) + \sigma T(r, f^{(m)})$$

holds on $\mathbb{R}^+$ except possibly a set E_1 of finite measure.

We distinguish two subcases.

(i). If there exist a number $\gamma_0, 0 < \gamma_0 \leq 1$, and a set S_2 in $\mathbb{R}^+$ with infinite measure such that

$$N(r, f) + \gamma_0 T(r, f^{(n)}) \leq n\overline{N}(r, f) + m(r, \frac{1}{f^{(n)}}),$$

then from Lemma 3, we have

$$\overline{N}(r, f) \leq N(r, \frac{1}{f^{(n+1)}}) - (1 + \varepsilon)\gamma_0 T(r, f^{(n)}) + (1 + \varepsilon)m(r, \frac{1}{f^{(n)}})$$
$$+ n\varepsilon\overline{N}(r, f) + S(r, f^{(n+1)}). \qquad (r \in S_2)$$

Hence for $r \in S_2$

$$\overline{N}(r, f) \leq N(r, \frac{1}{f^{(n+1)}}) - \gamma_0 T(r, f^{(n)}) + m(r, \frac{1}{f^{(n)}}) + 2\varepsilon T(r, f^{(n)}) + S(r, f^{(n)}).$$

Therefore for any finite set $\mathbb{C}_2$ in $\mathbb{C}\backslash\{0\}$,

$$m(r, \frac{1}{f^{(n)}}) + \sum_{a \in \mathbb{C}_2} m(r, \frac{1}{f^{(n)} - a}) \leq m(r, \frac{1}{f^{(n+1)}}) + S(r, f^{(n)})$$

$$\leq T(r, f^{(n)}) + \overline{N}(r, f) - N((r, \frac{1}{f^{(n+1)}}) + S(r, f^{(n)})$$

$$\leq (1 - \gamma_0 + 2\varepsilon)T(r, f^{(n)}) + m(r, \frac{1}{f^{(n)}}) + S(r, f^{(n)}), \quad (r \in S_2)$$

This yields

$$\triangle_n \leq 1 - \gamma_0 + 2\varepsilon.$$

Since ε is arbitrary, we have $\triangle_n \leq 1 - d_n$ by letting $\varepsilon \to 0$ and $d_n = \gamma_0$.

(ii) For arbitrary small positive number γ, the inequality

$$(3.2) \qquad n\overline{N}(r, f) + m(r, \frac{1}{f^{(n)}}) \leq N(r, f) + \gamma\, T(r, f^{(n)})$$

holds on $\mathbb{R}^+$ except possibly a set E_2 of finite measure.

We now apply Lemma 2 (2.4) to $\frac{1}{f^{(n)}(z)} = f_1(z)$, say. For $k = 2$, we have

$$N(r, \frac{1}{f_1}) - 2\overline{N}(r, \frac{1}{f_1}) \leq N(r, \frac{1}{f_1''}) + \varepsilon T(r, f_1) + S(r, f_1).$$

That is

$$(3.3) \qquad N(r, f^{(n)}) - 2\overline{N}(r, f^{(n)}) \leq N(r, \frac{1}{f_1''}) + \varepsilon T(r, f_1) + S(r, f_1),$$

From (3.1), (3.2) and (3.3) we have for any finite set $\mathbb{C}_3$ in $\mathbb{C}$,

$$\sum_{a \in \mathbb{C}_3} m(r, \frac{1}{f_1 - a}) \leq m(r, \frac{1}{f_1''}) + S(r, f_1)$$

$$\leq T(r, f_1) + 2\overline{N}(r, f_1) - N(r, \frac{1}{f_1''}) + S(r, f_1)$$

$$\leq T(r, f_1) + 2(n - m)\overline{N}(r, f) + 2\sigma T(r, f^{(m)})$$
$$-[N(r, f) + (n - 2)\overline{(r, f)}] + \varepsilon T(r, f_1) + S(r, f_1), \quad (r\overline{\in} E_1)$$
$$\leq T(r, f_1) + 2(1 - m)\overline{N}(r, f) + 2\sigma T(r, f^{(m)}) + \gamma T(r, f^{(n)}) -$$
$$-m(r, \frac{1}{f^{(n)}}) + \varepsilon T(r, f_1) + S(r, f_1). \quad (r\overline{\in} E_1 \cup E_2)$$

Giving a fixed $K > 1$, from Lemma 4, we have for $r \geq r_0, r\overline{\in} M(K)$

$$T(r, f^{(m)}) \leq 4eKT(r, f^{(n)}).$$

Let $E = E_0 \cup E_1 \cup E_2 \cup M(K) \cup [0, r_0]$. Therefore we get for $r\overline{\in} E$

$$\sum_{a \in \mathbb{C}_3} m(r, \frac{1}{f_1 - a}) + m(r, f_1) \leq (1 + 8eK\sigma + \gamma + \varepsilon)T(r, f_1) + S(r, f_1).$$

Since $T(r, f^{(n)}) = T(r, f_1) + O(1)$, we have

$$\sum_{a \in \mathbb{C}_3} \delta(a, f_1) + \delta_0 \leq \lim_{\substack{r \to \infty \\ r\overline{\in} E_0}} \frac{\sum_{a \in C_3} m(r, \frac{1}{f_1 - a}) + m(r, f_1)}{T(r, f_1)}$$

$$\leq \lim_{\substack{r \to \infty \\ r\overline{\in} E}} \frac{(1 + 8eK\sigma + \gamma + \varepsilon)T(r, f_1) + S(r, f_1)}{T(r, f_1)}$$

$$= 1 + 8eK\sigma + \gamma + \varepsilon.$$

Since σ, γ and ε are arbitrary, we finally get $\sum_{a \in \mathbb{C}} \delta(a, f_1) + \delta_0 \leq 1$. This means

$$\sum_{a \in \overline{\mathbb{C}} \backslash \{0\}} \delta(a, f^{(n)}) \leq 1 - \delta_0.$$

and hence we have $\triangle_n \leq 1 - d_n$ by letting $d_n = \delta_0$.

The proof of Theorem 1 is thus completed.

4   Proof of Theorem 2

Without loss of generality, we discuss the family F in the unit disk U. Suppose that F is not normal in U. We choose $\beta = \frac{k}{n}$. Then we have by Lemma 5, there exist (i) a real number r_0, $0 < r_0 < 1$; (ii) complex numbers $z_j, |z_j| < r_0$; (iii) functions $f_j \in F$; and (iv) positive numbers ρ_j, $\lim_{j \to \infty} \rho_j = 0$ such that $g_j(\zeta) = \rho_j^{-\frac{k}{n}} f_j(z_j + \rho_j \zeta)$ converges uniformly on any compact subset of $\mathbb{C}$ to a nonconstant meromorphic function $g(\zeta)$. We distinguish two cases.

(I) If $(g^n)^{(k)}(\zeta) \neq 1$, then by the result in *[23]*, $g(\zeta]$ must be a constant. It is a contradiction.

(II) If there exists ζ_0, such that

$$(g^n)^{(k)}(\zeta_0) - 1 = 0.$$

So $g(\zeta)$ is analytic at ζ_0. Then there exists a positive number δ, such that $g(\zeta), g_j(\zeta)$ are analytic in $D_{2\delta} = \{\zeta : |\zeta - \zeta_0| < 2\delta\}$ for sufficient large j. Thus $(g_j^n)^{(k)}$ converges uniformaly on $\overline{D}_\delta = \{\zeta : |\zeta - \zeta_0| \leq \delta\}$ to $(g^n)^{(k)}$. Since $(g_j^n(\zeta))^{(k)} = (\rho_j^{-k} f^n(z_j + \rho_j \zeta))^{(k)} = (f^n(z_j + \rho_j \zeta))^{(k)}|_{z_j + \rho_j \zeta} \neq 1.(\zeta \in \overline{D}_\delta, j \geq j_0)$, we have from Hurwitz's theorem *[20]*,

$$(g^n(\zeta))^{(k)} \equiv 1, \quad \zeta \in \mathbb{C}.$$

Therefore

$$(4.1) \qquad\qquad g^n(\zeta) = P_k(\zeta)$$

where $P_k(\zeta)$ is a polynomial with degree k.

Now we suppose that ζ_1 is a zero of $P_k(\zeta)$. Then ζ_1 is also a zero of $g(\zeta)$. Hence in a neighborhood of ζ_1, we have

$$(4.2) \qquad\qquad P_k(\zeta) = (\zeta - \zeta_1)^l P_0(\zeta), (l \leq k)$$

and

$$(4.3) \qquad\qquad g(\zeta) = (\zeta - \zeta_1)^m h(\zeta), (m \geq 1)$$

where $P_0(\zeta)$ is a polynomial with degree $k - l$, $P_0(\zeta_1) \neq 0$ and $h(\zeta)$ is an analytic function in the neighborhood. (4.1), (4.2) and (4.3) give

$$(4.4) \qquad\qquad (\zeta - \zeta_1)^{nm-l} h^n(\zeta) = P_0(\zeta).$$

Since $nm - l \geq n - l \geq n - k \geq 1$, it leads to a contradiction when $\zeta \to \zeta_1$ in (4.4).

So we get the conclusion of Theorem 2.

5 Proof of Theorem 3

We prove the theorem by using Lemma 1 and Lemma 2. (2.3) and (2.5) give

$$(5.1) \qquad m(r, \frac{f'}{f}) = O(\overline{N}(r, \frac{1}{f}) + N(r, \frac{1}{f^{(k)}})) + S(r, \frac{f'}{f}),$$

and

$$(5.2) \quad (k-2-\varepsilon)\overline{N}(r, f) \le (1+\varepsilon)\overline{N}(r, \frac{1}{f}) + N(r, \frac{1}{f^{(k)}})) + O(m(r, \frac{f'}{f})) + S(r, \frac{f'}{f}),$$

On the other hand, we have

$$(5.3) \qquad T(r, \frac{f'}{f}) = \overline{N}(r, f) + \overline{N}(r, \frac{1}{f}) + m(r, \frac{f'}{f}),$$

and

$$(5.4) \qquad S(r, \frac{f'}{f}) = O(\log T(r, \frac{f'}{f}) + \log r), \quad (r \overline{\in} E)$$

Therefore (1.4) follows from (5.1), (5.2), (5.3) and (5.4). We complete the proof of the theorem.

Note: Acknowledgements The author is indebted to Professor Lo Yang for his valuable suggestions and kind guidance. He is also grateful to Professor Günter Frank for some very helpful comments.

References

[1] Frank, G., Eine Vermutung von Hayman Über Nullstellen Meromorphe Funcktionen, Math. Z. 149(1976), 29-36.

[2] Frank G. and Hellerstein, S., On the meromorphic sloutions of non-homogeneous linear differential equations with polynomial coefficients, Proc. London Math. Soc. (3), 53(1986), 407-428.

[3] Frank G. and Hennkemper, W., Einige Ergebnisse Über die Werteverteilung meromorphe Funktionen und ihrer Ableitungen, Result. Math., 4(1981), 39-54.

[4] Frank, G., Hennkemper, W. and Polloczek, G., Über die Nullstellen meromorphe Funktionen und ihrer Ableitungen, Math. Ann. 255 (1971), 145-154.

[5] Frank, G. and Weissenbern, G., Rational deficient functions of meromorphic functions, Bull. London Math. Soc., 18 (1986), 29-33.

[6] Hayman, W. K., Picard values of meromorphic functions and their derivatives, Ann. of Math., 70 (1959), 9-42.

[7] Hayman, W. K., "Meromorphic functions", Clarendon Press, Oxford, 1964.

[8] Hayman, W. K., " Research problems in function theory", Athlone Press, 1967.

[9] Hayman, W. K. and Miles, J., On the growth of a meromorphic function and its derivatives, Complex Variables, 12 (1989), 245-260.

[10] Hennekemper, G., Über Fragen der Werteverteilung von Differtntiaploynomen meromorpher Funktionen, Dissertation Hagen, 1979.

[11]   Langley, J. K., The Tsuji characteristic and zeros of linear differential polynomials, Analysis, 9(1989), 269-282.

[12]   Langley, J. K., Proof of a conjecture of Hayman concerning f and f'', Preprint.

[13]   Lo Yang, " Value distribution theory and its new research", The Scientific Press, Beijing, 1982.

[14]   Lo Yang, Precise estimate of total deficiency of meromorphic derivatives, J. d'Analyse Math. 55 (1990), 287-296.

[15]   Lo Yang and Yuefei Wang, Drasin's problems and Mues' conjecture, Science in China (Series A), (10) 35 (1992), 1180-1190.

[16]   Mues, E., Über eine Defekt und Verzweigungsration für die Ableitung Meromorpher Funktionen, Manuscripta Math., 5 (1971), 275-297.

[17]   Mues, E., Über ein problem von Hayman, Math. Z., 164 (1979), 239-259.

[18]   Schwick, W., Normality criteria for families of meromorphic functions, J. d'Ana-lyse Math. 52 (1989), 241-289.

[19]   Steinmetz, N., On the zeros of $(f^{(p)} + a_{p-1}f^{(p-1)} + \cdots + a_0 f)f$, Analysis, 7 (1987), 375-389.

[20]   Titchmarch, E. C., " The theory of functions", 2nd edition, Oxford, 1939.

[21]   Xuecheng Pang, Bloch's principle and normal criterion, Science in China (Series A), (7) 32 (1989), 782-791.

[22]   Yongxing Gu, On normal criterion of meromorphic functions, Scientia Sinica, 4 (1978), 373-384.

[23]   Yuefei Wang, On Mues conjecture and Picard values, Science in China (Series A), (1) 36 (1993), 28-35.

[24]   Zalcman, L., A Heuristic principle in complex function theory, Amer. Math. Monthly, 82 (1975), 813-817.

Keywords. meromorphic function, defect, normal criterion.
1980 Mathematics *Mathematics subject classifications:* 30D35

Estimates For The Logarithmic Derivative Of A Meromorphic Function In An Angle, And Their Application

Shengjian Wu

Department of Mathematics, Peking University
Beijing 100871, P. R. China

Abstract: Let f be meromorphic on an angular region with finite order. Then for integers $n > m \geq 0$,

$$\left| \frac{f^{(n)}(z)}{f^{(m)}(z)} \right| < |z|^M$$

for $z \in G$, where G is some particular set of points and M is a finite constant. These estimates are used to study the growth of solutions of linear differential equations in an angular region.

1 Introduction

Let $f(z)$ be meromorphic in $\mathbb{C}$ with finite order ρ. Then [3] for every pair of integers n and m with $n > m \geq 0$

$$(1.1) \qquad \left| \frac{f^{(n)}(z)}{f^{(m)}(z)} \right| < |z|^{(n-m)(\rho+\varepsilon)}$$

for all $z \in \mathbb{C}$ except for a $\mathbb{R}$-set D, that is, a set of disks whose radii have a finite sum. The property that $\frac{f^{(n)}(z)}{f^{(m)}(z)}$ is polynomial-like in $\mathbb{C} \backslash D$ is useful in the study of the analytic theory of linear differential equations.

If $f(z)$ is meromorphic only on an angular region, Mokhon'ko recently obtained, among other things, the following result.

Theorem A: *[4] If $z = r \exp(i\varphi) \in \Omega(\alpha, \beta) = \{\alpha \leq \arg z \leq \beta\} \cap \{|z| \geq r_0\}$ and $f(z)$ is meromorphic on $\overline{\Omega}(\alpha, \beta)$ with finite order ρ, then for every $\varepsilon > 0$ there exists $K > 0$ depending only on f and ε such that*

$$\left| \frac{f'(z)}{f(z)} \right| < K|z|^{k+2\rho+1+\varepsilon}(\sin k(\varphi - \alpha))^{-2}$$

for all $z \in \Omega(\alpha, \beta)$ except for a $\mathbb{R}$-set, where $k = \frac{\pi}{\beta - \alpha}$.

Proceedings of the International Conference on Complex
Analysis at the Nankai Institute of Mathematics, 1992 pp. 235-241
©INTERNATIONAL PRESS 1994

For analytic functions on an angular region, we have extended Theorem A and obtained an analogy of (1.1) in *[5]*. By using these estimates we have also obtained some further results on the oscilation theory of linear differential equations. In this paper, we consider similar estimates of (1.1) of a meromorphic function on an angular region. In what follows, we shall adopt the usual notation of Nevanlinna theory (see *[2]*).

Theorem 1: *Let f be meromorphic on the angular region $\overline{\Omega}(\alpha, \beta)$ with finite order ρ, let $\Gamma = \{(n_1, m_1), (n_2, m_2), \cdots, (n_j, m_j)\}$ denote a finite set of distinct pair of integers which satisfy $n_i > m_i \geq 0$ for $i = 1, 2, \cdots, j$, and let $\varepsilon > 0$ and $\delta > 0$ be given constants. Then there exists $K > 0$ depending only on f, ε and δ such that*

$$(1.2) \qquad \left| \frac{f^{(n)}(z)}{f^{(m)}(z)} \right| < K|z|^{(n-m)(k_\delta + 2\rho + 1 + \varepsilon)} (\sin k_\delta(\varphi - \alpha - \delta))^{-2^{(n-m)}},$$

for all $(n, m) \in \Gamma$ and all $z = re^{i\varphi} \in \Omega(\alpha + \delta, \beta - \delta)$ except for a $\mathbb{R}$-set, where $k_\delta = \frac{\pi}{\beta - \alpha - 2\delta}$.

Remark. We note that Goldberg *[1]* has constructed an example which shows that the Nevanlinna's lemma concerning the logarithmic derivative is not true for meromorphic functions in halfplane. It seems difficult to get (1.2) without introducing the positive constant δ. Since it is impossible to obtain an estimate of the logarithmic derivative which is uniformly with respect to $\arg z$, there would not raise any difficulty in the application of Theorem 1. The interesting open problem is that whether or not a function which is meromorphic on an angular region and its derivatives have the same order.

2   Proof of Theorem 1

The proof of Theorem 1 will be presented only for the case in which $\Gamma = \{(2, 0)\}$ for the purpose of making the expression less cumbersome. In the general case the reasonings may be conducted precisely in the same manner for $f^{(i)}(z)(i = 1, 2, \cdots)$. In the following we set $k_\delta = \frac{\pi}{\beta - \alpha - 2\delta}$, and denote the order of the Nevanlinna function $S_{\alpha + \delta, \beta - \delta}(r, f)(0 < \delta < \frac{\beta - \alpha}{2})$ by $\rho_\delta(f)$.

We first show that $\rho_\delta(f') \leq \rho$ for any $\delta > 0$. From *[2]*(p. 43), there exists a nondecreasing function $\overset{\circ}{S}_{\alpha\beta}(r, f)$ defined by

$$(2.1) \;\; \overset{\circ}{S}_{\alpha\beta}(r, f) = \frac{k}{\pi} \int_1^r \int_\alpha^\beta \left(\frac{1}{t^k} - \frac{t^k}{r^{2k}} \right) \left(\frac{|f'(te^{i\varphi})|}{1 + |f(te^{i\varphi})|^2} \right)^2 \sin k(\varphi - \alpha) t \, d\varphi \, dt,$$

where $k = \frac{\pi}{\beta - \alpha}$, such that

$$(2.2) \qquad \overset{\circ}{S}_{\alpha\beta}(r, f) = S_{\alpha\beta}(r, f) + O(1)$$

for all $r \in (1, \infty)$.

Since $\frac{1}{2t^k} \leq \frac{1}{t^k} - \frac{t^k}{r^{2k}}$ for $1 \leq t \leq \frac{1}{2^{\frac{1}{2k}}} r$,

$$(2.3) \qquad \overset{\circ}{S}_{\alpha\beta}(r,f) \geq \frac{k}{2\pi} \int_1^{\frac{r}{2^{\frac{1}{2k}}}} \int_{\alpha+\delta}^{\beta-\delta} \frac{1}{t^{k-1}} \left(\frac{|f'(te^{i\varphi})|}{1+|f(te^{i\varphi})|^2} \right)^2 \sin k(\varphi-\alpha)\,d\varphi dt$$

$$\geq \frac{k\sin k\delta}{2\pi} \int_1^{\frac{r}{2^{\frac{1}{2k}}}} \int_{\alpha+\delta}^{\beta-\delta} \frac{1}{t^{k-}}$$

Thus

$$(2.4) \qquad \overset{\circ}{S}_{\alpha+\delta,\beta-\delta}\left(\frac{r}{2^{\frac{1}{2k}}}, f \right) \leq \frac{k}{\pi} \int_1^{\frac{r}{2^{\frac{1}{2k}}}} \int_{\alpha+\delta}^{\beta-\delta} \frac{1}{t^{k\delta-1}} \left(\frac{|f'(te^{i\varphi})|}{1+|f(te^{i\varphi})|^2} \right)^2 d\varphi dt$$

$$\leq \frac{2}{\sin k\delta} \overset{\circ}{S}_{\alpha,\beta}(r,f).$$

It follows from (2.4) that $\rho_\delta(f) \leq \rho$. To prove $\rho_\delta(f') \leq \rho$, we use Theorem A. Suppose that D is the $\mathbb{R}$-set in Theorem A. Since the set of the values θ such that the ray $\arg z = \theta$ meets infinitely many disks in D has linear measure zero, there exists $\delta'(0 < \delta' < \delta)$ such that both of the rays $\arg z = \alpha + \delta'$ and $\arg z = \beta - \delta'$ meet only a finite number of disks in D. In view of Theorem A, if $\varphi = \alpha + \delta'$ or $\beta - \delta'$, then

$$(2.5) \qquad \left| \frac{f'(re^{i\varphi})}{f(re^{i\varphi})} \right| < r^M$$

for all sufficiently large r, where $M(> 0)$ is a constant not depending on r. It is obvious that if $(|z| = r) \cap D = \emptyset$, then (2.5) also holds for all $\varphi \in (\alpha+\delta', \beta-\delta')$. Let E denote the set of r satisfying $(|z| = r) \cap D \neq \emptyset$. We deduce from (2.5) that

$$(2.6) \qquad A_{\alpha+\delta',\beta-\delta'}\left(r, \frac{f'}{f}\right) + B_{\alpha+\delta',\beta-\delta'}\left(r, \frac{f'}{f}\right) = O(1)$$

for all $r \overline{\in} E$.

In view of the Nevanlinna's first fundamental theorem and (2.6) we deduce that

$$
\begin{aligned}
(2.7) \qquad S_{\alpha+\delta',\beta-\delta'}(r,f') &\leq S_{\alpha+\delta',\beta-\delta'}(r,f) + S_{\alpha+\delta',\beta-\delta'}\left(r, \frac{f'}{f}\right) \\
&\leq S_{\alpha+\delta',\beta-\delta'}(r,f) + C_{\alpha+\delta',\beta-\delta'}\left(r, \frac{f'}{f}\right) + O(1) \\
&\leq S_{\alpha+\delta',\beta-\delta'}(r,f) + C_{\alpha+\delta',\beta-\delta'}(r,f) \\
&\quad + C_{\alpha+\delta',\beta-\delta'}\left(r, \frac{1}{f}\right) + O(1) \\
&\leq 3S_{\alpha+\delta',\beta-\delta'}(r,f) + O(1)
\end{aligned}
$$

for all $r \overline{\in} E$.

From (2.2) and (2.7) we have

$$\overset{\circ}{S}_{\alpha+\delta',\beta-\delta'}(r,f') \le 3\overset{\circ}{S}_{\alpha+\delta',\beta-\delta'}(r,f) + O(1)$$

for all $r\overline{\in}E$.

Since the linear measure of E is finite, we deduce that $[2]$(p. 360)

$$(2.8) \qquad \overset{\circ}{S}_{\alpha+\delta',\beta-\delta'}(r,f') \le 3\overset{\circ}{S}_{\alpha+\delta',\beta-\delta'}(2r,f) + O(1)$$

for all sufficiently large r. (2.8) together with the fact that $\rho_\delta(f') \le \rho_{\delta'}(f')$ implies $\rho_\delta(f') \le \rho$ and this proves our assertion.

By applying Theorem A to f and f' in $\Omega(\alpha + \delta, \beta - \delta)$, there exists $K > 0$ such that

$$(2.9) \qquad \left|\frac{f'(re^{i\varphi})}{f(re^{i\varphi})}\right| < Kr^{k_\delta+2\rho+1+\varepsilon}(\sin k_\delta(\varphi - \alpha - \delta))^{-2},$$

$$(2.10) \qquad \left|\frac{f''(re^{i\varphi})}{f'(re^{i\varphi})}\right| < Kr^{k_\delta+2\rho+1+\varepsilon}(\sin k_\delta(\varphi - \alpha - \delta))^{-2},$$

for all $z = re^{i\varphi} \in \Omega(\alpha + \delta, \beta - \delta) \cap (|z| > r_0)$ except for a $\mathbb{R}$-set.

Noting that the union of a finite number of $\mathbb{R}$-sets is also a $\mathbb{R}$-set, Theorem 1 follows from (2.9) and (2.10). The proof of Theorem 1 is completed.

3   Application of Theorem 1

In this section, we shall use Theorem 1 to investigate the growth of solutions of

$$(3.1) \qquad f'' + Af' + Bf = 0,$$

where A and B are meromorphic functions.

Let f be meromorphic in $\mathbb{C}$ with order $\rho(0 \le \rho \le +\infty)$ and let $\Omega(\alpha,\beta)(0 < \beta - \alpha \le 2\pi)$ be an angular region. Since

$$\overset{\circ}{S}_{\alpha\beta}(r,f) \le \frac{k}{\pi}\int_0^r\int_0^{2\pi}\left(\frac{|f'(te^{i\varphi})|}{1+|f(te^{i\varphi})|^2}\right)^2 td\varphi dt$$
$$\equiv kA(r,f)$$

where $k = \frac{\pi}{\beta-\alpha}$, and $A(r,f)$ has the same order as $T(r,f)$, we conclude that $\rho_{\alpha\beta}(f) \le \rho$. Especially if $\rho_{\alpha\beta}(f) = +\infty$, then $\rho = +\infty$.

Theorem 2: *Let A and B be meromorphic in $\mathbb{C}$ with $\rho(A) < \rho(B)$ and $\delta(\infty, B) > 0$, then every nontrival meromorphic solution f of (3.1) has infinite order. Furthermore if $\rho(B) \le \frac{1}{2}$ and $\delta(\infty, B) = 1$, then $\rho_{\alpha\beta}(f) = +\infty$ for every angular region $\Omega(\alpha,\beta)$.*

Proof: The first assertion of the theorem follows immediately from the generalized Nevanlinna's lemma comcerning the logarithmic derivatives (see *[6]*). To prove the second conclusion of the theorem, we choose a fixed $\varepsilon = \frac{\rho(B)-\rho(A)}{5}$ and note that from a result of Yang *[7]*, there exists a sequence $(r_n) \to +\infty$ as $n \to \infty$ such that

$$(3.2) \qquad \lim_{n\to\infty} \mathrm{mes} E_n = \lim_{n\to\infty} \mathrm{mes} E\{\theta; \log|B(r_n e^{i\theta})| > r_n^{\rho(B)-\varepsilon}\}$$

$$\geq \min\left\{2\pi, \frac{4}{\rho(B)} \arcsin\sqrt{\frac{\delta(\infty, B)}{2}}\right\} = 2\pi.$$

We next suppose for the contrary that there exist a nontrival meromorphic solution f of (3.1) and an angular region $\Omega(\alpha, \beta)$ such that $\rho_{\alpha\beta}(f) < +\infty$. In view of Theorem 1, there exists a constant $M > 0$ not depending on z such that

$$(3.3) \qquad \left|\frac{f'(z)}{f(z)}\right| < |z|^M$$

and

$$(3.4) \qquad \left|\frac{f''(z)}{f(z)}\right| < |z|^M$$

for all $z \in \Omega(\alpha + \delta, \beta - \delta)(\delta = \frac{\beta-\alpha}{4})$, except for a $\mathbb{R}$-set D. It is obvious that

$$(3.5) \qquad \lim_{n\to\infty} \mathrm{mes} E_n' = \lim_{n\to\infty} \mathrm{mes} E_n\{\theta, r_n e^{i\theta} \in D\}$$

$$= 0.$$

Set $E_{n,\delta} = (E_n \cap (\alpha + \delta, \beta - \delta))\backslash E_n'$. We deduce from (3.1)-(3.5) that

$$\delta r_n^{\rho(B)-2\varepsilon} \leq \int_{E_{n,\delta}} \log|B(r_n e^{i\theta})|d\theta$$

$$\leq \int_{E_{n,\delta}} \left(\log^+\left|\frac{f''(r_n e^{i\theta})}{f(r_n e^{i\theta})}\right| + \log^+\left|\frac{f'(r_n e^{i\theta})}{f(r_n e^{i\theta})}\right| +\right.$$

$$\left.\log^+|A(r_n e^{i\theta})|\right) d\theta \quad \leq r_n^{\rho(A)+2\varepsilon}$$

for all sufficiently large n.

Thus we obtain a contradiction which completes the proof of Theorem 2. ∎

Acknowledgement: The author would like to thank Professor Yang Lo for his helpful suggestions.

References

[1]   A. A. Goldberg, Nevanlinna's lemma on the logarithmic derivative of a meromorphic function, Mat. Zametki, 17: 4 (1975), 525-529.

[2]   A. A. Goldberg and I. V. Ostrovskii, The Distribution of values of meromorphic functions (in Russion), Nauka, Moscow (1970).

[3]   G. G. Gundersen, Estimates for the logarithemic derivative of a meromorphic function, plus similar estimates, J. London Math. Soc. (2) 37 (1988), 88-104.

[4]   A. Z. Mokhon'ko, An estimate of the modulus of the logarithmic derivative of a function which is meromorphic in an angular region, and its application, Ukrainskii Math. Zhurnal, 41 (1989) 839-843.

[5]   S. J. Wu, On the location of zeros of solutions of $f'' + Af = 0$ where A is entire, *Math. Scand.*, (to appear).

[6]   K. L. Xiong, Sur les fonctions holomorphes dont less dérivées admettent une valeur exceptionnelle, Ann. École Norm. Sup., 72 (1955), 165-197.

[7]   L. Yang, Borel directions of meromorphic functions in an angular domain, Sci. Sinica, 1979 Special Issue II, 149-164.

Keywords. logarithmic derivative, estimate, growth.

1980 Mathematics subject classification (1985 Revision) 30D35.

The Dynamics of the Family $z + 1/z + b$

Yongcheng Yin

Institute of Mathematics, Zhejiang University
Hangzhou, 310027, P. R. China.

Abstract: The purpose of this note is to give some dynamical properties of the family $z + 1/z + b$. We describe the topology of the parameter space and the mapping class group for all most b.

1 Introduction

Let $f : \overline{\mathbb{C}} \to \overline{\mathbb{C}}$ be a rational function with degree greater than one. f^n denote the n-th iterates of f. z is stable point of f if there exists a neighborhood U of z such that $\{f^n|_U\}$ is a normal family. The stable set $F(f)$ is the set of stable points of f. Its complement, $J(f)$ is called the Julia set of f. $J(f)$ is perfect, completely invariant and never empty. D.Sullivan ($[5, 6]$) proved that every component of $F(f)$ is eventually periodic, and that periodic stable components can be classified into four types – attractive basin, parabolic basin, Siegel disk and Herman ring.

In $[2]$, Goldberg and Keen given a description of the topology of the family $a(z + 1/z + b)$, where $|a| > 1$ and $b \in \mathbb{C}$.

Combine main theorem in $[7]$ with ideas of $[2]$, we discuss the dynamics of the family $z + 1/z + b$ in this note.

2 The Parameter Space of $z + 1/z + b$

Let $f_b(z) = z + 1/z + b$ with critical points ± 1 and $A_b(\infty)$ be the parabolic basin corresponding to ∞. There exist two attractive basins for $b = 0$, $A_b(\infty)$ has two components. There exists only one attractive basin for $b \neq 0$, $A_b(\infty)$ is connected. The main theorem in $[7]$ implies that $J(f_b)$ is connected if and only if $A_b(\infty)$ contains only one critical point for $b \neq 0$, and $J(f_0)$ is connected.

Set

$$U = \{b \in \mathbb{C} | J(f_b) \text{ is a Cantor set}\}$$

and

$$O = \{b \in \mathbb{C} | \text{ there are integers } m, n \geq 0$$
$$\text{such that } f_b^m(+1) = f_b^n(-1)\},$$

the set of orbit relation.

Proceedings of the International Conference on Complex
Analysis at the Nankai Institute of Mathematics, 1992 pp. 241-245
©INTERNATIONAL PRESS 1994

The Mandelbrot set $M_P = \{b|J(f_b)$ is connected$\} = \mathbb{C} - U$. M_P is symmetric about the imaginary. U contains the imaginary$-\{0\}$ and $O \subset U$.

It is easy to check that $M_P \subset \{b = u + iv| -2 \leq u, v \leq 2\}$ and $[-2, 2] \subset M_P$.

The polynomial $Q(z) = z^2 + z$ has a single parabolic fixed point at 0 with multiplier 1, a critical point at $-1/2$ and a critical value $-1/4$. By the Flower Theorem, there is an analytic map

$$\varphi : \text{Int}(K) \to \mathbb{C}$$

conjugates $Q(z)$ to $w \to w + 1$ for z near 0.

Figure 2.1: Julia set for $Q(z)$, with the curves
$\varphi(z) \in \mathbb{Z} + i\mathbb{R}$.

We denote by l_0 the leaf $\text{Re}\varphi(z) = \text{Re}\varphi(-1/2)$. l_0 is a figure eight curve through 0 and -1. $Q(l_0)$is a simple closed curve passing through 0 and $-1/4$. $Q(l_0)$ bounds a closed topological disk D.

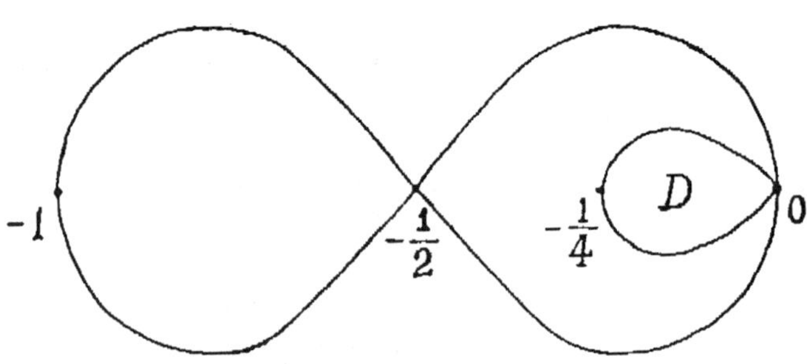

Figure 2.2

Let R be the right-half b-plane. We state the main result of this note in the following

Theorem 1: *There is a homeomorphism*

$$E : R - M_P \to Int(K) - D.$$

This theorem says that M_P is full. From computer pictures, $R \cap M_P$ seems to be a homeomorphic copy of the standard Mandelbrot set (J. Milnor noticed this also *[3]*).

3   The Mapping Class Group

For a rational map, we denote by $M(f)$, the space of $PSL(2,\mathbb{C})$ conjugacy classed of rational maps which are qc-conjugate to f. Let $Q(f)$ be the space of qc-homeomorphisms $h : \mathbb{C} \to \mathbb{C}$ for which $h \circ f \circ h^{-1}$ is again a rational map. Homeomorphisms h_0, h_1 are equivalent if there is $g \in \mathrm{PSL}(1,\mathbb{C})$ and an isotopy between $g \circ h_0$ and h_1 through elements of $Q(f)$. The quotient space of $Q(f)$ by this equivalence relation is the Teichmüller space of f we denote it $Teich(f)$. $Teich(f)$ is a complex manifold *[6]*.

Let $Q_0(f) \subset Q(f)$ be the subgroup of qc-homeomorphisms which commute with f. The mapping class group $MCG(f)$ is the quotient of $Q_0(f)$. It is often infinite and quite complicated.

$$M(f) \cong Teich(f)/MCG(f)$$

This follows that $MCG(f) = \pi_1(M(f))$.

Let $Aut(f)$ be the group of Möbius transformations commuting with f. It is a finite group.

For any $f_b(z) = z + 1/z + b$, $b \neq 0$. It is easy to check that $Aut(f_b) = id$. $Aut(f_0) = \mathbb{Z}/2 = \{z, -z\}$.

Theorem 2: *The mapping class group $MCG(f_0)$ has no nontrivial finite subgroup for any $b \neq 0$.*

Proof: Let G be a finite subgroup of $MCG(f_b)$. There exists a rational map g, qc-conjugate to f, such that G is realized as a subgroup of $Aut(g)$ *[4]*. g is $PSL(2,\mathbb{C})$ conjugate to f'_b, for some $b' \in \mathbb{C} - \{0\}$.

$$Aut(g) = Aut(f_{b'}) = id$$

Therefore, $G = id$.   QED.   ◾

The theorem 1 gives a conformal homeomorphism E between $R - M_P$ and $Int(K) - D$.

We conclude that the $b's$ satisfying orbit relations are isolated points in $\mathbb{C} - M_P$ which accumulated on M_P. Therefore $\pi_1(\mathbb{C} - M_P \cup O)$ is an infinitely generated free group, which is generated by loops enclosing orbit relations and one loop enclosing M_P.

We describe the mapping class group of the family $f_b(z) = z + 1/z + b$ in the following theorem

Theorem 3: *The mapping class group $MCG(f_b)$ of f_b is*

(1) $MCG(f_b) = \pi_1(\mathbb{C} - M_P \cup O)$ is infinitely generated for $b \in \mathbb{C} - M_P \cup O$ and $MCG(f_b) = id$ for $b \in O$;

(2) Let W be a hyperbolic component of M_P. Then $MCG(f_b) = \mathbb{Z}$ for $b \in W - \{the\ center\}$ and $MCG(f_b) = id$ for b being the center of W;

(3) $MCG(f_b) = id$ if f_b has an indifferent cycle or b is a Misurewicz point.

Acknowledgement: The author would like to thank professor Fuyao Ren for his instructions.

References

[1]   A. Douady and J. Hubbard, Étude dynamique des polynômes complexes, Publ. Math. d'Orsay, 1984-85.

[2]   L. Goldberg and L. Keen, The mapping class group of a generic quadratic rational map and automorphisms of the 2-shift, *Invent. Math.*, **101**(1990), 335-372.

[3]   J. Milnor, Remark on quadratic rational maps, Stony Brook IMS preprint 1992 /14.

[4]   C. McMullen, Automorphisms of rational maps, in *Holomorphic functions and moduli*, edit. Drasin et al., Springer-Verlag, 1988.

[5]   D. Sullivan, Quasiconformal homeomorphisms and dynamics I: Solution of the Fatou-Julia problem on wandering domains, *Ann. Math.*, **122**(1985), 401-418.

[6]   D. Sullivan, Quasiconformal homeomorphisms and dynamics III: Topological conjugacy classes and analytic endomorphisms, Preprint.

[7]   Yongcheng Yin, On the Julia sets of quadratic rational maps, *Complex Variables*, **18** (1992), 141-147.

Keywords: Julia set, Mandelbrot set, mapping class group.

1990 Mathematical Subject Classification: 30D05, 58F08.

On Iterates Of Holomorphic Maps In Strongly Pseudoconvex Domains

Wenjun Zhang[1]
Department of Mathematics
Henan University
Kaifeng, Hena, 475001 P.R. China

Fuyao Ren and Jainyong Qiao[2]
Institute of Mathematics
Fudan University
Shanghai, 200433 P.R. China

Abstract: In this paper the authors obtained the Denjoy-Wolff type theorem on bounded strongly pseudoconvex domains with C^2 boundaries in $\mathbf{C}^n$. It was proved that for any holomorphic self-map f of D, either the whole iterate sequence f^m is convergent to a fixed point in $\overline{D}$ or there is a holomorphic retraction R_f such that every subsequence limit g can be represented as $g = T \circ R_f$, where $T \in \mathrm{Aut}(R_f(D))$. Moreover, they gave an example to show that the results in this paper can not be improved.

1 Introduction

Let D_1, D_2 be two domains in the complex Euclidean space $\mathbf{C}^n$, and $H(D_1, D_2)$ denote the space of all holomorphic maps from D_1 to D_2. $H(D, \mathbf{C}^n)$ is endued with the compact open topology , and $H(D, D)$, the holomorphic self-map space, is given the induced topology.

For a map $f \in H(D, D)$, define by induction the m-th iterate f^m of f as:

$$f^1 = f, f^m = f \circ f^{m-1}, \qquad m = 2, 3, \dots$$

It is known that f^m is a normal family when $D \subset\subset \mathbf{C}^n$ is a strongly pseudo-convex domain with C^2 boundary (cf. *[21]*). So a natural object to study is the asymptotic behavior of the sequence of the iterates $\{f^m\}$.

Let $\Gamma(f)$ be the closure of the sequence $\{f^m\}$ in $H(D, \mathbf{C}^n)$ with respect to the compact open topology and $\Gamma'(f)$ be the set of all subsequential limits

---

[1] The first author is supported by the National Natural Science Fund of China.
[2] The other authors are supported in part by Tian-yuan fund of China.

Proceedings of the International Conference on Complex
Analysis at the Nankai Institute of Mathematics, 1992 pp. 245-254

of $\{f^m\}$. Then $\Gamma(f) \subset H(D,\overline{D})$ is an Abelian toplogical semigroup. If f is a holomorphic retraction of D, that is, $f = f^2$, then $\Gamma(f) = \Gamma'(f) = \{f\}$.

Denote by $\text{Fix}(f)$ the fixed point set of $f \in H(D,D)$ in D, then $\text{Fix}(f) = f(D)$ provided that f is a holomorphic retraction.

The first result about the asymptotic behavior of f^m was given by A. Denjoy [5] and J. Wolff [20] in 1926 and can be stated as follows:

Theorem 1.1: *Let $\Delta \subset \mathbf{C}$ be the unit disc and $f \in H(\Delta, \Delta)$.*

(I) If $\text{Fix}(f) = \emptyset$ (the empty set), then there exists a $\xi \in \partial\Delta$ such that $f^m \to \xi$;

(II) If $\text{Fix}(f) \neq \emptyset$, then either $f^m \to a \in \Delta$, or f is an automorphism of Δ with exactly one fixed point.

In 1941, M. H. Heins [7] extended this theorem to more general domains in $\mathbf{C}$.

In the case of $\mathbf{C}^n$, M. Hervé [8] in 1963, B. MacCluer [15] and Y. Kubota [11] in 1983, and G. N. Chen [4] in 1984 gave respectively the corresponding result in the unit ball of $\mathbf{C}^n$; More recently, M. Suzuki [17, 18] in 1987 and 1989, M. Abate [1] in 1988, and T. Kuczumow and A. Stachura [12] in 1990 extended the result to bounded convex domains with C^2 boundaries in $\mathbf{C}^n$. Their results can be formulated in the following:

Theorem 1.2: *Let $D \subset\subset \mathbf{C}^n$ be a convex domain with C^2 boundary, $f \in H(D,D)$*
(I) If $\text{Fix}(f) = \emptyset$, then there exists a $\xi \in \partial D$ such that $f^m \to \xi$;
(II) If $\text{Fix}(f) \neq \emptyset$, then $\exists R_f \in \Gamma'(f), R_f$ is a holomorphic retraction of D.

In the proof of the above theorem, the use of Brouwer's fixed point theorem is fundamental. With the same idea, Daowei Ma [14] in 1991 got a similar theorem for contractible strongly pseudoconvex domains with smooth boundaries in $\mathbf{C}^2$ (i. e., the continuous self map space $C(D,D)$ is arcwise connected with respect to the compact open topology.) His method can not yield the result in general $\mathbf{C}^n (n > 2)$. The theorem reads:

Theorem 1.3: *Let $D \subset\subset \mathbf{C}^2$ be a contractible strongly pseudoconvex domain with smooth boundary, $f \in H(D,D)$. If $\text{Fix}(f) = \emptyset$, then there exists a $\xi \in \partial D$ such that $f^m \to \xi$;*

The purpose of this paper is to give a complete description of the situation for arbitrary C^2 bounded strongly pseudoconvex domains . The methods used here are somewhat different from theirs and are clearly more simple.

2   Some Preliminary Results

When the domains in question are the unit ball, the convex domains or the contractible strongly pseudoconvex domains, the use of the so called horosphere defined by M. Abate in [1] is essential.

Definition: Let $D \subset\subset \mathbf{C}^n$ be a domain, and choose $a \in D, x \in \partial D$ and $R > 0$. Then the small horosphere $E_a(x, R)$ and the big horosphere $F_a(x, R)$ of center x, pole a and radius R are defined by

$$E_a(x, R) = \{z \in D \,|\, \limsup_{w \to x}[K_D(z, w) - K_D(a, w)] < \frac{1}{2}\log R\}$$

$$(2.1)$$

$$F_a(x, R) = \{z \in D \,|\, \liminf_{w \to x}[K_D(z, w) - K_D(a, w)] < \frac{1}{2}\log R\},$$

where $K_D(\cdot, \cdot)$ is the Kobayashi distance on D.

The key result of horospheres on strongly pseudoconvex domains which we will use in this paper is the following.

Theorem 2.1: *([1] Th.1.7) Let $D \subset\subset \mathbf{C}^n$ be a strongly pseudoconvex C^2 domain. Then for every $a \in D, x \in \partial D$ and $R > 0$*

$$\overline{F_a(x, R)} \cap \partial D = \{x\}. \tag{2.2}$$

By the holomorphic contractibility of the unit ball, the convex domains or the topological contractibility of the contractible strongly pseudoconvex domains, it was proved in *[1]* and *[14]* that :

Theorem 2.2: *Let $D \subset\subset \mathbf{C}^n$ be a convex domain or a contractible C^3 strongly pseudoconvex domain, and $f \in H(D, D)$, $Fix(f) = \emptyset$. Then there exists a $x \in \partial D$ such that for any $a \in D, R > 0, k = 1, 2, ...,$ it holds*

$$f^k(E_a(x, R)) \subset F_a(x, R). \tag{2.3}$$

(2.3) is extremely important in the study of $\{f^m\}$ on the above mentioned domains. Unfortunately, the following example shows that it is no longer true for an arbitrary strongly pseudoconvex domain D, even if D is of smooth boundary.

Example 2.3 Define on $\mathbf{C}^n$ the real valued function $\rho(z)$ by

$$\rho(z) = \|z\|^2 - 2|z_1| + 1 - r^2,$$

where $0 < r < \frac{1}{2}$ is a given number. Let

$$\Omega = \{z \in \mathbf{C}^n \,|\, \rho(z) < 0\}.$$

We claim that Ω is a bounded strongly pseudoconvex domain with smooth boundary.

In fact, it is not difficult to verify that when $P \in \partial\Omega$, then $|P_1| \geq 1 - r > \frac{1}{2}$ and the Levi form of Ω at $P \in \partial\Omega$ is

$$\sum_{j,k=1}^{n} \frac{\partial^2 \rho}{\partial z_j \partial \overline{z}_k}(P)w_j\overline{w}_k = (1 - \frac{1}{2|P_1|})|w_1|^2 + \sum_{j=2}^{n}|w_j|^2$$

$$> (1 - \frac{1}{2|P_1|})\|w_1\|^2, \qquad \forall w \in \mathbf{C}^n - \{0\}.$$

So Ω is a strongly pseudoconvex domain, the boundedness and smoothness are obvious. Now consider the map $f(z) = -z$, clearly $f \in H(\Omega, \Omega)$ and as a map from $\mathbf{C}^n$ to $\mathbf{C}^n$, it has only one fixed point 0 and 0 is not in Ω. So $\mathrm{Fix}(f) = \emptyset$. But $f^{2k+1}(z) = -z$, and so

$$f^{2k+1}(E_a(x, R)) = E_{-a}(-x, R) \subset F_{-a}(-x, R).$$

When a is sufficiently close to the boundary point x, and R is small enough, it is easily seen that

$$F_{-a}(-x, R) \cap F_a(x, R) = \emptyset,$$

and consequently $f^{2k+1}(E_a(x, R)) \not\subset F_a(x, R)$. This shows that Theorem 2.2 is not   true
for Ω.

In fact, note that when D is a C^2 convex domain, the condition $\mathrm{Fix}(f) = \emptyset$ is equivalent to the statement that every convergent subsequence tends to a point in the boundary. So it is reasonable to extend Theorem 2.2 to the following more general form:

Theorem 2.4: *Let $D \subset\subset \mathbf{C}^n$ be any domain, and K_D be the Kobayashi distance, $f : D \to D$ be a K_D-nonexpansive mapping (i. e., $K_D(f(z), f(w)) \leq K_D(z, w), \forall z, w \in D$). If $\{f^m\}$ is normal on D and every convergent subsequence tends to a point in the boundary, then for any $a \in D, \exists x \in \partial D$, such that $f^k(E_a(x, R)) \subset F_a(x, R)$ holds for all $R > 0$ and $k \in \mathbf{Z}^+$.*

Proof: Given $a \in D$, since all limit points of $\{f^k\}$ lie in the boundary. So we have

$$\lim_{m \to \infty} K_D(a, f^m(a)) = \infty.$$

With no difficulty, one can choose a subsequence $\{m_j\}, m_j \to \infty$ and

$$K_D(a, f^{m_j}(a)) < K_D(a, f^{m_j+k}(a)), \qquad \forall k \in \mathbf{Z}^+. \tag{2.4}$$

By the normality of $\{f^m\}$, taking a subsequence if necessary, we can assume that f^{m_j} is convergent. Let $f^{m_j} \to x \in \partial D$, then

$$\begin{aligned}
\lim_{j \to \infty} f^{m_j}(a) &= x, \\
\lim_{j \to \infty} f^{m_j+k}(a) &= \lim_{j \to \infty} f^{m_j}(f^k(a)) = x, \qquad \forall k = 1, 2, \ldots
\end{aligned} \tag{2.5}$$

Now given any $z \in E_a(x, R)$, for $k = 1, 2, \ldots$, by (2.4), (2.5) and the nonexpansivity of f, we have

$$\liminf_{w \to x}[K_D(f^k(z), w) - K_D(a, w)]$$

$$\leq \liminf_{j\to\infty}[K_D(f^k(z), f^{m_j+k}(a)) - K_D(a, f^{m_j+k}(a))]$$

$$\leq \liminf_{j\to\infty}[K_D(z, f^{m_j}(a)) - K_D(a, f^{m_j+k}(a))]$$

$$\leq \liminf_{j\to\infty}[K_D(z, f^{m_j}(a)) - K_D(a, f^{m_j}(a))]$$

$$\leq \limsup_{w\to x}[K_D(z, w) - K_D(a, w)]$$

$$< \frac{1}{2}\log R.$$

That is, $f^k(z) \in F_a(x, R)$, and the proof is completed. ∎

It is well-known that any holomorphic map is K_D -nonexpansive, and if $D \subset \mathbf{C}^n$ is taut (cf. *[21]*), then $H(D, D)$ is normal, and so is $\{f^m\}$ for any $f \in H(D, D)$. Hence we get

Corollary 2.5: *Let $D \subset\subset \mathbf{C}^n$ be taut, and $f \in H(D, D)$. If every convergent subsequence tends to a boundary point, then for any $a \in D$, there exists a $x \in \partial D$ such that $f^k(E_a(x, R)) \subset F_a(x, R)$ holds for all $R > 0$ and $k \in \mathbf{Z}^+$.*

Remark2.6 A large class of domains which are taut is the class of all (weakly) pseudoconvex domains with C^1 boundaries (cf. *[9]*). Hence Corollary 2.5 is valid for such domains, and a priority, for C^2 strongly pseudoconvex domains.

In the conclusion of Corollary 2.5, $x \in \partial D$ may depend on the varying choice of $a \in D$. We do not know whether or not that x is uniquely determined by f and D for an arbitrary domain in $\mathbf{C}^n$. But we do have an affirmative answer for C^2 strongly pseudoconvex domains. In fact, we have

Corollary 2.7: *Let $D \subset\subset \mathbf{C}^n$ be a C^2 strongly pseudoconvex domain, and $f \in H(D, D)$. If every convergent subsequence tends to a boundary point, then there exists a unique $x \in \partial D$ such that*
(I) The sequence $\{f^m\}$ converges to x uniformly on compact subset of D ;
(II) For any $a \in D, R > 0$ and $k \in \mathbf{Z}^+$, we have

$$f^k(E_a(x, R)) \subset F_a(x, R). \tag{2.6}$$

Proof: By Corollary 2.5, for any given $a \in D$, there exists a point $x \in \partial D$ with property (2.6).

Now fix such points $a \in D$ and $x \in \partial D$. Given any convergent subsequence $\{f^{m_j}\}$. Let $f^{m_j} \to x_1 \in \partial D$. Then from (2.6) we know $x_1 \in \overline{F_a(x, R)}$. So (2.2) implies $x_1 \in \overline{F_a(x, R)} \cap \partial D = \{x\}$, and this means $x_1 = x$, and $f^m \to x \in \partial D$. Since $\{f^m\}$ is normal, the convergence is uniform on each compact subset of D, and (I) is proved.

From (I), recall the proof of Theorem 2.4, we know that x is unique and (II) holds. ∎

The following lemmas will be used to prove our main theorem in Section 3.

Lemma 2.8: *Let $D \subset\subset \mathbf{C}^n$ be a C^2 strongly pseudoconvex domain, and $f \in H(D, \overline{D})$. Then either $f \in H(D, D)$ or $f(z) \equiv \xi \in \partial D$.*

Proof: Suffice it to show that : if $f(a) = \xi \in \partial D$ for some $a \in D$, then $f(z) \equiv \xi$.

In fact, it is well-known that every boundary point of a C^2 strongly pseudoconvex domain D is a peak point (see, for example, *[16]*). So if we denote by $A(D)$ the set of all functions continuous on $\overline{D}$ and holomorphic in D , then for $\xi = f(a) \in \partial D$, one can find a $g \in A(D)$ with $g(\xi) = 1$, and $|g(z)| < 1, \forall z \in \overline{D} - \{\xi\}$.

Consider the holomorphic function $h = g \circ f$, then $h(a) = g(f(a)) = g(\xi) = 1$, and $|h(z)| \leq 1, \forall z \in \overline{D}$. So the Maximum Modulus Theorem implies $h(z) \equiv 1$, or $g(f(z)) \equiv 1$, this means $f(z) \equiv \xi$ by the choice of g . ∎

Lemma 2.9:   *Let $D \subset\subset \mathbf{C}^n$ be a C^2 strongly pseudoconvex domain, and $f \in H(D, D)$. If $\Gamma'(f) \bigcap H(D, D) \neq \emptyset$, then $\Gamma(f) \subset H(D, D)$.*

Proof: Let $g \in \Gamma'(f) \bigcap H(D, D)$ and seek a subsequence $f^{m_j} \to g$. Then $\{f^{m_j}(a)\} \subset\subset D$ for $a \in D$. If we can prove that D is finitely totally bounded under the Kobayashi metric (i. e., each bounded subset of D is totally bounded, cf. *[10]*), then Theorem 5.6 of A. Calka *[3]* implies that for any $z \in D$ we have $f^m(z) \subset\subset D$, and so $\Gamma(f) \subset H(D, D)$.

Now since the Kobayashi metric is complete in strongly pseudoconvex domains (cf. *[6]*). The result in *[13]* shows that $\overline{B(z, r)}$ is compact for any $z \in D$ and $r > 0$, where $B(z, r)$ is the ball with center z and radius r under the Kobayashi distance. The above discussion means that the closure of any bounded subset of D is compact, so a priority, every bounded subset of D is totally bounded, as desired. ∎

3   The Main Results and Proofs

Now we are in a position to state our main theorem:

Theorem 3.1: *Let $D \subset\subset \mathbf{C}^n$ be a C^2 strongly pseudoconvex domain, and $f \in H(D, D)$. Then one of the following holds:*

(I) The sequence $\{f^m\}$ converges to a point $\xi \in \overline{D}$ uniformly on compact subset of D ;

(II) There exists a unique holomorphic retraction $R_f \in \Gamma'(f)$, such that for any $g \in \Gamma'(f), \exists T \in Aut(V)$ with $g = T \circ R_f$, and $V = R_f(D)$ is a submanifold of D.

Proof: We break it into two cases:

Case 1 Every convergent subsequence f^{m_j} converges to a constant map.

If each $\{f^{m_j}\}$ tends to the boundary, then Corollary 2.7 gives the result with $\xi \in \partial D$.

So we suppose that there exists a subsequence $f^{m_j} \to a \in D$, then

$$f(a) = \lim_{j\to\infty} f(f^{m_j}(z)) = \lim_{j\to\infty} f^{m_j}(f(z)) = a,$$

that is, $f(a) = a$. Now for any convergent subsequence f^{k_j}, we have

$$\lim_{j\to\infty} f^{k_j}(z) = \lim_{j\to\infty} f^{k_j}(a) = a,$$

so

$$f^m \to a \in D.$$

Anyway, we get $f^m \to \xi \in \overline{D}$. That the convergence is uniformly on compact subsets follows from the normality of f^m .

Case 2 There exists a nonconstant map $g \in \Gamma'(f) \subset H(D,\overline{D})$.

Lemma2.8 shows $g \in H(D,D)$, and so $\Gamma'(f) \bigcap H(D,D) \neq \emptyset$, consequently $\Gamma(f) \subset H(D,D)$ by Lemma 2.9. Hence $\Gamma(f)$ is a compact Abelian semigroup. By the theory of semigroup (cf. [19]), there is a unique idempotent map $R_f \in \Gamma(f), R_f = R_f^2, R_f$ is nothing but a holomorphic retraction. The result in [2] tells us that $V = R_f(D) = \text{Fix}(R_f)$ is a submanifold of D.

Now given any $g \in \Gamma'(f) \subset H(D,D)$, seek a subsequence $f^{m_j} \to g$. Taking a subsequence if necessary, we may assume that

$$k_j = m_{j+1} - m_j \to \infty,$$

and

$$l_j = k_j - m_j = m_{j+1} - 2m_j \to \infty.$$

Again taking the subsequences if necessary, we may assume that both subsequences $\{f^{k_j}\}$ and $\{f^{l_j}\}$ are convergent. Let

$$f^{k_j} \to R \in \Gamma'(f) \subset H(D,D),$$

$$f^{l_j} \to h \in \Gamma'(f) \subset H(D,D),$$

By the relation $f^{m_{j+1}} = f^{m_j} \circ f^{k_j} = f^{k_j} \circ f^{m_j}$, passing to the limit as $j \to \infty$ we get

$$g = g \circ R = R \circ g. \tag{3.1}$$

The same idea used for $f^{k_j} = f^{l_j} \circ f^{m_j} = f^{m_j} \circ f^{l_j}$ gives

$$R = g \circ h = h \circ g. \tag{3.2}$$

Now from (3.1) and (3.2) we know

$$R^2 = h \circ g \circ h \circ g = h \circ R \circ g = h \circ g = R$$

so $R \in \Gamma(f)$ is an idempotent map, and the uniqueness of idempotent map yields
$R = R_f$.

Now (3.1) can be rewritten as

$$g = g \circ R_f.$$

The remaining thing is to prove that $T = g|_V \in \mathrm{Aut}(V)$.

First note that $R_f^2 = R_f$ implies $R_f|_V = id$. Next, since $h \in \Gamma'(f)$, repeat the above argument with h in place of g , one gets

$$h = h \circ R = R \circ h. \tag{3.3}$$

Now (3.1) shows $g(V) = R(g(V)) \subset R(D) = V$, and so $g \in H(V,V)$. (3.3) shows $h \in H(V,V)$. Finally, (3.2) implies $h \circ g|_V = g \circ h|_V = R_f|_V = id$. This means $g|_V \in \mathrm{Aut}(V)$, the proof is completed. ∎

Remark 3.2 One can not expect, for general C^2 strongly pseudoconvex domains, to strengthen Theorem 3.1 to the form as in Theorem 1.2 or Theorem 1.3, that is to say, conclusion (II) can not generally ensure that $\mathrm{Fix}(f) \neq \emptyset$.

For example, let Ω be the domain given in Example 2.3, and

$$f(z) = (-z_1, \frac{1}{2}z_2, ..., \frac{1}{2}z_n).$$

Then $f \in H(\Omega, \Omega)$, and $f^{2k}(z) \to R_f(z) = (z_1, 0, ..., 0)$, but $\mathrm{Fix}(f) = \emptyset$.

Remark3.3 In conclusion (I) of Theorem 3.1, $\xi \in \overline{D}$ is necessarilly the fixed point of f. If $\xi \in D$, then $\mathrm{Fix}(f) = \{\xi\}$. But

(I) If $\xi \in \partial D$, we can not say that f has a unique fixed point $\xi \in \overline{D}$.

(II) If $a \in D$, and $\mathrm{Fix}(f) = \{a\}$, one can not expect that $f^m \to a$, even if f is not an automorphism. In fact, Let $D = B_n$ be the unit ball of $\mathbf{C}^n$, $f(z)$ be as in the remark 3.2, we do have $\mathrm{Fix}(f) = \{0\}$, but $f^m \nrightarrow 0$.

Remark 3.4 Theorem 3.1 gives a description of $\Gamma'(f)$, that is, either

$$\Gamma'(f) = \{g(z) \equiv \xi \in \overline{D}\}$$

or

$$\Gamma'(f) = \{T \circ R_f| \ R_f \text{ is a holomorphic retraction, and } T \in \mathrm{Aut}(R_f(D))\}.$$

References

[1] Abate, M.,*Math. Z.* **198**(1988), 225–238.

[2] Bedford, E. ,*Math. Ann.* **266** (1983) , 215–227 .

[3] Calka, A. ,*Colloq. Math.* , **48** (1984), 219–227.

[4] Chen, G. N. ,*J. Math. Anal. Appl.*, **98**(1984)305–313 .

[5] Denjoy, A. ,*C. R. Acad. Sci., Paris, SerI.*, **182**(1926), 255–257 .

[6] Graham, I. , *Trans. Am. Math. Soc.*, **207**(1975) , 219–240 .

[7] Heins, M. H. ,*Am. J. Math.* , **63**(1940) , 461–480.

[8] Hervé, M. , *J. Math. Pure. Appl.*, **(9)42**(1963), 117–147 .

[9] Kerzman, N. ,*Notices Am. Math. Soc.*, **16**(1969) , 675 .

[10] Kobayashi, S. , *Bull. Am. Math. Soc.*, **82**(1982), 357–416.

[11] Kubota, Y. , *Proc. Am. Math. Soc.*, **88**(1983), 476–480 .

[12] Kuczumow, T., Stachura, A. ,*Adv. Math.*, **81**(1990) , 90–98 .

[13] Lang, S. ,*Introduction to Complex Hyperbolic Spaces.* Springer-Verlag, 1987.

[14] Ma , D., *Math. Z.* , **207**(1991), 417–428 .

[15] MacCluer, B. D., *Mich. Math. J.* , **30**(1983), 97–106 .

[16] Rudin, W. , *Pac. J. Math.* , **75**(1978), 267–277.

[17] Suzuki, M. , *Kodai Math. J.*, **10**(1987), 298–306.

[18] Suzuki, M. , *Kobe J. Math.*, **6**(1989), 229–232 .

[19] Wallace, A. , *Bull. Am. Math. Soc.*, **61**(1955), 95–112 .

[20] Wolff, J. , *C. R. Acad. Sci. Paris.* , **182**(1926), 42–43.

[21] Wu, H. ,*Acta Math.*, **119**(1967), 193–233 .

A John-Nirenberg Type Theorem for BMOA on Riemann surfaces [1]

Ruhan Zhao

Wuhan Institute of Mathematical Sciences, Academia Sinica,
Wuhan 430071,P.R.China

Abstract: A kind of John-Nirenberg type theorem for BMOA on general open Riemann surfaces is given. As an application a new characterization of BMOA on Riemann surfaces is given too.

In $[4]$, T.A.Metzger asked if the John-Nirenberg theorem for BMOA on the unit disk is true on Riemann surfaces. We have given a positive answer for compact bordered Rimann surfaces in $[1]$. Using exhaustions we give a kind of John-Nirenberg theorem for BMOA on general open Riemann surfaces in $[6]$. But the exhaustion sequences in $[6]$ depend on the reference point a. In this paper we will give another kind of John-Nirenberg type theorem for BMOA on general open Riemann surfaces which the exhaustion sequences in this theorem do not depend on the reference point a.

Let R be a Riemann surface which possesses a Green's function, i.e., $R \notin O_G$, then there exists an exhaustion $R_1 \subset R_2 \subset \cdots R_k \nearrow R$, where R_k are compact bordered Riemann surfaces whose borders consist of finite numbers of analytic Jordan curves $(1 \leq k < \infty)$. We say R_k is a regular exhaustion of R. Let $G_R(w, a)$ be the Green's function on R with logarithmic singularity at $a \in R$. In $[3]$ T.A.Metzger gave the definition of BMOA(R) as follows: Let F be an analytic function on R. We say $F \in BMOA(R)$, if

$$B_R^2(F) = sup_{a \in R} \frac{2}{\pi} \int\!\!\int_R |F'(w)|^2 G_R(w, a) dw d\overline{w} < \infty.$$

We first give the next lemma:

Lemma 1: *Let R be a Riemann surface, $R \notin O_G$, R_k be a regular exhaustion of R. If we denote*

$$B_k^2(F) = sup_{a \in R_k} \frac{2}{\pi} \int\!\!\int_{R_k} |F'(w)|^2 G_k(w, a) dw d\overline{w},$$

where $G_k(w, a)$ is the Green's function on R_k with logarithmic singularity at $a \in R_k$. Then $\{B_k^2(F)\}$ is an increasing sequence, and for every $k \geq 1$, $B_k^2(F) \leq B_R^2(F)$.

---

[1]Project supported by the National Natural Science Foundation of China.

Proceedings of the International Conference on Complex
Analysis at the Nankai Institute of Mathematics, 1992 pp. 254-260
©INTERNATIONAL PRESS 1994

Proof: For convenience, we denote $R_\infty = R$ and $B_\infty(F) = B_R(F)$. Let $1 \le k_1 < k_2 \le \infty$. For $w \in R_{k_1}$, let

$$G(w) = G_{k_2}(w, a) - G_{k_1}(w, a).$$

It is easy to verify that $G(w)$ is a harmonic function on R, and

$$G(w)|_{\partial R_{k_1}} = G_{k_2}(w, a)|_{\partial R_{k_1}} \ge 0.$$

By the maximum principle of harmonic function we know that $G(w) \ge 0$ for every $w \in R_{k_1}$, i.e.,

$$G_{k_1}(w, a) \le G_{k_2}(w, a).$$

Thus

$$\frac{2}{\pi} \iint_{R_{k_1}} |F'(w)|^2 G_{k_1}(w, a) dw d\bar{w} \le \frac{2}{\pi} \iint_{R_{k_2}} |F'(w)|^2 G_{k_2}(w, a) dw d\bar{w}.$$

Taking the supremum we have

$$B_{k_1}^2(F) \le B_{k_2}^2(F).$$

This is the conclusion of the lemma.

Next we give an equivalent definition of BMOA on Riemann surfaces by a regular exhaustion.

Theorem 1: *Let R be a Riemann surface, $R \notin O_G$, $\{R_k\}$ be a regular exhaustion of R. F is an analytic function on R. Then*

$$B_R^2(F) = \lim_{k \to \infty} B_k^2(F).$$

Proof: Because

$$B_R^2(F) = \sup_{a \in R} \frac{2}{\pi} \iint_R |F'(w)|^2 G_R(w, a) dw d\bar{w},$$

we know that there is a point esquence $\{a_n\}$ such that

$$B_R^2(F) = \lim_{n \to \infty} \frac{2}{\pi} \iint_R |F'(w)|^2 G_R(w, a_n) dw d\bar{w}.$$

For every $a \in R$ and every $k \ge 1$, Let

$$\tilde{G}_k(w, a) = \begin{cases} G_k(w, a), & w \in R_k \\ 0, & \in R/R_k. \end{cases}$$

Then from

$$\lim_{k \to \infty} G_k(w, a) = G_R(w, a),$$

we know

$$lim_{k\to\infty}\tilde{G}_k(w,a) = G_R(w,a).$$

Let n be an arbitrary positive integer. Because R_k is an exhaustion of R, there is a k_n such that $a_n \in R_{k_n}$. So for every $k \geq k_n$, $a_n \in R_{k_n} \subset R_k$. Thus

$$\frac{2}{\pi}\int\int_R |F'(w)|^2\tilde{G}_k(w,a_n)dwd\bar{w} \leq sup_{a\in R_k}\frac{2}{\pi}\int\int_R |F'(w)|^2\tilde{G}_k(w,a)dwd\bar{w}.$$

Then

$$\frac{2}{\pi}\int\int_R |F'(w)|^2 G_R(w,a_n)dwd\bar{w} = \frac{2}{\pi}\int\int_R |F'(w)|^2 lim_{k\to\infty}\tilde{G}_k(w,a_n)dwd\bar{w}$$

$$\leq lim_{k\to\infty}\frac{2}{\pi}\int\int_R |F'(w)|^2\tilde{G}_k(w,a_n)dwd\bar{w}$$

$$\leq lim_{k\to\infty}sup_{a\in R_k}\frac{2}{\pi}\int\int_R |F'(w)|^2\tilde{G}_k(w,a)dwd\bar{w}$$

$$= lim_{k\to\infty}sup_{a\in R_k}\frac{2}{\pi}\int\int_{R_k} |F'(w)|^2 G_k(w,a)dwd\bar{w}$$

$$= lim_{k\to\infty}B_k^2(F).$$

Because the right side is independent on n, we have

$$B_R^2(F) = lim_{n\to\infty}\frac{2}{\pi}\int\int_R |F'(w)|^2 G_R(w,a_n)dwd\bar{w} \tag{1}$$
$$\leq lim_{k\to\infty}B_k^2(F).$$

By Lemma 1, for every $k \geq 1$,

$$B_R^2(F) \geq B_k^2(F).$$

So

$$B_R^2(F) \geq lim_{k\to\infty}B_k^2(F) \tag{2}$$

Combining (1) and (2) we have

$$B_R^2(F) = lim_{k\to\infty}B_k^2(F)$$

This completes the proof of Theorem 1. ∎

Remark: Theorem 1 says that $F \in BMOA(R)$ if and only if $sup_{k\geq 1} B_k(F) \leq M < \infty$, i.e., $B_k(F)$ are uniformly bounded ($1 \leq k < \infty$).

The next example shows that from $F \in BMOA(R_k)$ only we can not infer that $F \in BMOA(R)$. Let $D_r = \{z, |z| < r < 1\}$ and $D = \{z, |z| < 1\}$, then $\{D_r\}$ is a regular exhaustion of D. Let

$$f(z) = \frac{1}{1-z}, \qquad z \in D,$$

then it is obvious that $f(z)$ is bounded on $\partial D_r = \{z \in D, |z| = r\}$. So $f \in H^\infty(D_r)$ and then $f \in BMOA(D_r)$. But f has Taylor series $f(z) = \sum_{n=0}^{\infty} z^n$, whose Taylor coefficients are $a_n = 1$, $n = 0, 1, 2, \ldots$. Thus

$$\sum_{n=0}^{\infty} |a_n|^2 = \infty,$$

Therefore $f \notin H^2(D)$. From $BMOA(D) \subset H^2(D)$ we know that $f \notin BMOA(D)$. Now let us give a kind of John-Nirenberg type theorem on general Riemann surfaces.

Theorem 2: *Let R be a Riemann surface, $R \notin O_G$, $\{R_k\}$ be a regular exhaustion of R, F be an analytic function on R, then $F \in BMOA(R)$ if and only if for every $k \geq 1$ and every $a \in R_k$,*

$$\mu_{a,k}(E_{a,k,\lambda}) = \frac{1}{2\pi} \int_{E_{a,k,\lambda}} \frac{\partial G_k(w, a)}{\partial n} ds \leq Ke^{-\beta\lambda}, \tag{3}$$

where $E_{a,k,\lambda} = \{w \in \partial R_k, |F(w) - F(a)| > \lambda\}$, $\frac{\partial}{\partial n}$ is the inner normal derivative with respect to R_k, K, β are constants. When $F \in BMOA(R)$, $\beta = c/B_R(F)$, c is another constant.

Proof: Let $F \in BMOA(R)$. By Lemma 1 we know $F \in BMOA(R_k)$ for every $k \geq 1$. By Theorem 1 in *[1]* we have

$$\mu_{a,k}(E_{a,k,\lambda}) \leq Ke^{-\beta_k\lambda}, \tag{4}$$

where K is a constant, $\beta_k = c/B_k(F)$. c is another constant. By Lemma 1 we know that $B_k(F) \leq B_R(F)$. Thus $\beta_k \geq \beta = c/B_R(F)$. So $\mu_{a,k}(E_{a,k,\lambda}) \leq Ke^{-\beta\lambda}$.

Conversely, suppose (3) is true. Let $H_{a,k}(w)$ be the least harmonic majorant of the subharmonic function $|F(w) - F(a)|^2$ on R_k, then

$$H_{a,k}(a) = \frac{1}{2\pi} \int_{\partial R_k} |F(w) - F(a)|^2 \frac{\partial G_k(w, a)}{\partial n} ds.$$

Let

$$\Lambda_{a,k}(\lambda) = \mu_{a,k}(E_{a,k,\lambda}),$$

Then

$$H_{a,k}(a) = 2 \int_0^{\infty} \lambda\Lambda(\lambda)d\lambda \leq 2K \int_0^{\infty} \lambda e^{-\beta\lambda}d\lambda = \frac{2K}{\beta^2} < \infty.$$

Thus

$$\sup_{a \in R_k} H_{a,k}(a) \leq \frac{2K}{\beta^2} < \infty \tag{5}$$

By lemma 1 of *[2]*,
$$B_k^2(F) = sup_{a \in R_k} H_{a,k}(a),$$

Then (5) is
$$B_k^2(F) \leq \frac{2K}{\beta^2} < \infty.$$

So by Therem 1,
$$B_R^2(F) = lim_{k \to \infty} B_k^2(F) \leq \frac{2K}{\beta^2} < \infty.$$

Hence $F \in BMOA(R)$. The proof is completed. ∎

As an application of Theorem 2, we give a characterization of $BMOA$ on Riemann surfaces.

Corollary 0.5: *Let R be a Riemann surface, $R \notin O_G$, $\{R_k\}$ be a regular exhaustion of R, then $F \in BMOA(R)$ if and only if there is a constant $\tau = \tau_F > 0$ such that*

$$sup_{k \geq 1} sup_{a \in R_k} \frac{1}{2\pi} \int_{\partial R_k} exp(\tau |F(w) - F(a)|) \frac{\partial G_k(w,a)}{\partial n} ds < \infty. \tag{6}$$

Proof: Taking $\Lambda_{a,k}(\lambda) = \frac{1}{2\pi} \int_{E_{a,k,\lambda}} \frac{\partial G_k(w,a)}{\partial n} ds$, where $E_{a,k,\lambda} = \{w \in \partial R_k,\ |F(w) - F(a)| > \lambda\}$. Then by Theorem 2, if $F \in BMOA(R)$,

$$\Lambda_{a,k}(\lambda) \leq K e^{-\beta \lambda},$$

where K is a constant, $\beta = c/B_R(F)$, c is another constant. Taking τ so that $0 < \tau < \beta$, By Lemma 3 of *[5]* we have

$$\frac{1}{2\pi} \int_{\partial R_k} exp(\tau |F(w) - F(a)|) \frac{\partial G_k(w,a)}{\partial n} ds$$

$$\leq 1 + \tau \int_0^\infty e^{\tau \lambda} \Lambda_{a,k}(\lambda) d\lambda$$

$$\leq 1 + K\tau \int_0^\infty e^{(\tau - \beta)\lambda} d\lambda \leq 1 + \frac{K\tau}{\beta - \tau}.$$

Because the right side is independent on a and k, we have

$$sup_{k \geq 1} sup_{a \in R_k} \frac{1}{2\pi} \int_{\partial R_k} exp(\tau |F(w) - F(a)|) \frac{\partial G_k(w,a)}{\partial n} ds \leq 1 + \frac{K\tau}{\beta - \tau} < \infty.$$

Conversely, let (6) be true. Because for every $x > 0$, $x^2 < e^x$, by Theorem 1 and Lemma 1 of *[2]* we have

$$B_R^2(F) = sup_{k \geq 1} sup_{a \in R_k} \frac{1}{2\pi} \int_{\partial R_k} |F(w) - F(a)|^2 \frac{\partial G_k(w,a)}{\partial n} ds$$

$$\leq \frac{1}{\tau^2} sup_{k \geq 1} sup_{a \in R_k} \frac{1}{2\pi} \int_{\partial R_k} exp(\tau |F(w) - F(a)|) \frac{\partial G_k(w,a)}{\partial n} ds < \infty.$$

Therefore $F \in BMOA(R)$. This completes the proof. ∎

Acknowledgment: The auther wishes to thank Professor He Yuzan and Professor Ouyang Caiheng for their enthusiatic guidance.

References

[1]   He Yu-Zan and Zhao Ru-Han, BMOA and Ba spaces on compact bordered Riemann surfaces, Chinese Sci. Bull., 36 (1991), 20, 1677-1682.

[2]   S.Kobayashi, Range sets and BMO norms of analytic functions, Can. J.Math., 36(1984), 745-755.

[3]   T.A.Metzger, On BMOA for Riemann surfaces, Can. J.Math., 33(1981). 1255-1260.

[4]   T.A.Metzger, Bounded mean oscillation and Riemann surfaces, Bounded mean oscillation in complex analysis, Joensuu, 1989, 79-99.

[5]   Ouyang Caiheng, Some classes of functions with exponential decay in the unit ball of C^n, *Publ. RIMS, Kyoto Univ.*, 25 (1989), 263-277.

[6]   Zhao Ru-Han, The charateristics of BMOA on Riemann surfaces, Kodai Math. J., 15(1992), 2, 221-229.

On the zeros of $f(g(z)) - \alpha(z)$

Chung-chun Yang

1. Introduction and main results

It was conjectured by Gross [12] that if f and g are transcendental entire functions, then f(g) has infinitely many fix-points, i.e. $f(g(z)) - z = 0$ has infintely many zeros. The case that f(g) is of finite order follows from either of the more general results of Gol'dberg and Prokopovich [9], Goldstein [10], Gross and Yang [16], and Mues [21] obtained earlier by essentially different arguments. Only recently Bergweiler completely confirmed the conjecture. In fact, it is shown in [2] that $f(g(z)) - p(z)$ has infintely many zeros for any non-constant polynomial p(z); namely $n(r) \to \infty$, where n(r) denotes the number of zeros of $f(g(z)) - p(z)$ in $|z| \le r$. The following two questions was raised in [1].

Question 1. Can some lower bound for n(r) be estimated?

Question 2. Whether the function $f(g(z)) - \alpha(z)$ always has infinitely many zeros, if α is a transcendental entire function satisfying $T(r, \alpha) = o(1)T(r, g)$?

It was then proved by Laine and his students [19] that the answer to Question 2 is affirmative when both f and g are of finite order. More specifically: Let f be a non-linear entire function of finite order, g a transcendental entire function of finite order, α a non-constant entire function such that $T(r, \alpha) = S(r, g)$. If π denotes the canonical product formed with zeros of $f(g) - \alpha$, then $T(r, \pi) \ne S(r, g)$. It is remarked at the end of [14], that the assertion $T(r, \pi) \ne S(r, g)$ holds if α is a non-constant meromorphic function such that $T(r, \alpha) = S(r, g)$, and π denotes the quotient of cannonical products formed by the zeros of poles and $f(g) - \alpha$. respecitively. However, $T(r, \pi)$ is not small does not imply that π must have infinitely many zeros. This question has been resolved in [20] as Theorem 3 of the present paper.

The above results lead one naturally to raise the following question:

Question 3. Will the conclusion of Question 2 holds if f and α are meromorphic functions?

In this talk, we shall present an affirmative answer to Question 3 if the order of f(g) is finite, and provide a quantitative measure about the number of zeros in Question 2.

Theorem 1 [5]. Let f be meromorphic and transcendental, g transcendental entire, and R a non-constant rational function. Then the function $f(g) - R$ has infinitely many zeros.

Remark. It is proved in [3] when $R(z) \equiv z$ or any bilinear form, then Theorem 1 is true regardless the order of f(g).

Theorem 2 [28]. If f(z) and g(z) are two transcendental entire functions and $\alpha(z)$ is a non-constant polynomial, then there exists a set I of infinite linear measure such that for some d with $0 < d < 1$ and $r \in I$,

$$N(r, \frac{1}{f(g) - \alpha}) > dT(r^{1/3}, g).$$

We note that if f is entire, then the conclusion of theorem 1 remains valid even R is constant (cf. [22]). It is, however, easy to show that this is not the case if f is a non-entire meromorphic function. We shall present sketch proofs of Theorems 1 and 2. The techniques involved here are essentially based on a powerful result of Steinmetz and some modifications of Bergweiler's arguments used in [2]. Also we note in proving Gross' conjecture, Bergweiler [2] has utilized, besides the Steinmetz's result, Wiman-Valiron theory, Landau's and Schottky's types of theorems. The latter two results are not applicable to meromorphic functions. Thus different arguments are employed here in proving our Theorem 1. We shall assume that the reader is familiar with the basic results of Nevanlinna theory (cf[17]) as well as its standard notations such as $T(r, f)$, $N(r, f)$, and $S(r, f)$.

2. __Lemmas__

Lemma 1 (Steinmetz [23, Satz 1, Korollar 1]). Let F_0, F_1, ..., F_m be not identifically vanishing meromorphic functions and let h_0, h_1, ..., h_m be meromorphic functions that do not all vanish identically. Let g be a nonconstant entire function and suppose that there exists a positive constant K such that

$$\sum_{j=0}^{m} T(r, h_j) \leq KT(r, g) + S(r, g),$$

where $S(r, g) = o(T(r, g))$ as $r \to \infty$ possibly outside some exceptional set of finite measure. Suppose also that
$$F_0(g)h_0 + F_1(g)h_1 + \ldots F_m(g)h_m = 0.$$
Then there exist polynomials P_0, P_1, ..., P_m that do not all vanish identically such that
$$P_0(g)h_0 + P_1(g)h_1 + \ldots P_m(g)h_m = 0$$
and there exist polynomials Q_0, Q_1, ..., Q_m that do not all vanish identically such that
$$Q_0 F_0 + Q_1 F_1 + \ldots + Q_m F_m = 0.$$

Generalizations and different proofs of this result have been given by Brownawell [6] and Gross and Osgood [14, 15]. For more applications of this lemma, we refer the reader to [13, 19, 20, 23, 25, 28].

Lemma 2 Let $\{a_n\}$ be a sequence of complex numbers and $|a_n| \to \infty$ $(n \to \infty)$ and let $r_n = |a_n|$ with $0 < r_1 \leq r_2 \leq \ldots \leq r_n \leq \ldots$. Define

$$\sigma_n = \frac{\log n}{\log r_n}, \quad n = r_n^{\sigma_n}$$

and

$$\rho(r) = \log n(r)/\log r, \quad n(r_n) \geq n,$$

where $n(r)$ denotes the number of $\{a_r\}$ in $|z| \leq r$.

Define $p_n = [2\rho(r_n)]$, where $[x]$ denotes the largest integer no greater than x, and

$$P(z) = \prod_{n=1}^{\infty} E(\frac{z}{z_n}, p_n),$$

where $E(z, p) = (1 - z) \exp(z + z^2/2 + \ldots + z^p/p)$.
Then, for all sufficiently large r,

$$\log|P(z)| < n(r)^{2\log r} + n(r^{2+\varepsilon}/2)(3 + \log n(r^{2+\varepsilon}/2)). \tag{1}$$

And further if

$$\limsup_{r\to\infty} \frac{\log n(r, 1/P)}{\log r} = \infty,$$

then there exists a set I of infinite linear measure such taht for $r \in I$,

$$\log|P(z)| < n(r^{2+\varepsilon}/2)^{7/2}.$$

Lemma 3 ([24] or cf. [2]). Let g be entire and transcendental and assume that $d > 0$, $\varepsilon > 0$, $\eta > 0$ and $\gamma > \frac{1}{2}$. Suppose that $|z_0| = r$, $|g(z_0)| = \eta M(r, g)$ and $|\tau| < dv(r, g)^{-\gamma}$, where $v(r, g)$ denotes the central index of g. Then

$$g(z_0 e^\tau) \sim g(z_0)e^{v(r, g)\tau}, \qquad (r \notin E); \tag{2}$$

$$g(z_0 e^\tau) \sim (v(r, g)/z_0 e^\tau)g(z_0)e^{v(r, g)\tau}, \qquad (r \notin E); \tag{3}$$

$$v(r, g) \le (\log \mu(r, g))^{1+\varepsilon} \le (\log M(r, g))^{1+\varepsilon}, \qquad (r \notin E); \tag{4}$$

$$\log M(r, g) \le (1+o(1))\log \mu(r, g) \le (1 + o(1))v(r, g)\log r, \qquad (r \notin E); \tag{5}$$

By suitably modifying a proof of Bergweiler [2, Lemma 3], one can easily derive the following:

Lemma 4 ([28]). Let $g(z)$ be entire and transcendental and assume that $c > 0$, $1 \ge \gamma > \frac{1}{2}$ and $\eta > 0$. If $r \notin E$, $|z_0| = r$ and $|g(z_0)| \ge \eta M(r, g)$, then there exists a function $\tau(z)$ analytic in $|z - z_0| \le crv(r, g)^{-\gamma}$ satisfying $|\tau(z)v(r, g) - 2\pi i| = o(1)$,

$$g(ze^{\tau(z)}) = g(z).$$

Lemma 5. ([28].). If f and g are two transcendental entire functions and $\alpha(z)$ a non-constant polynomial. By $P(z)$ we denote as the Weierstrass product of the zeros of $f(g) - \alpha$ formed in the same manner as $P(z)$ in Lemma 2. Let $f(g(z)) = \alpha(z) + P(z)e^{D(z)}$, where D is an entire function.

If

$$N(r, 1/P(z)) < dT(r^{1/3}, g), \qquad (r \notin E), \tag{6}$$

where d is a given number, $0 < d < 1$, then there exists a set I of infinite linear measure such that for $r \in I$,

$$v(r, g) < v(r, D)^{1+\delta}, \tag{7}$$

where ε is a given positve number.

Lemma 6. ([27, p. 68]). If $f(z)$ is meromorphic in $|z| < R$, with $f(0) \neq 0, \infty$ and $f'(0) \neq 0$, then for $0 < r < R$,

$$T(r,f) < 2\{N(R, f) + N(R, \tfrac{1}{f}) + N(R, \tfrac{1}{f-1})\} + 191 + 4\log^+ |f(0)|$$

$$+ 2\log^+ |f'(0)R|^{-1} + 12\log^+ R/(R-r).$$

3. Proofs of the Theorems 1 and 2.

3.1   Sketch Proof of Theorem 1.

First we note that f has order zero and g has finite order by a result of Edrie and Fuchs [8, Corollary 1.1]. Hence $f = \dfrac{f_1}{f_2}$, where f_1 and f_2 denote the canonical products of the zeros and poles of f., respectively. Then f_1 and f_2 have order zero.

Suppose now that $f(g(z)) - R(z)$ has only finitely many zeros, which means that $f_1(g(z)) - R(z)f_2(g(z))$ has only finitely many zeros and poles. Hence

$$f_1(g(z)) - R(z)f_2(g(z)) = q(z)e^{\alpha(z)} \tag{8}$$

for a rational function q and an entire function α.

First we show that the left side of (8) has finite lower order. To do this, by noting that g is of finite order, we then can apply a result of Edrie and Fuchs [8, Theorem 1] and Valiron [24] and deduce that

$$n(M(r, g), \tfrac{1}{f}) \leq n(r^{1+\varepsilon}, \tfrac{1}{f(g)}) + O(1)$$

and

$$n(M(r, g), f) \leq n(r^{1+\varepsilon}, f(g)) + O(1)$$

By defining $N(r) = N(r, \tfrac{1}{f_1}) + N(r, \tfrac{1}{f_2})$, that is, $N(r) = N(r, \tfrac{1}{f}) + N(r, f)$, and noting f is of order zero, we conclude that there exist arbitrary large r such that

$$\frac{N(t)}{\sqrt{t}} \leq \frac{N(M(r, g))}{\sqrt{M(r, g)}}$$

for $t \geq M(r, g)$. From a well-known estimates of canonical products (cf. [17, p. 102]) we have

$$\log M(M(r, g), f_1) + \log M(M(r, g), f_2)$$

$$\leq M(r, g) \int_{M(r, g)}^{\infty} \frac{N(t)}{t^2} \, dt$$

$$\leq N(M(r, g)) \sqrt{M(r, g)} \int_{M(r, g)}^{\infty} \frac{1}{t^{3/2}} \, dt$$

$$= 2N(M(r, g))$$

for these r. Since $N(r, \tfrac{1}{f_i}) \leq n(r, \tfrac{1}{f_i}) \log r + O(1)$ and $M(r, f_j(g)) \leq M(M(r, g), f_i)$ for $j = 1, 2$, it follows that

$$\log M(r, f_1(g)) + \log M(r, f_2(g))$$

$$\leq 2(n(M(r, g), \frac{1}{f}) + n(M(r, g), f)) \log M(r, g) + O(1)$$

$$\leq r^{(1+\varepsilon)(\rho(f(g))+\varepsilon)+\rho(g)+\varepsilon}$$

for a sequence of r values, where $\rho(f(g))$ and $\rho(g)$ denote the order of $f(g)$ and g, respectively. We deduce that the left side of (8) has finite lower order.

It follows that e^α has finite lower order which implies that α is a polynomial. Differentiating (8) and then by eliminating e^α, we have

$$f_1'(g)g' - f_2'(g)g'R - f_1(g)(\frac{q'}{q}+\alpha') + f_2(g)((\frac{q'}{q}+\alpha')R-R') = 0.$$

We deduce that $g'(z) = P(g(z), z)$ for some function P which is rational in $g(z)$ and z. (Here we have used the hypothesis that R is nonconstant.) A theorem of Malmquist (cf. [18], p. 170]) implies that this differential equation for g is actually a Riccati equation. Furthermore, because g is entire, we can deduce that P has the form $P(g(z), z) = a(z)g(z) + b(z)$, where a and b are rational functions. It follows that if A is an antiderivative of a, then

$$g(z) = e^{A(z)}(\int_{z_0 z} b(t)e^{-A(t)}dt + c)$$

for suitable constants c and z_0. Suppose that $A(z) = A_0 e^{i\theta}z^d + o(|z|^d)$ as $|z| \to \infty$ where $A_0 > 0$. Then g has order d, in fact, we have $T(r, g) \sim \gamma r^d$ for some positive γ as $r \to \infty$. Moreover, we can deduce from the fact that f_1 and f_2 are of zero order, expression of g above and that

$$\mathrm{Re}\, \alpha(re^{i\varphi}) = \log |f_1(g(re^{i\varphi}) - R(re^{i\varphi})f_2(g(re^{i\varphi})|$$

that α is either constant or $\alpha(z) = \alpha_0 e^{i\theta}z^d + o(|z|^d)$ for some $\alpha_0 > 0$ as $|z| \to \infty$. In any case, we have $T(r, e^\alpha) = O(T(r, g))$ as $r \to \infty$. Hence Steinmetz's theorem can be applied to the equation (1). It follows that there exists polynomials $Q_1, Q_2,$ and Q_3 that do not all vanish identically such that

$$Q_1 f_1 + Q_2 f_2 = Q_3 . \tag{9}$$

Suppose first Q_2 does not vanish identically. Solving (9) for f_2, substituting in (8), and solving the resulting equation for $f_1(g)$ we find that

$$f_1(g) = \frac{qe^\alpha Q_2(g) + RQ_3(g)}{Q_2(g) + RQ_1(g)} .$$

But if Q_2 does vanish identically, then this equation follows directly from (9).
From this and a result of Clunie [7, Theorem 1] we can conclude that both f_1 and f_2 are polynomials, contradicting the hypothesis. This also completes the proof.

3.2 Sketch Proof of Theorem 2.

First we write

$$f(g(z)) = \alpha(z) + P(z)e^{D(z)},$$

where $D(z)$ is an entire function and $P(z)$ the Weierstrass product formed by the zeros of $f(g) - \alpha(z)$ as expressed in Lemma 2.

We need treat two cases (I) and (II), separately.

(I) There exists a set I of infinite linear measure such taht for $r \in$ I,

$$40 \log M(r, P(z)) \geq \log M(\frac{r}{2}, f(g)).$$

Then (4) of Lemma 2 implies that

$$\log M(\frac{r}{2}, f(g)) < 40n(r)^{2\log r} + 40n(r^{5/2}) \log \log n(r^{5/2}) < 41n(r^{5/2})^{2\log r},$$

and further

$$\log M(r^{1/3}, f(g)) < 41n(r)^{(2/3)\log r} < n(r)^{2 \log r}$$

The assertion follows.

(II) $\quad 40\log M(r, P(z)) < \log M(\frac{r}{2}, f(g)), \ r \notin$ E. $\hspace{2cm}$ (10)

Suppose now that the theroem is false so that

$$N(r, \frac{1}{f(g)} - \alpha(z))) < dT(r^{1/3}, g), \quad (0 < d < 1), \quad r \notin E. \hspace{1cm} (11)$$

We choose a point z_0 with $|z_0| = r$ such that

$$|f(g(z_0))| = M(r, f(g)) = M((1 + o(1))M((r, g), f),$$

then

$$|g(z_0)| \geq (1 - o(1))M(r, g),$$

and

$$\begin{aligned}
|D(z_0)| &> \text{Re } D(z_0) = \log |\exp(D(z_0))| = \log |f(g(z_0)) - \alpha(z_0))/P(z_0)| \\
&\geq \log |f(g(z_0))| - \log^+ |P(z_0)| - \log 2 \\
&\geq \log M(r, f(g)) - \log M(r, \alpha(z)) - \log M(r, P(z)) - \log 2 \\
&\geq \log M(r, \exp(D(z))) - 2\{\log M(r, \alpha(z)) + \log M(r, P(z)) + \log 2\} \\
&\geq (\frac{7}{8})(1 - o(1))\log M(r, \exp(D(z))) > (\frac{2}{3})M(r, D). \ r \notin E. \hspace{0.5cm} (12)
\end{aligned}$$

From Lemma 4, we can construct an analytic function $\tau(z)$ in $|z - z_0| \leq 12crv(r, g)^{-2/3}$,

$c = \text{arctg } 2 + \frac{\pi}{2}, r \notin$ E, satisfying

$$|\tau(z)v(r, g) - 2\pi i| = o(1), \text{ and } g(ze^{\tau(z)}) = g(z).$$

Set $k(z) = ze^{\tau(z)}$ and

$$h(z) = \frac{f(g(z)) - \alpha(z)}{\alpha(k(z)) - \alpha(z)}.$$

Since

$$\alpha(k(z)) - \alpha(z) \sim (e^{q\tau(z)} - 1)\alpha(z) \sim q\tau(z)\alpha(z) \sim (2\pi qi/v(r, g)) \alpha(z), \text{ as } r \to \infty,$$

where q is the degree of $\alpha(z)$. It follows that $h(z)$ is analytic in $|z - z_0| \leq 12crv(r, g)^{-2/3}$.

Now defing $z_1 = z_0e^{i\theta}$ where $\theta \in$ R is chosed such that $\text{Re}D(z_1) = 0$ and $|v(r, D)\theta$

$-\frac{\pi}{2}\bigm| = \bigm|\arg D(z_0) + o(1)\bigm| = \bigm|\mathrm{arctg}(\mathrm{Im}D(z_0)/\mathrm{Re}D(z_0)) + o(1)\bigm| < \mathrm{arctg}\, 2$, we can do this according to Lemma 3. Then

$$|z_1 - z_0| = r\bigm|e^{i\theta} - 1\bigm| \sim r|\theta| < rv(r, D)^{-1}.$$

In Lemma 5 by taking $\varepsilon = \frac{1}{2}$, we have

$$v(r, g) < v(r, D)^{3/2}, \quad \text{i.e. } v(r, D)^{-1} < v(r, g)^{-2/3}, \qquad r \notin E.$$

and hence $|z_1 - z_0| < (1 + o(1))crv(r, g)^{-2/3} < 2crv(r, g)^{-2/3}$.

It follows that $h(z)$ may be defined for $|z - z_1| < (1 + o(1))crv(r, g)^{-2/3}$ and, setting $R = 2crv(r, g)^{-2/3}$,
we now estimate

$$N(R) \stackrel{\bullet}{=} N(5R, z_1, h) + N(5R, z_1, 1/h) + N(5R, z_1, \frac{1}{(h-1)}). \tag{13}$$

For this, consider the disk $|z - z_1| < R$. Then

$$|z| < |z_1| + 5R < |z_0| + 6R = r(1 + 12\, crv(r, g)^{-2/3})$$

$$|k(z)| = |ze^{\tau(z)}| < (1 + 12\, crv(r, g)^{-2/3})re^{|\tau(z)|} < (1 + 12\, crv(r, g)^{-2/3})(1 +$$

$7crv(r, g)^{-1})r < (1 + 20\, crv(r, g)^{-2/3})r$, $r \notin E$, since $|\tau(z)v(r, g) - 2\pi i| = o(1)$. Define $t = (1 + 20\, crv(r, g)^{-2/3})r$. Thus it clearly follows from $g(k(z)) = g(z)$ and the form of $h(z)$ that

$$N(R) < 2N(t, \frac{1}{P}) < 2dT(r^{1/2}, g), r \notin E.$$

Now we estimate $\log|h'(z_1)|$ and $\log^+|h(z_1)|$.

$$\log^+|h(z_1)| = \log^+|P(z_1)\exp D(z_1)/(\alpha(k(z_1)) - A(z_1))| <$$

$$\log^+|P(z_1)| + \log^+|\alpha(z_1)| + \log v(r, g) + O(1) < \log M(r, P(z)) +$$

$$O(\log rv(r, g)) < (\frac{1}{40})\log M(r, f(z)) + O(\log rv(r, g)). \tag{14}$$

$$h'(z) = \frac{f'(g(z))g'(z) - \alpha'(z)}{\alpha(k(z)) - A(z)} - \frac{(f(g(z)) - \alpha(z))(\alpha'(k(z))k'(z) - \alpha'(z))}{(\alpha(k(z)) - \alpha(z))^2}$$

$$= \left[\frac{P'(z) + P(z)D'(z)}{\alpha(k(z)) - \alpha(z)} - \frac{P(z)(\alpha'(k(z))k'(z) - \alpha'(z)}{(\alpha(k(z)) - \alpha(z))^2}\right]e^{D(z)}.$$

Then, from this and (2) we can get the estimation:

and

$$\log|h'(z_1)| > -\frac{1}{5}\log M(r, f(g))), r \notin E. \tag{15}$$

Now it follows from Lemma 6 that

$$T(4R, z_1, h) < 2N(R) + 191 + 12\log^+\frac{5R}{(5R{-}4R)} + 2\log^+\frac{1}{R} + 4\log^+|h(z_1)| + 2\log^+\frac{1}{|h'(z_1)|}$$

$$< 4dT(r^{1/2}, g) + \frac{1}{2} \log M(r, f(g)) + O(\log rv(r, g)). \qquad (16)$$

On the other hand, from

$$\left| h(z_0) \right| = \left| \frac{f(g(z_0)) - \alpha(z_0)}{\alpha(\kappa(z_0)) - \alpha(z_0)} \right| \sim \left| \frac{v(r, g)}{2\pi q \alpha(z_0)} \right| \left| f(g(z_0)) - \alpha(z_0) \right|,$$

we have, $\log \left| h(z_0) \right| > \log M(r, f(g)) + \log v(r, g) - O(\log r)$. $\qquad (17)$
Note

$$\log \left| h(z_0) \right| < \log M(r, z_1, h) \le \frac{5}{3} T(4R, z_1, h). \qquad (18)$$

Thus by combining (16)m (17) and (18), we have

$$\log M(r, f(g)) + \log v(r, g) < 8dT(r^{1/2}, g) + O(\log rv(r, g)) + \frac{5}{6} \log M(r, f(g)),$$

and hence

$$T(r, f(g)) < \log M(r, f(g)) < 48dT(r, g) + O(\log rv(r, g)), \qquad r \notin E.$$

this is absurd and thus Theorem 2 follows.

4.   Remarks and a Conjecture

As a further study of the zeros of $f(g) - R$ and by arguments similar to the ones used in [19] the folllowing quite general result has been obtained by Laine and his students.

Theorem 3[20]. Let f be a rational function of degree ≥ 2 or a transcendental meromorphic function of finite order, g be a transcendental entire function of finite lower order $\mu(g)$, and Q be a non-constant meromorphic function of order $\lambda(Q) < \mu(g)$. Let π denote the canonical product formed with the zeros of $f(g) - Q$. Then the exponent of convergence of zeros $\sigma(\pi) \ge \mu(g)$. In particular $f(g) - Q$ has infinitely many zeors.

We note that the above theorem is, in a sense, extension of Theorem 1 to the case $f(g)$ being of infinite order but its argument fails to include Theorem 1. It was indicated in [19] that if Q is a non-constant rational function, then Theorem 3 holds if the lower order of g is zero.

During the Conference, X.H. Hua reported the following result:
Let $f(z)$, $g(z)$ be transcendental entire functions. Then there exists a set I of infinite logarithmic meausre such that
$$T(r,g) = oN(r, f(g)), r \in I.$$

Finally, we would like to conclude the paper with the following strong conjecture:

Conjecture. Let f be a transcendental meromorphic function, g a transcendental entire function, and $\alpha(z)$ a non-constant mermorphic function such that $T(r, \alpha) = o(1) T(r, g)$. Then

$$N(r, \frac{1}{f(g) - \alpha}) \ne S(r, f(g)).$$

References

[1]   W. Bergweiler,On Factorization of Certain Entire Functions, to appear.

[2]   W. Bergweiler, Proof of a conjectue of Gross concerning fix-points, Math. Z. 204 (1990), 381-390.

[3]   W. Bergweiler, On the existence of fixpoints of composite meromorphic functions, Proc. Amer. Math. Soc., to appear.

[4]   W. Bergweiler, On factorization of certain functions, in "The Mathematical Heritage of Carl Friedrich Gauß", World Scientic, Singapore, to appear.

[5]   W. Bergweiler and C.C. Yang, On the value distribution of composite meromorphic functions, to appear in Bull. London Math. Soc.

[6]   W. D. Brownawell, On the factorization of partial differential equations, Can. J. Math. 39 (1987), 825-834.

[7]   J. Clunie, The composition of entire and mermorphic functions, in "Mathematical essays dedicated to A. H. Macintyre", Ohio University Press, Athens, Ohio, 1970, 75-92.

[8]   A. Edrei and W.H. Fuchs, On the zeros of $f(g(z))$ where f and g are entire functions. J. Analyse Math. 12 (1964), 243-255.

[9]   A.A. Gol'dberg and S. Prokopovich, On the simplicity of certain entire functions, Ukrain. Math. J. 22 (1970), 701-704 (translation from Ukrain. Mat. Zh. 22, 813-817).

[10]   R. Goldstein, On factorisation of certain entire functions, II, Proc. London Math. Soc. 22 (1971), 483-506.

[11]   F. Gross, On factorization of meromorphic functions, Trans. Amer. Math. Soc. 131 (1968), 215-222.

[12]   F. Gross, Factorization of meromorphic functions, U.S.: Government Printing Office, Washington, D.C., 1972.

[13]   F. Gross and C.F. Osgood, On fixed points of composite entire functions, J. London Math. Soc. (2) 28 (1983), 57-61.

[14]   F. Gross and C.F. Osgood, A simpler proof of a theorem of Steinmetz, J. Math. Anal. Appl. 143 (1989), 490-496.

[15]   F. Gross and C.F. Osgood, An extension of a theorem of Steinmetz, J. Math. Anal. Appl. 156 (1991), 290-294.

[16] Gross and C.C. Yang, Further results on prime entire functions, Trans. Amer. Math. Soc. 142 (1974), 347-355.

[17] K. Hayman, Meromorphic Functions, Clarendon Press, Oxford 1964.

[18] G. Jank and L. Volkmann, Einführung in die Theorie der ganzen und meromorphen Funktionen mit Anwendungen auf Differentialgleichungen, Birkhäuser, Basel - Boston - Stuttgart 1985.

[19] K. Katajamäki, L. Kinnunen, and I. Laine, On the value distribution of composite entire functions, preprint.

[20] ______________, On the vlaue distribution of some composite meromorphic function, preprint.

[21] E. Mues, Über faktorisierbare Lösungen Riccastischer Differentialgeichungen, Math. Z. 121 (1971), 145-156.

[22] M. Ozawa, On the solution of the functional equation f o g(z) = F(z), I, Kodai Math. Sem. Rep. 20(1968), 159-162.

[23] N. Steinmetz, Über faktorisierbare Lösungen gewöhnlicher Differential-gleichungen, Math. Z. 170 (1980), 169-180.

[24] G. Valiron, Sur un théorème de M. Fatou, Bull. Sci. Math. 46 (1922), 200-208.

[25] C.C. Yang, Further results on the fix-points of composite transcendentla functions, J. Math. Anal. Appl. 90 (1982), 259-269.

[26] C.C. Yang, On the fix-points of composite transcendental entire functions, J. Math. Anal. Appl. 108 (1985), 366-370.

[27] L. Yang, New researches on value distribution theory, Science Press, 1982 (in Chinese).

[28] Z. H. Zheng and C.C. Yang, Further results on fixpoints and zeros of entire functions, to appear in Trans. Amer. Math. Soc.

Department of Mathematics
The Hong Kong Unviersity of Science & Technology
Kowloon
Hong Kong